Birkhäuser

Advances in Mechanics and Mathematics

Volume 57

Mechanics and mathematics have been complementary partners since Newton's time, and the history of science shows much evidence of the beneficial influence of these disciplines on each other. Driven by increasingly elaborate modern technological applications, the symbolic relationship between mathematics and mechanics is continually growing. Mechanics is understood here in the most general sense of the word, including - besides its standard interpretation - relevant physical, biological, and information-theoretical phenomena, such as electromagnetic, thermal, and quantum effects; bio- and nano-mechanics; information geometry; multiscale modelling; neural networks; and computation, optimization, and control of complex systems. Topics with multi-disciplinary range, such as duality, complementarity, and symmetry in sciences and mathematics, are of particular interest.

Advances in Mechanics and Mathematics (AMMA) is a series intended to promote multidisciplinary study by providing a platform for the publication of interdisciplinary content with rapid dissemination of monographs, graduate texts, handbooks, and edited volumes on the state-of-the-art research in the broad area of modern mechanics and applied mathematics. All contributions are peer reviewed to guarantee the highest scientific standards. Monographs place an emphasis on creativity, novelty, and innovativeness in the field; handbooks and edited volumes provide comprehensive surveys of the state-of-the-art in particular subjects; graduate texts may feature a combination of exposition and electronic supplementary material.

The series is addressed to applied mathematicians, engineers, and scientists, including advanced students at universities and in industry.

All manuscripts are peer-reviewed to meet the highest standards of scientific literature:

- Series editors provide their final decision for inclusion in the series after manuscripts have been fully peer-reviewed.
- For contributed volumes, volume editors conduct the anonymous peer-review process before final submission.
- The peer-review process is anonymous and confidential in order to ensure reviewers provide impartial opinions.

Kaïs Ammari • Boumediène Chentouf

Stabilization of Wave-Type Equations with Delay or Memory

Birkhäuser

Kaïs Ammari
Department of Mathematics, Faculty of Sciences of Monastir
University of Monastir
Monastir, Tunisia

Boumediène Chentouf
Department of Mathematics
Kuwait University
Kuwait City, Kuwait

ISSN 1571-8689 ISSN 1876-9896 (electronic)
Advances in Mechanics and Mathematics
ISBN 978-3-032-28495-2 ISBN 978-3-032-28496-9 (eBook)
https://doi.org/10.1007/978-3-032-28496-9

Mathematics Subject Classification: 35L05, 35B35, 93D15

This book is published under the imprint Birkhäuser, www.birkhauser-science.com by the registered company Springer Nature Switzerland AG
The registered company address is: Gewerbestrasse 11, 6330 Cham, Switzerland

Preface

The study of wave equations and their stabilization properties has been a central theme in control theory and applied mathematics, with wide-ranging applications in engineering, physics, and mechanical systems. This book provides a comprehensive exploration of the stabilization of the multidimensional wave equation, covering various aspects of boundary and distributed delayed or memory control mechanisms. In particular, it delves into different stabilization notions—polynomial stabilization, exponential stabilization, and asymptotic stabilization—under miscellaneous damping and control settings. It is our intent to illustrate, in a simple manner, how to eliminate any negative impact of the occurrence of any delay or memory phenomenon on systems' stability or to nullify their presence. Our approach is direct, which makes the exposition accessible to a wide audience.

A significant portion of this book is dedicated to practical applications, with a special emphasis on the stabilization of overhead cranes, a system widely used in industrial operations and freight handling. The techniques presented in the book are not only mathematically rigorous but also applicable to real-world control problems. It is also worth mentioning that the methodology adopted in this book consists in tackling all problems, except one (see Sect. 3.7), independently, while detailed proofs are provided in each case. This is contrary to the abstract approach in which a class of problems is set in an abstract way, and under general assumptions, the proof of the results is concluded. Although the proofs in some of the chapter's sections appear to be slightly repetitive, we believe that this will allow the reader to follow each section with a certain level of independence. Thereby, this monograph is not meant to be read in a linear manner that obliges the reader to follow the sections' order. Indeed, the reader interested in a specific section may start immediately with it. Meanwhile, in order to have a smooth transition, we tried to highlight the link between the sections of each chapter.

We begin by examining the fundamentals of wave equation stabilization, followed by specific cases incorporating local damping, Kelvin–Voigt damping, and memory-type control. These approaches introduce dissipative mechanisms that play a crucial role in ensuring system stabilization. Special attention is given to boundary stabilization via a single control input, demonstrating how minimal intervention can still achieve desirable stability properties.

Furthermore, the role of delayed control is investigated, including time-dependent delays in boundary stabilization. Delays are inherent in many control systems and their effects on stability require careful mathematical treatment. The book also extends its scope to nonlinear boundary stabilization, offering insights into how nonlinearity affects the long-term behavior of wave equations. Unless otherwise specified, all of the material contained in this book has resulted from the authors' research over the last 10 years, together with the collaboration of their colleagues. We would like to take this opportunity to thank all of them for their precious contributions to this work: Serge Nicaise, Cristina Pignotti, Sabeur Mansouri, Stefan Gerbi, Baowei Feng, Aissa Guesmia, and Zhong–Jie Han.

Each chapter presents theoretical results, proofs, and illustrative examples to guide readers through the intricate aspects of wave equation stabilization. Whether the reader is interested in theory, numerical simulations, or real-world applications, this book serves as a valuable resource for researchers, graduate students, and engineers working in the fields of control theory, partial differential equations (PDEs), and mechanical system stabilization. Implementation of designed controls and experiments is not addressed in this book.

By the end of this book, readers will have developed a deep understanding of stabilization techniques for wave equations and overhead cranes despite the presence of a delay or memory, equipping them with the necessary tools to tackle more complex control problems in dynamical systems.

Monastir, Tunisia Kaïs Ammari
Kuwait, Kuwait Boumediène Chentouf
December 2025

Contents

Chapter 1
Introduction

The study of wave equations and their stabilization properties has been a central theme in control theory and applied mathematics, with broad applications across engineering, physics, and mechanical systems. From structural mechanics to fluid dynamics and acoustic wave propagation, the need to control and stabilize wave-type systems is crucial to ensure the efficiency, safety, and reliability of various technologies. This book provides a comprehensive exploration of stabilization techniques for wave-type equations with a particular focus on delay and memory effects, two key factors that significantly impact the system's behavior and control performance.

In many real-world systems, delays arise naturally due to signal transmission, computational processing, or actuator response times. Similarly, memory effects, often modeled through integral terms, are essential to describe materials with viscoelastic properties, hereditary effects, or long-term interactions. Understanding the interplay between these factors and stabilization properties is vital for designing effective control strategies. This book thoroughly examines boundary and interior (also called distributed) control mechanisms subject to delay and memory effect, offering rigorous mathematical formulations, recent developments on this topic, and practical insights into their application. It is noteworthy that in practice, the implementation of an interior control requires relatively more actuators and sensors than in the case of boundary controls. The latter option is often considered as more economical (say less costly) method to stabilize systems.

A significant portion of this book is devoted to different notions of stabilization, including logarithmic stabilization, polynomial stabilization, exponential stabilization, and asymptotic stabilization under several damping and control settings. We explore stabilization techniques incorporating local damping, Kelvin-Voigt damping, and memory-type control, all of which introduce dissipative mechanisms that play a fundamental role in ensuring system stability. In addition, special attention is given to boundary stabilization via a single control input, demonstrating how minimal intervention can achieve robust stability properties. Furthermore, the

K. Ammari, B. Chentouf, *Stabilization of Wave-Type Equations with Delay or Memory*, Advances in Mechanics and Mathematics 57,
https://doi.org/10.1007/978-3-032-28496-9_1

book investigates the effects of time-dependent delays in boundary or internal stabilization, a critical aspect in many control applications where delay parameters evolve over time. The challenges associated with nonlinear boundary stabilization are also addressed, providing insights into how nonlinearity influences the long-term behavior of wave equations and control performance. Beyond theoretical advancements, this book also emphasizes practical applications, particularly in the stabilization of overhead cranes—a crucial system in industrial operations that requires precise control to minimize oscillations and ensure safe transportation of loads. The methodologies discussed extend to other mechanical and engineering applications where wave-type dynamics, delays, and memory effects are present. Each chapter presents a balanced combination of theoretical results, rigorous proofs, and a few illustrative examples, guiding the reader through the intricate aspects of wave equation stabilization. The book is designed to be accessible to a broad audience, from researchers and graduate students to engineers and practitioners working in control theory, partial differential equations (PDEs), mechanical system stabilization, and applied mathematics. In addition to being a specialized reference text, the book is sufficiently self-contained to be considered as a graduate text in control of PDEs.

As stated earlier, this book is mainly concerned with the stabilization problem of either the wave equation or the overhead crane system. In each situation, the input control will be specified. This book contains one chapter for mathematical preliminaries and two chapters for the recent advances and developments in the stabilization of wave-type equations with delay or memory. To be more precise, the book starts with a brief introduction to the mathematical tools used throughout the book, together with proofs of many results. A considerable portion of the mathematical background necessary to understand the topic is provided in Chap. 2.

The second chapter addresses the stabilization problem of the multidimensional wave equation subject to a time delay or memory, in either the boundary or the interior of the domain. In addition to a conclusion section, this chapter consists of six main sections, where each section ends with concluding remarks and open questions:

1. **Static Boundary Delay: Logarithmic and Exponential Convergence**
   In the first section, we analyze the case where the delay occurs at the boundary of the domain. We begin by introducing the problem and defining the boundary conditions in Sect. 3.2.1. Next, Sect. 3.2.2 provides the necessary preliminaries and sets up the problem. The existence and uniqueness of solutions for our system are established in Sect. 3.2.3. In Sect. 3.2.4, we investigate the asymptotic behavior of solutions, proving their convergence to a steady equilibrium state. Furthermore, Sect. 3.2.5 demonstrates that this convergence is logarithmic [15], even in the absence of the Bardos-Lebeau-Rauch (BLR) geometric condition [32, 133]. Finally, in Sect. 3.2.6, we establish an exponential convergence result [16] under the assumption that the BLR condition holds.

2. **Dynamic Boundary Delay: Exponential Stabilization**
   The second section focuses on the stabilization of the wave equation with dynamic boundary delay. We establish stabilization results based on an appropriate choice of the damping operator. The proof of the main stability result relies on the frequency-domain method (also called the resolvent method) and combines a contradiction argument with the multiplier technique to perform a detailed resolvent analysis [18].
3. **3D Wave Equation with Trapped Ray and Interior Delay: Polynomial Stabilization**
   In the third section, we investigate the long-time behavior of a wave equation in a three-dimensional domain containing a trapped ray, which leads to a violation of the geometric control condition. More specifically, we consider the delayed wave equation with an internal (distributed) localized damping and analyze the resulting stability properties [16].
4. **Interior Memory Versus Interior Delay: Exponential Stabilization**
   The fourth section is dedicated to the asymptotic behavior analysis of solutions to a space-dependent wave equation with distributed memory or time-delay term. The history approach is used for the distributed memory case. The solutions are shown to converge exponentially to their stationary state by utilizing the resolvent method. The same approach is also used for diverse physical systems [62].
5. **Interior Delay: Boundary Exponential Stabilization**
   In the fifth section, we consider a boundary stabilization problem for the wave equation with interior delay. We propose a boundary delay control to prove an exponential stability result under some Lions geometric condition [20]. Indeed, a uniform decay estimate is proved via a suitable Lyapunov functional.
6. **Abstract-Wave Equation With Delay and a Kelvin-Voigt Damping: Exponential Stabilization**
   In this last section, an abstract-wave equation with delay and a Kelvin-Voigt damping is considered. Using the resolvent method, an exponential stability result is shown for appropriate damping coefficients [22]. This result is then applied to a concrete example of a delayed wave equation with a Kelvin-Voigt damping.

In the third chapter, we study the 2D overhead cranes that are widely used in industrial and construction environments for transporting suspended loads from one point to another. Typically, these cranes consist of a hoisting mechanism—including a hoisting cable and a hook—and a support mechanism, such as a girder or trolley [3]. Due to the inherent flexibility of cables, they can only sustain tension forces [3]. This natural characteristic leads to deflections in the transverse direction of the cable and contributes to load oscillations, which can arise from different sources. As a result, the dynamics of an overhead crane system with a flexible cable could be highly complex (see [3] for further details).

In this chapter, we analyze one typical example of an overhead crane system consisting of a motorized platform with a mass moving along a horizontal rail. A flexible cable is attached to the platform. Furthermore, a payload of a certain mass is suspended at the end of the cable. The system is supposed to suffer from the presence of a delay in one of the end points of the cable. The chapter is structured into nine sections (plus a concluding section), each addressing different aspects of stabilization and control, and presenting future research directions:

1. **Boundary Delay: Polynomial Stabilization**
   This section establishes polynomial convergence rates for the crane system via a boundary delayed control of velocity type. First, the convergence of solutions of the system to the equilibrium state is shown by applying LaSalle principle. Then, thanks to the resolvent method, it is demonstrated that the convergence is, in fact, polynomial [12]. This provides foundational results that will be extended in subsequent sections.
2. **Boundary Delay: Exponential Stabilization via a Global Interior Control**
   Building upon the results of Sect. 4.2, we introduce an additional distributed (interior) control to enhance the convergence rate, achieving exponential convergence [17]. Note that the control is of type velocity and acts in the whole spatial domain.
3. **Boundary Delay: Exponential Stabilization via a Localized Interior Control**
   We examine the energy decay of the flexible crane system under localized interior damping. Our results demonstrate that the localized damping is sufficient to ensure exponential decay of the energy of the system [64], thereby improving upon the findings of Sect. 4.3, where damping was applied over the entire spatial domain.
4. **Boundary Delay: Exponential Stabilization via Kelvin-Voigt Damping Control**
   This section investigates the stabilization of the delayed crane system through a higher-order nondelayed boundary control. Unlike previous sections, the delayed and nondelayed boundary terms considered here are of different types. To address this scenario, we employ the Kelvin-Voigt damping mechanism [63].
5. **Boundary Delay: Exponential Boundary Stabilization Via a Single Control**
   Motivated by the analysis in Sect. 4.5, we aim to relax the stabilization conditions previously imposed on the proposed control. Specifically, we show that the high-order interior control in Sect. 4.5 acts as a strong damping due to the analyticity of the semigroup governing the system [141]. Additionally, we explore how the implementation of a secondary boundary control at the lower end of the cable influences the cost and feasibility of the proposed feedback law [65].
6. **Boundary Time-Dependent Delay: Exponential Stabilization**
   Here, we consider the presence of a time-dependent delay at one boundary of the crane system. Our primary objective is to stabilize both the cable's dynamics and the coupled dynamics of the cart and suspended load. We begin by proving that the system is well-posed in the semigroup sense, provided certain conditions on the delay are met. Subsequently, we construct an appropriate Lyapunov function, leading to the exponential decay of the energy of the system [66]. Moreover, an estimate of the decay rate is provided by means of the approach used in [155, 156].

7. **Nonlinear Boundary Delay: Nonlinear Boundary Stabilization**
   We investigate the stabilization of a variant of the crane system in which the mass m is assumed to be sufficiently small, allowing its dynamics to be neglected. Under this assumption, two nonlinear effects emerge: one associated with the velocity term and another arising from the delayed term (see [167]). Under specific conditions on the nonlinearities, several decay rates of the energy of the system are established [59].
8. **Boundary Memory-Type Control: Exponential Stabilization**
   This section addresses the boundary stabilization of an overhead crane system that incorporates infinite memory effects. We establish the well-posedness of the problem within the semigroup framework, under suitable assumptions on the memory kernel. Additionally, we prove a uniform exponential stability result of the solutions of our system using the frequency-domain method [67].
9. **Interior and Boundary Memory-Type Controls: Exponential Stabilization**
   This section is consecrated to the stabilization problem of the crane structure under the effect of an infinite memory term of the same type as in the previous section. The feedback law consists of a localized interior control and a boundary control. However, the latter has lower-order derivatives (unlike the previous one) and hence simpler. Well-posedness and exponential stability results are also obtained under reasonable conditions on the memory kernel [68].

Chapter 2
Preliminaries

This chapter is intended to provide the necessary material, from functional analysis, that is heavily used in this book. We strike a balance between detailed proofs of some results and others. Since we are driven by concrete wave problems whose setting is in Hilbert spaces, we shall often confine ourselves to reviewing concepts and results in Hilbert spaces. Notwithstanding, we begin by introducing general spaces and assure a smooth transition between the "most" general ones to special ones. The emphasis will be on semigroups theory of operators.

2.1 Inner Product Spaces and Related Topics

Definition 2.1.1 Given a set $\mathscr{X}$, a *metric space* is a pair $(\mathscr{X}, d)$, where d is a *metric* (or a distance) on $\mathscr{X}$ defined by

$$d : \mathscr{X} \times \mathscr{X} \to \mathbb{R}$$

such that for every $x, y, z \in \mathscr{X}$, we have:

1. $d(x, y) \geq 0$ and $d(x, y) = 0$ if and only if $x = y$.
2. $d(x, y) = d(y, x)$.
3. $d(x, y) \leq d(x, z) + d(z, y)$.

Example 2.1 A typical example of a metric space is $\mathscr{X} = \mathbb{R}^n$ (also $\mathbb{C}^n$), $n \in \mathbb{N}$, with a metric d defined by (among others)

$$d(x, y) = \sum_{i=1}^{n} |x_i - y_i|,$$

where $x = (x_1, x_2, \cdots, x_n)$ and $y = (y_1, y_2, \cdots, y_n)$.

K. Ammari, B. Chentouf, *Stabilization of Wave-Type Equations with Delay or Memory*, Advances in Mechanics and Mathematics 57,
https://doi.org/10.1007/978-3-032-28496-9_2

Definition 2.1.2 Let $\mathscr{X}$ be a linear space over a field $\mathbb{F}$, where $F = \mathbb{R}$ or $\mathbb{C}$. A *norm* on $\mathscr{X}$, denoted by $\|\cdot\|$, is a map from $\mathscr{X}$ to $\mathbb{R}$ such that for every $x, y \in \mathscr{X}$ and for each scalar $a \in F$, the following properties hold:

1. $\|x\| \geq 0$ and $\|x\| = 0$ if and only if $x = 0$.
2. $\|ax\| = |a|\|x\|$.
3. $\|x + y\| \leq \|x\| + \|y\|$.

A *normed space* is a linear space $\mathscr{X}$ endowed with a norm $\|\cdot\|$. In such a situation, one is able to "measure the size" of every element x of $\mathscr{X}$. Indeed, the "distance" between two vectors x and y of $\mathscr{X}$ is given by

$$d(x, y) = \|x - y\|,$$

which makes $\mathscr{X}$ a metric space (one can easily verify that d is a distance according to Definition 2.1.1).

Example 2.2 Let Ω be an open bounded (to be simple) domain of $\mathbb{R}^n$. Given a measurable function $f : \Omega \to \mathbb{C}$, we define $L^1(\Omega)$ as the space of integrable functions from Ω into $\mathbb{C}$. This space is a normed space with the norm

$$\|f\|_{L^1(\Omega)} = \int_\Omega |f(x)|\, dx.$$

Now, if $1 < p < \infty$, we define

$$L^p(\Omega) = \left\{: \Omega \to \mathbb{C};\ f \text{ is measurable and} |f|^p \in L^1(\Omega)\right\}.$$

Moreover, $L^p(\Omega)$ endowed with

$$\|f\|_{L^p(\Omega)} = \left(\int_\Omega |f(x)|^p\, dx\right)^{1/p}$$

is a normed space.

Finally, we define

$$L^\infty(\Omega) = \{f : \Omega \to \mathbb{C};\ f \text{ is Lebesgue measurable and there exists}$$
$$\text{a positive constant } C \text{ such that; } |f(x)| \leq C, \quad \text{a.e. on } \Omega\}.$$

This space equipped with

$$\|f\|_{L^\infty(\Omega)} = \inf\{C;\ |f(x)| \leq C, \quad \text{a.e. on } \Omega\}$$

is also a normed space.

Definition 2.1.3 Let $\mathscr{X}$ be a vector space over a field $\mathbb{F}$, where $\mathbb{F} = \mathbb{R}$ or $\mathbb{C}$. The *inner product* is a map from $\mathscr{X} \times \mathscr{X}$ to $\mathbb{F}$ satisfying:

1. $\langle x, x\rangle \geq 0, \quad \forall x \in \mathscr{X}$ and $\langle x, x\rangle = 0$ if and only if $x = 0$.
2. $\langle x, y\rangle = \overline{\langle y, x\rangle}, \quad \forall x, y \in \mathscr{X}$.
3. $\langle ax + by, z\rangle = a\langle x, z\rangle + b\langle y, x\rangle, \quad \forall x, y, z \in \mathscr{X}$ and $\forall a, b \in \mathbb{F}$.

The norm on an inner product space is defined by $\|x\| = \sqrt{\langle x, x\rangle}$:

Definition 2.1.4

1. A sequence (x_n) in a normed space $\mathscr{X}$ is *convergent* if there exists $x \in \mathscr{X}$ such that $\lim_{n\to\infty} \|x_n - x\| = 0$.
2. A sequence (x_n) in a normed space $\mathscr{X}$ is a *Cauchy sequence* if for any $\varepsilon > 0$, there exists a positive integer N such that $\|x_n - x_m\| \leq \varepsilon, \quad \forall n, m \geq N$.
3. A normed space $\mathscr{X}$ is *complete* if every Cauchy sequence $\{x_n\}$ converges to an element x in $\mathscr{X}$.
   A complete normed space is a *Banach space*.
4. If $\mathscr{X}$ is a complete inner product space, then it is called a *Hilbert space*.

Example 2.3

- The vector space $\mathbb{C}^n$ is a Hilbert space with the inner product

$$\langle z, w\rangle = \sum_{k=1}^{n} z_k \overline{w_k},$$

  where $z = (z_1, ..., z_n)$ and $w = (w_1, ..., w_n)$.
- The vector space

$$l^2 = \left\{ (a_n)_{n\in\mathbb{N}} \, ; \, a_n \in \mathbb{C} \text{ and } \sum_{n=1}^{\infty} |a_n|^2 < \infty \right\},$$

  is a Hilbert space with the inner product

$$\langle a, b\rangle = \sum_{n=1}^{\infty} a_n \overline{b_n},$$

  where $a = (a_n)_{n\in\mathbb{N}} \in l^2$ and $b = (b_n)_{n\in\mathbb{N}} \in l^2$.
- For each $p \in [1, \infty]$, the space $L^p(\Omega)$ (see Example 2.2) equipped with its corresponding norm is a Banach space.
  The most favorable case is $p = 2$ in which $L^2(\Omega)$ is a Hilbert space with the inner product defined by

$$|\langle f, g\rangle| = \int_{\Omega} f(x)\overline{g(x)}\, dx.$$

Definition 2.1.5 A metric space $\mathscr{X}$ is *compact* if every sequence in $\mathscr{X}$ has a convergent subsequence.

A subset B of $\mathscr{X}$ is compact if it is compact as a subspace of $\mathscr{X}$, that is, every sequence in B has a convergent subsequence whose limit belongs to B.

A subset B of $\mathscr{X}$ is precompact if its closure is compact.

When studying PDEs, the regularity of solutions is required to be as minimal as possible. This is motivated by the fact that numerous natural models have nonsmooth solutions such as in quantum mechanics, fluid dynamics, and heat transfer. In particular, the concept of differentiability in the traditional sense is no longer valid, and hence, a wider mathematical setting arises. This is one of the motivations for introducing Sobolev spaces (a solid introduction about this topic can be found in [4, 40]).

2.1.1 *Sobolev Spaces*

Let $L^1_{\text{loc}}(\Omega)$ be the space of all measurable functions $f : \Omega \to \mathbb{C}$ such that $\int_K |f(x)|\, dx < \infty$ for every compact $K \subset \Omega$. As the reader may know, we do not distinguish between functions that are equal almost everywhere. In other words, two functions $f, g \in L^1_{\text{loc}}(\Omega)$ are equal almost everywhere if and only if $\int_K |f(x) - g(x)|\, dx = 0$, for every compact $K \subset \Omega$ (this is also true for $L^p(\Omega)$ spaces introduced earlier).

We denote $\mathbb{N}_0 = \mathbb{N} \cup \{0\}$. For $\alpha = (\alpha_1, \ldots, \alpha_n) \in \mathbb{N}_0^n$ and $x = (x_1, \ldots, x_n) \in \mathbb{R}^n$, the Euclidean norm of x is denoted by $|x| = \left((x_1)^2 + (x_2)^2 + \cdots + (x_n)^2\right)^{\frac{1}{2}}$.

Whereupon, we set

$$x^\alpha = x_1^{\alpha_1} \cdots x_n^{\alpha_n}, \qquad \text{and } |\alpha| = \sum_{i=1}^{n} \alpha_j.$$

In order to define the Sobolev spaces, we shall need a number of classical spaces:

For any $m \in \mathbb{N}_0$, the set $C^m(\Omega)$ is the space of all functions $f : \Omega \to \mathbb{C}$ for which all the partial derivatives up to order m exist and are continuous. In turn, $C^0(\Omega)$ will be denoted by $C(\Omega)$. Furthermore, $C^\infty(\Omega)$ is the intersection of all spaces $C^m(\Omega)(m \in \mathbb{N})$.

Now, let $\alpha \in \mathbb{N}_0^n$ and $f \in C^m(\Omega)$ with $|\alpha| \leqslant m$. We denote

$$\partial^\alpha y = \frac{\partial^{|\alpha|} y}{\partial x_1^{\alpha_1} \cdots \partial x_n^{\alpha_n}} = \frac{\partial^{\alpha_1}}{\partial x_1^{\alpha_1}} \cdots \frac{\partial^{\alpha_n}}{\partial x_n^{\alpha_n}} y.$$

Now, let K be the closure of an open subset of $\mathbb{R}^n$, and let $m \in \mathbb{N}_0$ or $m = \infty$.

We denote by $C^m(K)$ the space of all the restrictions to K of functions in $C^m(\mathbb{R}^n)$. Additionally, we denote by $C^\infty(K)$ the intersection of all the spaces $C^m(K), m \in \mathbb{N}$.

Assume that K is also compact. Then, given $m \in \mathbb{N}$ and $\varphi \in C^m(K)$, we can define sup-norm

$$\|\varphi\|_{C^m(K)} = \sup_{x \in K, |\alpha| \leqslant m} \left| \left(\partial^\alpha \varphi\right)(x) \right|.$$

$C^m(K)$ equipped with this norm is a Banach space.

Definition 2.1.6 If $\varphi \in C(\Omega)$, the support of φ is the closure (in $\mathbb{R}^n$) of $\{x \in \Omega \mid \varphi(x) \neq 0\}$. The support of φ is denoted by $\operatorname{supp} \varphi$.

We denote by $\mathscr{D}(\Omega)$ the set of all $\varphi \in C^\infty(\Omega)$, which have compact support contained in Ω. These functions are called *test functions*. For a compact $K \subset \Omega$, we denote by $\mathscr{D}_K(\Omega)$ the set of all $\varphi \in \mathscr{D}(\Omega)$ with $\operatorname{supp} \varphi \subset K$.

Let Ω be an open set of $\mathbb{R}^n$ and let $p \in \mathbb{R}$ with $1 \leq p \leq \infty$.

Definition 2.1.7 The *Sobolev space* $W^{1,p}(\Omega)$ is defined by

$$W^{1,p}(\Omega) = \left\{ y \in L^p(\Omega); \begin{array}{l} \text{there exist } f_1, f_2, \ldots, f_n \in L^p(\Omega) \text{ such that} \\ \displaystyle\int_\Omega y \frac{\partial \varphi}{\partial x_i}\, dx = -\int_\Omega f_i \varphi\, dx \quad \text{for all } \varphi \in \mathscr{D}(\Omega), \\ \text{and for all } i = 1, 2, \ldots, n \end{array} \right\}.$$

For the special case $p = 2$, we denote

$$H^1(\Omega) = W^{1,2}(\Omega).$$

As a direct consequence of the above definition, if $y \in W^{1,p}(\Omega)$, we define the partial derivatives of y in the $W^{1,p}(\Omega)$ sense as follows (also called weak derivatives): $\dfrac{\partial y}{\partial x_i} = f_i$. Moreover, we write

$$\nabla y = \left(\frac{\partial y}{\partial x_1}, \frac{\partial y}{\partial x_2}, \ldots, \frac{\partial y}{\partial x_n} \right).$$

Example 2.4 Let $\Omega = (0, 1)$ and $y(x) = \begin{cases} x, & \text{if } x \in (0, 1/2), \\ 1/2, & \text{if } x \in [1/2, 1) \end{cases}$. Next, let $\varphi \in \mathscr{D}(0, 1)$. We have the following:

$$\int_0^1 y\varphi'\, dx = \int_0^{1/2} x\varphi'\, dx + \int_{1/2}^1 \varphi'/2\, dx = -\int_0^{1/2} \varphi\, dx = -\int_0^1 f\varphi\, dx,$$

where $f(x) = \begin{cases} 1, & \text{if } x \in (0, 1/2), \\ 0, & \text{if } x \in (1/2, 1) \end{cases}$. Hence, $y' = f$ and $y \in W^{1,p}(0, 1)$, for every $p \in [1, \infty]$.

Note that the space $W^{1,p}(\Omega)$ is equipped with the norm

$$\|y\|_{W^{1,p}(\Omega)} = \|y\|_{L^p(\Omega)} + \sum_{i=1}^{n} \left\| \frac{\partial y}{\partial x_i} \right\|_{L^p(\Omega)}.$$

In turn, if $1 \leq p < \infty$, then an equivalent norm could also be used:

$$\left(\|y\|_{L^p(\Omega)}^p + \sum_{i=1}^{n} \left\| \frac{\partial y}{\partial x_i} \right\|_{L^p(\Omega)}^p \right)^{1/p}.$$

Concerning the space $H^1(\Omega)$, the latter is equipped with the scalar product

$$\langle y, z \rangle_{H^1(\Omega)} = \langle y, z \rangle_{L^2(\Omega)} + \sum_{i=1}^{n} \left\langle \frac{\partial y}{\partial x_i}, \frac{\partial z}{\partial x_i} \right\rangle_{L^2(\Omega)} = \int_{\Omega} \left(yz + \sum_{i=1}^{n} \frac{\partial y}{\partial x_i} \frac{\partial z}{\partial x_i} \right) dx.$$

Clearly, the associated norm

$$\|y\|_{H^1(\Omega)} = \left(\|y\|_{L^2(\Omega)}^2 + \sum_{i=1}^{n} \left\| \frac{\partial y}{\partial x_i} \right\|_{L^2(\Omega)}^2 \right)^{1/2}$$

is equivalent to that of $W^{1,2}(\Omega)$.

We have the following result:

Proposition 2.1.1 *$W^{1,p}(\Omega)$ is a Banach space for every $1 \leq p \leq \infty$, while $H^1(\Omega)$ is a separable Hilbert space.*

Remark 2.1.1 It is worth noting that if $y \in C^1(\Omega) \cap L^p(\Omega)$ and the usual partial derivative of y, namely, $\frac{\partial y}{\partial x_i}$, belong to $L^p(\Omega)$ for all $i = 1, 2, \ldots, n$, then $y \in W^{1,p}(\Omega)$. Additionally, the usual partial derivative of y coincides with the weak partial derivative, which means that the latter is an extension of the usual partial derivative. Thereby, if Ω is bounded, then $C^1(\bar{\Omega}) \subset W^{1,p}(\Omega)$ for all $1 \leq p \leq \infty$.

Conversely, if $y \in W^{1,p}(\Omega)$ and $\frac{\partial y}{\partial x_i} \in C(\Omega)$ for all $i = 1, 2, \ldots, n$ (here $\frac{\partial y}{\partial x_i}$ means the weak partial derivative), then $y \in C^1(\Omega)$.

2.1.1.1 The Spaces $W^{m,p}(\Omega)$

Let $m \geq 2$ be an integer and let p be a real number with $1 \leq p \leq \infty$. We define

$$W^{m,p}(\Omega) = \left\{ y \in L^p(\Omega); \begin{array}{c} \text{for all } \alpha \text{ with } |\alpha| \leq m, \text{ there exists } f_\alpha \in L^p(\Omega) \text{ such that} \\ \int_\Omega y \partial^\alpha \varphi \, dx = (-1)^{|\alpha|} \int_\Omega f_\alpha \varphi \, dx, \quad \forall \varphi \in \mathcal{D}(\Omega) \end{array} \right\},$$

where we use the multi-index notation introduced in the beginning of this subsection.

We set $\partial^\alpha y = f_\alpha$. The space $W^{m,p}(\Omega)$ equipped with the norm

$$\|y\|_{W^{m,p}} = \sum_{0 \leq |\alpha| \leq m} \left\| \partial^\alpha y \right\|_{L^p(\Omega)}$$

is a Banach space.

The space $H^m(\Omega) = W^{m,2}(\Omega)$ equipped with the scalar product

$$(y, v)_{H^m} = \sum_{0 \leq |\alpha| \leq m} \left(\partial^\alpha y, \partial^\alpha v \right)_{L^2(\Omega)}$$

is a Hilbert space.

Alternatively, these sets could also be defined by induction

$$W^{m,p}(\Omega) = \left\{ y \in W^{m-1,p}(\Omega); \frac{\partial y}{\partial x_i} \in W^{m-1,p}(\Omega) \quad \forall i = 1, 2, \ldots, n \right\}.$$

2.1.1.2 Sobolev Spaces: A Distributional Approach

It turns out that the Sobolev spaces defined above can also be introduced in the language of distributions. To be more precise, we say that $W^{1,p}(\Omega)$ is the set of functions $y \in L^p(\Omega)$ for which all the partial derivatives $\frac{\partial y}{\partial x_i}$, $1 \leq i \leq n$ (in the sense of distributions), belong to $L^p(\Omega)$. For sake of clarity, we shall elaborate on this point. To do so, we first denote any linear mapping $u : \mathcal{D}(\Omega) \to \mathbb{C}$ by $\langle u, \varphi \rangle$, and hence, the linearity is valid for both arguments.

Definition 2.1.8 A *distribution* is a linear map $u : \mathcal{D}(\Omega) \to \mathbb{C}$, which satisfies the following continuity assumption:

For every compact $K \subset \Omega$, there exist $m \in \mathbb{N}_0$ and a constant $C \geq 0$ (both may depend on K) so that

$$|\langle u, \varphi \rangle| \leqslant C \|\varphi\|_{C^m(K)}, \quad \forall \varphi \in \mathcal{D}_K(\Omega).$$

Obviously, the set of all distributions on Ω is a vector space that will be denoted by $\mathscr{D}'(\Omega)$.

Given $u \in \mathscr{D}'(\Omega)$, if m can be chosen independently of K, then the smallest value of m is called the order of u.

Example 2.5 Let $f \in L^1_{\mathrm{loc}}(\Omega)$. Then we can define $u_f : \mathscr{D}(\Omega) \to \mathbb{C}$ by

$$\langle u_f, \varphi \rangle = \int_\Omega f(x)\varphi(x)\,dx \quad \forall \varphi \in \mathscr{D}(\Omega).$$

Thereby, $u_f \in \mathscr{D}'(\Omega)$, and it is of order zero. Indeed, for every compact $K \subset \Omega$, we have

$$|\langle u_f, \varphi \rangle| \leqslant \left[\int_K |f(x)|\,dx \right] \|\varphi\|_{C(K)} \quad \forall \varphi \in \mathscr{D}(\Omega).$$

Proposition 2.1.2 *If $f \in L^1_{loc}(\Omega)$ is such that $u_f = 0$, then $f(x) = 0$ for almost every $x \in \Omega$.*

Based on the above discussion, we have $L^1_{\mathrm{loc}}(\Omega) \subset \mathscr{D}'(\Omega)$. Accordingly, distributions can be viewed as generalizations of L^1_{loc} functions. As a matter of fact, distributions are sometimes called generalized functions.

Example 2.6 Let $a \in \Omega$ and define a linear map $\delta_a : \mathscr{D}(\Omega) \to \mathbb{C}$ as follows:

$$\langle \delta_a, \varphi \rangle = \varphi(a) \quad \forall \varphi \in \mathscr{D}(\Omega).$$

This implies that if K is a compact subset of Ω, then it holds

$$|\langle \delta_a, \varphi \rangle| \leqslant \|\varphi\|_{C(K)} \quad \forall \varphi \in \mathscr{D}(\Omega).$$

Consequently, δ_a is a distribution of order zero on Ω, called the Dirac mass at a.

The following definition gives an insight on the set of test functions $\mathscr{D}(\Omega)$ as well as the set of distributions $\mathscr{D}'(\Omega)$.

Definition 2.1.9 Let $(\varphi_m)_m$ (resp. $(u_m)_m$) be a sequence in $\mathscr{D}(\Omega)$ (resp. $\mathscr{D}'(\Omega)$). We say that:

(i) $(\varphi_m)_m$ converges to $\varphi \in \mathscr{D}(\Omega)$ if there exists a compact $K \subset \Omega$ such that:

- $\operatorname{supp} \varphi_m \subset K$ for all $m \in \mathbb{N}$ and $\operatorname{supp} \varphi \subset K$.
- For all $k \in \mathbb{N}_0$, we have $\lim_{m\to\infty} \|\varphi_m - \varphi\|_{C^k(K)} = 0$.

(ii) The sequence $(u_m)_m$ converges to $u \in \mathscr{D}'(\Omega)$ if

$$\lim_{m\to\infty} \langle u_m, \varphi \rangle = \langle u, \varphi \rangle \quad \forall \varphi \in \mathscr{D}(\Omega).$$

Now, we are able to define the derivative in the sense of distribution.

Definition 2.1.10 Let $u \in \mathscr{D}'(\Omega)$ and let $j \in \{1, \dots, n\}$. The partial derivative of u with respect to x_j, denoted $\frac{\partial u}{\partial x_j}$, is the distribution defined by

$$\left\langle \frac{\partial u}{\partial x_j}, \varphi \right\rangle = -\left\langle u, \frac{\partial \varphi}{\partial x_j} \right\rangle \quad \forall \varphi \in \mathscr{D}(\Omega).$$

Higher-order partial derivatives are defined recursively, and hence, for all $\alpha \in \mathbb{N}_0^n$,

$$\langle \partial^\alpha u, \varphi \rangle = (-1)^{|\alpha|} \langle u, \partial^\alpha \varphi \rangle \quad \forall \varphi \in \mathscr{D}(\Omega).$$

Example 2.7 Consider the Heaviside function H on $\mathbb{R}$ defined by

$$H(x) = \begin{cases} 1, \text{ if } x \geq 0, \\ 0, \text{ if } x < 0 \end{cases}.$$

We have

$$\left\langle \frac{\mathrm{d}H}{\mathrm{d}x}, \varphi \right\rangle = -\left\langle H, \frac{\mathrm{d}\varphi}{\mathrm{d}x} \right\rangle = -\int_0^\infty \frac{\mathrm{d}\varphi}{\mathrm{d}x}\, dx = \varphi(0) \quad \forall \varphi \in \mathscr{D}(\mathbb{R}).$$

Hence, $\partial^1 H = \frac{\mathrm{d}H}{\mathrm{d}x} = \delta_0$ in $\mathscr{D}'(\mathbb{R})$.

Note that we can define distributions with values in $\mathbb{C}^m (m \in \mathbb{N})$, and spaces of such distributions are defined componentwise. The notation $\mathscr{D}'(\Omega, \mathbb{C}^m)$ is used for such distributions. Thereby, we can define differential operators

$$\operatorname{div} : \mathscr{D}'\left(\Omega; \mathbb{R}^n\right) \to \mathscr{D}'(\Omega),$$

and

$$\nabla : \mathscr{D}'(\Omega) \to \mathscr{D}'\left(\Omega; \mathbb{R}^n\right)$$

by

$$\operatorname{div} v = \frac{\partial v_1}{\partial x_1} + \cdots + \frac{\partial v_n}{\partial x_n}$$

and

$$\nabla \psi = \left(\frac{\partial \psi}{\partial x_1}, \dots, \frac{\partial \psi}{\partial x_n} \right).$$

We also denote $\Delta = \text{div}\, \nabla$, which is called the Laplacian. To be more precise, for every distribution $\psi \in \mathscr{D}'(\Omega)$, we have

$$\langle \Delta\psi, \varphi \rangle = -\langle \nabla\psi, \nabla\varphi \rangle = \langle \psi, \Delta\varphi \rangle \quad \forall \varphi \in \mathscr{D}(\Omega).$$

Lastly, the Sobolev space $W^{m,p}(\Omega)$ is formed by the distributions $f \in \mathscr{D}'(\Omega)$ having the property that $\partial^\alpha f \in L^p(\Omega)$ for every $\alpha \in \mathbb{N}_0^n$ with $|\alpha| \leqslant m$.

Thus, $H^m(\Omega)$ is formed by the distributions $f \in \mathscr{D}'(\Omega)$ such that $\partial^\alpha f \in L^2(\Omega)$ for every $\alpha \in \mathbb{N}_0^n$ with $|\alpha| \leqslant m$. Moreover, $H^m(\Omega)$ is a Hilbert space with the scalar product

$$\langle y, z \rangle_m = \sum_{|\alpha| \leqslant m} \int_\Omega (\partial^\alpha y)\, \overline{(\partial^\alpha z)}\, dx \quad \forall y, z \in H^m(\Omega).$$

2.1.1.3 Further Discussion on Sobolev Spaces

The Space $W_0^{m,p}(\Omega)$

Definition 2.1.11 Let $1 \leq p < \infty$; $W_0^{m,p}(\Omega)$ denote the closure of $\mathscr{D}(\Omega)$ in $W^{m,p}(\Omega)$.

We denote

$$H_0^m(\Omega) = W_0^{m,2}(\Omega).$$

The space $W_0^{m,p}(\Omega)$, equipped with the $W^{m,p}(\Omega)$ norm, is a separable Banach space. Moreover, $H_0^m(\Omega)$, equipped with the $H^m(\Omega)$ scalar product, is a Hilbert space.

One can roughly interpret $W_0^{m,p}(\Omega)$ as the set of functions $y \in W^{m,p}(\Omega)$ such that $\partial^\alpha y = 0$ on the boundary $\Gamma = \partial\Omega$, for all $|\alpha| \leq m-1$. This point will be made clearer in the following result. First, we need to define smooth domains:

Definition 2.1.12 We say that an open bounded set Ω is of class C^k if for every $x^0 \in \partial\Omega = \Gamma$, there exist $r > 0$ and a C^k function $\gamma : \mathbb{R}^n \to \mathbb{R}$ such that

$$\Omega \cap B(x^0, r) = \left\{ x \in B(x^0, r);\ x_n > \gamma(x_1, x_2, \cdots, , x_{n-1}) \right\}.$$

Theorem 2.1.1 *Suppose that Ω is of class C^1. Let*

$$u \in W^{1,p}(\Omega) \cap C(\bar{\Omega}) \text{ with } 1 \leq p < \infty.$$

Then the following properties are equivalent:

(i) $u = 0$ *on* Γ.
(ii) $u \in W_0^{1,p}(\Omega)$.

The Dual Space of $W_0^{1,p}(\Omega)$

Let $p \in [1, \infty]$. The conjugate exponent of p will be denoted by p', that is, $\frac{1}{p} + \frac{1}{p'} = 1$.

The dual space of $W_0^{1,p}(\Omega)$, $1 \leq p < \infty$ is denoted by $W^{-1,p'}(\Omega)$, while $H^{-1}(\Omega)$ is the dual of $H_0^1(\Omega)$.

It is commonly admitted to identify the dual of $L^2(\Omega)$ with itself, but we do not identify $H_0^1(\Omega)$ with its dual. We have the inclusions

$$H_0^1(\Omega) \subset L^2(\Omega) \subset H^{-1}(\Omega),$$

where these injections are continuous and dense.

As Ω is bounded, we have

$$W_0^{1,p}(\Omega) \subset L^2(\Omega) \subset W^{-1,p'}(\Omega) \quad \text{if} \quad 2n/(n+2) \leq p < \infty$$

with continuous and dense injections.

The elements of $W^{-1,p'}$ are fully described as follows:

Let $F \in W^{-1,p'}(\Omega)$. Then there exist functions $f_0, f_1, f_2, \ldots, f_n \in L^{p'}(\Omega)$ such that

$$\langle F, v\rangle = \int_\Omega (f_0 v)\, dx + \sum_{i=1}^{n} \int_\Omega (f_i \frac{\partial v}{\partial x_i})\, dx, \quad \forall v \in W_0^{1,p}(\Omega)$$

and

$$\|F\| = \max_{0 \leq i \leq n} \|f_i\|_{p'}.$$

If Ω is bounded, we can take $f_0 = 0$.

We end this part by recalling Green's identity:

$$\int_\Omega \Delta y z\, dx = \int_\Gamma \frac{\partial y}{\partial n} z\, d\sigma - \int_\Omega \nabla y \cdot \nabla z\, dx, \text{ for all } y, z \in H^2(\Omega). \tag{2.1.1}$$

Some Useful Sobolev Embedding Results and Inequalities

Theorem 2.1.2 (Rellich-Kondrachov Embedding) *Suppose that Ω is bounded and of class C^1. Then we have the following compact injections:*

$$\begin{array}{lll} W^{1,p}(\Omega) \subset L^q(\Omega) & \forall q \in [1, \tilde{p}), \quad \textit{where } \frac{1}{\tilde{p}} = \frac{1}{p} - \frac{1}{n}, & \textit{if } p < n, \\ W^{1,p}(\Omega) \subset L^q(\Omega) & \forall q \in [p, +\infty), & \textit{if } p = n, \\ W^{1,p}(\Omega) \subset C(\bar{\Omega}), & & \textit{if } p > n. \end{array}$$

In particular, $W^{1,p}(\Omega) \subset L^p(\Omega)$ with compact injection for all p and n.

A more general Sobolev embedding is given below.

Theorem 2.1.3 *Suppose that* Ω *is bounded, open, and of class* C^1. *Let* $y \in W^{m,p}(\Omega)$. *If* $m < \frac{n}{p}$, *then* $y \in L^q(\Omega)$, *for any* q *satisfying* $\frac{1}{q} = \frac{1}{p} - \frac{k}{n}$. *In this case, there exists a positive constant* C *depending on* m, p, n *and* Ω *such that the following estimate holds*

$$\|y\|_{L^q(\Omega)} \leq C\|y\|_{W^{m,p}(\Omega)}.$$

Theorem 2.1.4 (Gagliardo-Nirenberg Interpolation Inequalities)

- *Let* $u \in L^p(\Omega) \cap W^{2,r}(\Omega)$ *with* $1 \leq p \leq \infty$ *and* $1 \leq r \leq \infty$. *Then* $u \in W^{1,q}(\Omega)$, *where* q *is the harmonic mean of* p *and* r, *i.e.,* $\frac{1}{q} = \frac{1}{2}\left(\frac{1}{p} + \frac{1}{r}\right)$, *and*

$$\|\partial^1 u\|_{L^q(\Omega)} \leq C\|u\|^{1/2}_{W^{2,r}(\Omega)}\|u\|^{1/2}_{L^p(\Omega)}.$$

If $p = \infty$, *and thus* $q = 2r$, *we have*

$$\|\partial^1 u\|_{L^q(\Omega)} \leq C\|u\|^{1/2}_{W^{2,r}(\Omega)}\|u\|^{1/2}_{L^\infty(\Omega)}.$$

If $p = q = r$, *then it holds*

$$\|\partial^1 u\|_{L^p(\Omega)} \leq C\|u\|^{1/2}_{W^{2,p}(\Omega)}\|u\|^{1/2}_{L^p(\Omega)},$$

from which one deduces in particular that

$$\|\partial^u\|_{L^p(\Omega)} \leq \varepsilon \left\|\partial^2 u\right\|_{L^p(\Omega)} + C_\varepsilon\|u\|_{L^p(\Omega)} \quad \forall \varepsilon > 0.$$

- *Let* $1 \leq q \leq p < \infty$. *Then*

$$\|u\|_{L^p(\Omega)} \leq C\|u\|^{1-a}_{L^q(\Omega)}\|u\|^a_{W^{1,n}(\Omega)} \quad \forall u \in W^{1,n}(\Omega), \text{ where } a = 1 - (q/p).$$

We note a particular case that is frequently used.

$$n = 2, \quad p = 4, \quad q = 2, \quad \text{and} \quad a = 1/2,$$

that is to say,

$$\|u\|_{L^4(\Omega)} \leq C\|u\|^{1/2}_{L^2(\Omega)}\|u\|^{1/2}_{H^1(\Omega)}, \quad \forall u \in H^1(\Omega).$$

- *We also have*

$$\|u\|_{L^p(\Omega)} \leq \|u\|^{1-a}_{L^q(\Omega)}\|u\|^a_{L^\infty(\Omega)}, \quad \text{with} \quad a = 1 - (q/p).$$

- *Let* $1 \leq q \leq p \leq \infty$ *and* $r > n$. *Then*

$$\|u\|_{L^p(\Omega)} \leq C\|u\|_{L^q(\Omega)}^{1-a}\|u\|_{W^{1,r}(\Omega)}^{a} \quad \forall u \in W^{1,r}(\Omega),$$

where $a = \left(\frac{1}{q} - \frac{1}{p}\right)\left(\frac{1}{q} + \frac{1}{n} - \frac{1}{r}\right)$.

Lastly, the following inequalities will be frequently used:

- **(Young's Inequality)** Let $1 < p, q < \infty$ such that $\frac{1}{p} + \frac{1}{q} = 1$. Then

$$ab \leq \frac{a^p}{p} + \frac{b^q}{q}, \qquad \text{for } a, b \geq 0. \tag{2.1.2}$$

- Given x and y in an inner product space $\mathscr{X}$, we have:
  (i) **(Cauchy-Schwarz Inequality)**

$$|\langle x, y\rangle| \leq \|x\| \, \|y\| \, .$$

  (ii) **(Triangle Inequality)**

$$\|x + y\| \leq\| x \| + \| y \| \, .$$

- **(Hölder's Inequality)** Let $p \in [1, \infty]$. If $f \in L^p(\Omega)$ and $g \in L^{p'}(\Omega)$, then $fg \in L^1(\Omega)$ and

$$\int_\Omega |f(x)| \, |g(x)| \, dx \leq \|f\|_p \|g\|_{p'}.$$

- **(Wirtinger's Inequality)** [115, estimate (7.6.1) in p. 182]
  Let $y \in H^2(0, L)$ such that $y_x(0) = 0$. Then

$$\int_0^L y_x^2 \, dx \leq \frac{4L^2}{\pi^2} \int_0^L y_{xx}^2 \, dx.$$

- **(Poincaré's Inequality)**
  Suppose that $p \in [1, \infty)$ and Ω is a bounded open set of $\mathbb{R}^n$. Then there exists a constant $C > 0$ (depending on p and Ω) such that

$$\|y\|_{L^p(\Omega)} \leq C\|\nabla y\|_{L^p(\Omega)}, \quad \forall y \in W_0^{1,p}(\Omega).$$

Thereby, $\|\nabla y\|_{L^p(\Omega)}$ is a norm on $W_0^{1,p}(\Omega)$ and is equivalent to the usual norm $\|y\|_{W^{1,p}(\Omega)}$.

- **(Generalized Poincaré's Inequality)**
  There exists a constant $C > 0$ (depending on p and Ω) such that

$$\|y\|_{L^p(\Omega)} \leq C\|\nabla y\|_{L^p(\Omega)} + \left|\int_\Omega y\,dx\right|, \quad \forall y \in W^{1,p}(\Omega).$$

2.1.2 Linear Operators

Definition 2.1.13 A mapping from a normed space $\mathscr{X}$ into a normed space $\mathscr{Y}$ is called an *operator* .

Now, consider an operator $\mathscr{A}$, with domain $\mathscr{D}(\mathscr{A})$, where $\mathscr{D}(\mathscr{A})$ is assumed to be a subspace of $\mathscr{X}$. Then $\mathscr{A}$ is said to be *linear* if $\mathscr{A}(\alpha x + \beta y) = \alpha\mathscr{A}x + \beta\mathscr{A}y$, for all $x, y \in \mathscr{D}(\mathscr{A})$, and for all $\alpha,\ \beta \in \mathbb{F}$.

It is frequent to invoke a special type of norm when dealing with linear operators. In fact, if $\mathscr{A} : \mathscr{D}(\mathscr{A}) \subset \mathscr{X} \to \mathscr{X}$ is a linear operator, the *graph norm* denoted by $||\ ||_{\mathscr{D}(\mathscr{A})}$ is defined by

$$||x||_{\mathscr{D}(\mathscr{A})} = ||x|| + ||\mathscr{A}x||. \tag{2.1.3}$$

Definition 2.1.14 Let $\mathscr{X}$ and $\mathscr{Y}$ be two normed spaces and consider a linear operator $\mathscr{A} : \mathscr{D}(\mathscr{A}) \subseteq \mathscr{X} \longrightarrow Y$. Then $\mathscr{A}$ is *bounded* if $\mathscr{D}(\mathscr{A}) = \mathscr{X}$, and there exists a constant $C > 0$ such that

$$\|\mathscr{A}x\|_{\mathscr{Y}} \leq C\|x\|_{\mathscr{X}}, \quad \text{for all } x \in \mathscr{D}(\mathscr{A}) = \mathscr{X}.$$

Otherwise, the operator $\mathscr{A}$ is called unbounded.

Example 2.8

- Let $G = [0, 1]$, and consider the Banach space $C(G)$ endowed with the sup-norm $\|f\| = \sup_{x\in[0,1]} |f(x)|$ (see the beginning of Sect. 2.1.1). Then, let the operator $T : C(G) \to C(G)$ defined by $(Tf)(x) = \int_0^x f(s)\,ds$.
  T is linear and bounded with norm 1.
- Consider the operator $T : \mathscr{X} = C^1[0, 1] \to \mathscr{Y} = C[0, 1]$ such that $Tf(t) = f'(t)$, $t \in [0, 1]$ (the norm on X and Y is the sup- norm). Then T is linear but unbounded. The latter is due to the following counterexample: For $n \in \mathbb{N}$, we have $f_n(t) = t^n$, $t \in [0, 1]$. Thereby, $f_n \in \mathscr{X}$ and $Tf_n = nf_{n-1}$, for $n > 1$. Moreover $\|f_n\| = 1$, but $\|Tf_n\| = n$.

Let $\mathscr{L}(\mathscr{X},\mathscr{Y})$ be the set of linear bounded operators from $\mathscr{X}$ to $\mathscr{Y}$ endowed with the following norm:

$$\|\mathscr{B}\| = \sup_{x\neq 0}\frac{\|\mathscr{B}x\|}{\|x\|}, \qquad \text{for all } \mathscr{B} \in \mathscr{L}(\mathscr{X},\mathscr{Y}).$$

Then, $\mathscr{L}(\mathscr{X},\mathscr{Y})$ is a normed space and also a Banach space in the case where $\mathscr{Y}$ is a Banach space [40].

If $Y = \mathbb{F}$ (recall that $\mathbb{F} = \mathbb{R}$ or $\mathbb{C}$), then $\mathscr{L}(\mathscr{X},\mathbb{F})$ is called *dual space* of $\mathscr{X}$ and is denoted by $\mathscr{X}^* = \mathscr{L}(\mathscr{X},\mathbb{F})$.

Any element f of $\mathscr{L}(\mathscr{X},\mathbb{F})$ is called a *linear bounded functional* on $\mathscr{X}$.

Definition 2.1.15 Let $\mathscr{X}$ and $\mathscr{Y}$ be Banach spaces and $\mathscr{D}(\mathscr{A}) \subseteq \mathscr{X}$. Let $\mathscr{A} : \mathscr{D}(\mathscr{A}) \longrightarrow \mathscr{Y}$ be a linear operator:

- $\mathscr{A}$ is *densely defined* if $\overline{\mathscr{D}(\mathscr{A})} = \mathscr{X}$.
- $\mathscr{A}$ is *closed* if the graph of $\mathscr{A}$

$$\mathscr{G} = \{(x,y) \in \mathscr{X} \times \mathscr{Y};\ x \in \mathscr{D}(\mathscr{A}),\ y = \mathscr{A}x\}$$

is closed in $\mathscr{X} \times \mathscr{Y}$; in other words, $\mathscr{A}$ is closed if for any sequence (x_n) in $\mathscr{D}(\mathscr{A})$ such that

$$x_n \in \mathscr{D}(\mathscr{A}),\ x_n \to x \ \text{ and } \ \mathscr{A}x_n \to y,\ \text{ as } n \to \infty,$$

then we have

$$x \in \mathscr{D}(\mathscr{A}) \ \text{ and } \ \mathscr{A}x = y.$$

Now, let us recall the following result [40]:

Theorem 2.1.5 *A closed linear operator defined on all of a Banach space $\mathscr{X}$ on a Banach space $\mathscr{Y}$ must be bounded.*

Definition 2.1.16 Let $\mathscr{X}$ and $\mathscr{Y}$ be Banach spaces and $\mathscr{A} : \mathscr{X} \to \mathscr{Y}$ a linear operator. $\mathscr{A}$ is said to be *compact* if for every bounded subset B of $\mathscr{X}$, the image $\mathscr{A}(B)$ is relatively compact, that is, $\overline{\mathscr{A}(B)}$ is compact. In other words, whenever (x_n) is a bounded sequence in $\mathscr{X}$, then $(\mathscr{A}x_n)$ has a convergent subsequence in $\mathscr{Y}$.

2.1.3 Hilbert Adjoint Operators

Definition 2.1.17 Let $\mathscr{A} : \mathscr{D}(\mathscr{A}) \subseteq \mathscr{H} \longrightarrow \mathscr{H}$ be a densely defined linear operator on a Hilbert space $\mathscr{H}$. Then, the *adjoint operator* maps as follows:

$$\mathscr{A}^* : \mathscr{D}(\mathscr{A}^*) \longrightarrow \mathscr{H}$$

and

$$\mathscr{D}(\mathscr{A}^*) = \left\{y \in \mathscr{H};\ \exists\, y^* \in \mathscr{H} \text{ satisfying } \left\langle \mathscr{A}x, y\right\rangle = \langle x, y^*\rangle,\ \forall x \in \mathscr{D}(\mathscr{A})\right\}.$$

In this case, $\mathscr{A}^* y = y^*$.

Definition 2.1.18 A densely defined linear operator $\mathscr{A}$ is *self-adjoint* if $\mathscr{A} = \mathscr{A}^*$, that is, $\mathscr{D}(\mathscr{A}) = \mathscr{D}(\mathscr{A}^*)$ and $\mathscr{A}x = \mathscr{A}^* x$ for all $x \in \mathscr{D}(A)$.

A densely defined linear operator $\mathscr{A}$ is *skew adjoint* in a Hilbert space $\mathscr{H}$ if $\mathscr{A}^* = -\mathscr{A}$.

Definition 2.1.19 (Bilinear Form)

(1) Given a Hilbert space $\mathscr{H}$, consider a mapping $a : \mathscr{H} \times \mathscr{H} \to \mathbb{C}$. Then a is called a bilinear form if it is linear in the first argument and conjugate linear in the second one.
(2) A bilinear form $a : \mathscr{H} \times \mathscr{H} \to \mathbb{R}$ is said to be:

   (i) Continuous if there exists a positive constant C such that

$$|a(u, v)| \leq C\|u\|\|v\|, \quad \forall u, v \in H$$

   (ii) Coercive if there is a constant $K > 0$ such that

$$a(v, v) \geq K\|v\|^2, \quad \forall v \in \mathscr{H}$$

The following result will play a crucial role in the proof of the well-posedness of the problems that we are going to tackle in the next chapters:

Theorem 2.1.6 (Lax-Milgram) *Assume that $a(u, v)$ is a continuous coercive bilinear form on a Hilbert space$\mathscr{H}$. Then given any $\phi \in \mathscr{L}(\mathscr{H})$, there exists a unique element $u \in \mathscr{H}$ such that $a(u, v) = \langle \phi, v\rangle \in \mathscr{H}$.*

Another important result that will be frequently used is:

Theorem 2.1.7 (Banach-Steinhaus Theorem) *Let $|X$ and $\mathscr{Y}$ be two Banach spaces and let $(\mathscr{A}_i)_{i\in I}$ be a family (not necessarily countable) of continuous linear operators from $\mathscr{X}$ into $\mathscr{Y}$. Assume that*

$$\sup_{i\in I} \|\mathscr{A}_i x\|_{\mathscr{Y}} < \infty, \quad \text{for all } x \in \mathscr{X}.$$

Then

$$\sup_{i\in I} \|\mathscr{A}_i\|_{\mathscr{L}(\mathscr{X},\mathscr{Y})} < \infty.$$

2.1.4 Spectrum and Resolvent of Linear Operators

Consider a linear operator $\mathscr{A} : \mathscr{D}(\mathscr{A}) \subseteq \mathscr{X} \longrightarrow \mathscr{X}$, where $\mathscr{X}$ is a Banach space.

Definition 2.1.20 The *resolvent set* of $\mathscr{A}$, denoted by $\rho(\mathscr{A})$, is the set of all complex numbers λ such that $(\lambda I - \mathscr{A})$ is invertible (bijective) from $\mathscr{D}(\mathscr{A})$ to $\mathscr{X}$. In this case, $(\lambda I - \mathscr{A})^{-1}$ is called the *resolvent operator* and will be denoted by $R(\lambda, \mathscr{A})$.

Note that for a closed linear operator $\mathscr{A}$, if $\lambda \in \rho(\mathscr{A})$, then $(\lambda I - \mathscr{A})^{-1}$ is a bounded linear operator on $\mathscr{X}$. This is a direct consequence of the closed graph theorem.

Definition 2.1.21 The *spectrum* of the operator $\mathscr{A}$ is the complement of $\rho(\mathscr{A})$ in the complex plane, that is,

$$\sigma(\mathscr{A}) = \mathbb{C} - \rho(\mathscr{A}) = \{\lambda \in \mathbb{C} | (\lambda I - \mathscr{A}) \text{ is not invertible } \}.$$

Let $\mathscr{X} = \{f \in C[0, 1];\ f(1) = 0\}$ be the Banach space endowed with the sup-norm. Then, consider the operator $\mathscr{A} : \mathscr{D}(\mathscr{A}) = \left\{f \in C^1[0, 1];\ f(1) = f'(1) = 0\right\}$ such that $\mathscr{A} f = f'$ for any $f \in \mathscr{D}(\mathscr{A})$.

Given any $\lambda \in \mathbb{C}$ and $g \in \mathscr{X}$, the equation $(\lambda I - \mathscr{A}) f = g$ has a unique solution $f(x) = \int_x^1 e^{\lambda(x-r)} g(r)\, dr$. Thereby, $\rho(\mathscr{A}) = \mathbb{C}$ and $\sigma(\mathscr{A}) = \emptyset$.

Contrary to finite-dimensional operators, there are three types of spectrum of a linear unbounded operator $\mathscr{A}$, namely:

- Point spectrum (a set of eigenvalues):

$$\sigma_p(\mathscr{A}) = \{\lambda \in \mathbb{C};\ (\lambda I - \mathscr{A}) \text{ is not one to one } \}.$$

- Continuous spectrum:

$$\sigma_c(\mathscr{A}) \\ = \left\{\lambda \in \mathbb{C};\ (\lambda I - \mathscr{A}) \text{ is one to one but not onto and } \overline{\Im m(\lambda I - \mathscr{A})} = \mathscr{X}\right\},$$

  where $\Im m(P)$ denotes the range of P.
- Residual spectrum:

$$\sigma_r(\mathscr{A}) \\ = \left\{\lambda \in \mathbb{C};\ (\lambda I - \mathscr{A}) \text{ is one to one but not onto and } \overline{\Im m(\lambda I - \mathscr{A})} \neq \mathscr{X}\right\}.$$

The three sets are mutually disjoint subsets of $\mathbb{C}$, i.e.,

$$\sigma(A) = \sigma_p(A) \cup \sigma_r(A) \cup \sigma_c(A).$$

Remark 2.1.2 It turns out in numerous systems, the resolvent of a closed operator $\mathscr{A}$ has a compact resolvent $R(\lambda_0, \mathscr{A})$, for some λ_0 (and hence for all λ_0) in $\rho(\mathscr{A})$ (see Remark 2.3.1). For such a case, the spectrum of $\mathscr{A}$ consists entirely of a discrete set of eigenvalues with finite multiplicities [121]. This result will be frequently used.

2.2 Dissipative and m-Dissipative Linear Operators

In this section, $\mathscr{H}$ is a Hilbert space endowed with an inner product $<,>$, whose corresponding norm is $\| \ \|$. In turn, $\mathscr{A}$ is an unbounded linear operator with domain $\mathscr{D}(\mathscr{A}) \subset \mathscr{H}$. The range and kernel of $\mathscr{A}$ are respectively denoted by $\Im m\mathscr{A}$ and Ker $\mathscr{A}$. Finally, given $r \in \mathbb{R}$, $\mathbb{C}_r$ denotes $\{z \in \mathbb{C};\ \Re ez > 0\}$, where $\Re ez$ is the real part of the complex number z.

Definition 2.2.1 The operator A is *dissipative* if $\Re e < \mathscr{A}x, x > \leq 0$, for all x in $\mathscr{D}(\mathscr{A})$.

We have the following characterization of a dissipative operator:

Proposition 2.2.1 *The following statements are equivalent:*

(i) The operator A is dissipative.
(ii) For any x in $\mathscr{D}(\mathscr{A})$ and for any $\lambda \in \mathbb{C}_+ = \{\lambda \in \mathbb{C};\ \Re e\lambda > 0\}$, we have

$$\|(\lambda I - \mathscr{A})x\| \geq \Re e\lambda \|x\|.$$

(iii) For any x in $\mathscr{D}(\mathscr{A})$ and for any $\gamma > 0$, it holds

$$\|(\gamma I - \mathscr{A})x\| \geq \gamma \|x\|.$$

Proof (i) $\Rightarrow$ (ii): Let A be dissipative. Then, consider x in $\mathscr{D}(\mathscr{A})$ and $\lambda \in \mathbb{C}_+$. Then, a simple computation together with Definition 2.2.1 gives

$$\|(\lambda I - \mathscr{A})x\|^2 = |\lambda|^2\|x\|^2 - 2\Re e\lambda \Re e < \mathscr{A}x, x > \\ + \|\mathscr{A}x\|^2 \geq |\lambda|^2\|x\|^2 \geq (\Re e\lambda)^2 \|x\|^2.$$

(ii) $\Rightarrow$ (iii): This implication is trivial.
(iii) $\Rightarrow$ (i): Let $x \in \mathscr{D}(\mathscr{A})$ and $\gamma > 0$. We have

$$\|(\gamma I - \mathscr{A})x\|^2 = \gamma^2\|x\|^2 - 2\gamma \Re e < \mathscr{A}x, x > + \|\mathscr{A}x\|^2,$$

and hence, thanks to (iii), it follows that

$$\Re e\langle Ax, x\rangle = \frac{1}{2\gamma}\|Ax\|^2 + \frac{\gamma^2\|x\|^2 - \|(\gamma I - A)x\|^2}{2\gamma} \leqslant \frac{1}{2\gamma}\|Ax\|^2.$$

Taking $\gamma \to \infty$, we see that A is dissipative. □

Remark 2.2.1 If $\mathscr{A}$ is dissipative, then, owing to Proposition 2.2.1, we conclude that for each $\gamma > 0$, $\gamma I - \mathscr{A}$ is injective and

$$\left\|(\gamma I - \mathscr{A})^{-1} x\right\| \leq \frac{1}{\gamma}\|x\|,$$

for all x in $\Im m(\gamma I - \mathscr{A})$.

Proposition 2.2.2 *Let $\mathscr{A}$ be a closed and dissipative operator on the Hilbert space $\mathscr{H}$. Then for every $\lambda \in \mathbb{C}_+$, the subspace $\Im m(\lambda I - \mathscr{A})$ is closed in $\mathscr{H}$.*

Proof Let (y_n) be a sequence in $\Im m(\lambda I - \mathscr{A})$ such that $y_n \to y$ in $\mathscr{H}$. Hence, there exists a sequence (x_n) in $\mathscr{D}(\mathscr{A})$ such that

$$\lambda x_n - \mathscr{A}x_n = y_n \quad \text{for all } n \geqslant 1. \tag{2.2.1}$$

In view of Proposition 2.2.1 and the convergence of (y_n), we deduce that (x_n) is a Cauchy sequence in $\mathscr{X}$ and hence converges to some x in $\mathscr{H}$. Additionally, we infer from (2.2.1) that $\mathscr{A}x_n \to \lambda x - y$ in $\mathscr{H}$. Lastly, since $\mathscr{A}$ is closed, we conclude that $x \in \mathscr{D}(\mathscr{A})$ and $\mathscr{A}x = \lambda x - y$, that is, $y \in \Im m(\lambda I - \mathscr{A})$. □

Proposition 2.2.3 *If $\mathscr{A}$ is dissipative and $\Im m(\lambda I - \mathscr{A}) = \mathscr{H}$ for some $\lambda \in \mathbb{C}_+$, then $\mathscr{D}(\mathscr{A})$ is dense in $\mathscr{H}$.*

Proof Let $z \in \Im m(\lambda I - \mathscr{A}) = \mathscr{H}$ such that $< z, x >= 0$, for all $x \in \mathscr{D}(\mathscr{A})$. This implies that there exists $x_0 \in \mathscr{D}(\mathscr{A})$ such that $z = \lambda x_0 - \mathscr{A}x_0$ and $< z, x_0 >= 0$. Thereby, we have

$$\begin{aligned} 0 =& \Re e < z, x_0 >= \Re e < \lambda x_0 - \mathscr{A}x_0, x_0 > \\ &= \Re e\lambda\|x_0\|^2 - \Re e < \mathscr{A}x_0, x_0 >\geq \lambda\|x_0\|^2, \end{aligned}$$

as $\mathscr{A}$ is dissipative. Consequently, $x_0 = 0$ and, hence, $z = 0$, which leads to the desired result. □

Proposition 2.2.4 *Let $\mathscr{A}$ be dissipative. Then the following statements are equivalent:*

(i) $\lambda I - \mathscr{A}$ *is onto* $\mathscr{H}$ *for all* $\lambda > 0$.
(ii) $I - \mathscr{A}$ *is onto* $\mathscr{H}$.
(iii) $\lambda_0 I - \mathscr{A}$ *is onto* $\mathscr{H}$ *for some* $\lambda_0 > 0$.

Proof We shall prove that (ii) $\Rightarrow$ (i). The other implications are either similar or trivial.

By virtue of Remark 2.2.1 and the fact that $\Im m(I - \mathscr{A}) = \mathscr{H}$, it follows that $(I - \mathscr{A})^{-1}$ is a linear bounded operator in $\mathscr{H}$ and

$$\left\| (I - \mathscr{A})^{-1} \right\| \leq 1.$$

The task ahead is to show that given $\lambda > 0$ and $y \in \mathscr{H}$, we seek $x \in \mathscr{D}(\mathscr{A})$ such that

$$(I - \lambda A)x = y, \tag{2.2.2}$$

which takes the form

$$x = (I - \lambda A)^{-1} Bx,$$

where B is the operator with domain $\mathscr{D}(B) = \mathscr{D}(\mathscr{A})$ such that $Bx = \left(\frac{\lambda-1}{\lambda}\right) x + \frac{1}{\lambda} y$. One can readily check that $(I - \lambda A)^{-1} B$ is contractive if $\lambda > 1/2$, and hence, (2.2.2) has a unique solution. Repeating this reasoning, we obtain by induction that (2.2.2) has a unique solution for all $\lambda > 0$. Finally, changing λ to $\frac{1}{\lambda} > 0$, we deduce that (i) holds. □

Definition 2.2.2 A dissipative operator $\mathscr{A}$ is *maximal dissipative (m-dissipative)* if $\mathscr{A}$ is dissipative and one of the assertions of Proposition (2.2.4) holds.

Note that in light of Proposition (2.2.4), a dissipative operator $\mathscr{A}$ is m-dissipative if and only if all the assertions of Proposition (2.2.4) hold.

Theorem 2.2.1 *If $\mathscr{A}$ is m-dissipative, then we have:*

1. *$\mathscr{D}(\mathscr{A})$ is dense in $\mathscr{H}$.*
2. *For all $\lambda > 0$, the operator $\lambda I - \mathscr{A}$ is an isomorphism from $\mathscr{D}(\mathscr{A})$ onto $\mathscr{H}$. Furthermore, $R(\lambda, \mathscr{A})$ is a linear bounded operator that satisfies*

$$\lambda \left\| R(\lambda, \mathscr{A}) \right\|_{\mathscr{L}(\mathscr{H})} \leq 1. \tag{2.2.3}$$

3. *$\mathscr{A}$ is closed.*

Proof The first assertion follows from Proposition 2.2.3, while the second is a consequence of Proposition 2.2.4 and Remark 2.2.1. With regard to the third assertion, let (x_n) be a sequence in $\mathscr{D}(\mathscr{A})$ such that $x_n \to x$ and $\mathscr{A}x_n \to y$ in $\mathscr{H}$. Since $I - \mathscr{A}$ has a bounded inverse, we obtain $x_n = (I - \mathscr{A})^{-1} (x_n - \mathscr{A}x_n) \to (I - \mathscr{A})^{-1} (x - y)$. We infer from the uniqueness of limit that $x = (I - \mathscr{A})^{-1} (x - y)$, that is, $x \in \mathscr{D}(\mathscr{A})$ and $\mathscr{A}x = y$. □

2.3 Semigroups of Operators and Linear Evolution Equations

This section is consecrated to a short introduction to semigroups theory of linear operators. This topic is vast and has miscellaneous applications. In turn, we shall focus on the basic concepts and properties related to semigroups theory. The target is to prove two important results known in the literature as Hill-Yosida and Lumer-Phillips theorems. Furthermore, semigroups theory is used to solve abstract Cauchy problems (linear evolution equations in Hilbert spaces). This establishes a link between semigroups theory and linear evolution equations. Last but not least, the concept of stability of semigroups is discussed from different perspectives as well as the characterization of this concept.

2.3.1 Semigroups of Operators and Linear Evolution Equations

Definition 2.3.1 A one parameter family $(S(t))_{t\geq 0}$ of $\mathscr{L}(\mathscr{H})$ is a *semigroup of bounded linear operators* on $\mathscr{H}$ if:

(i) $S(0) = Id_{\mathscr{H}}$.
(ii) $S(t+s) = S(t)S(s)$, for all $t, s \geq 0$.

If, in addition to these two properties, $(S(t))_{t\geq 0}$ satisfies

$$\lim_{t\to 0^+} S(t)x = x, \text{ for all } x \in \mathscr{H}, \tag{2.3.1}$$

then the semigroup is said to be *strongly continuous or simply a C_0-semigroup* .

This definition is mainly motivated by the fact that the family $(S(t))_{t\geq 0}$ provides a complete idea about the time evolution of the system in the following sense: If x_0 is the initial state of the process at time $t = 0$, then $S(t)x_0$ is its state at time $t \geq 0$.

Furthermore, a typical example of a C_0-semigroup is

$$S(t) = e^{tM} = \sum_{n=0}^{\infty} \frac{(tM)^n}{n!},$$

for $M \in \mathscr{L}(\mathscr{H})$. Indeed, it is easy to check that the above series converges in $\mathscr{L}(\mathscr{H})$ and satisfies the definition. Actually, it defines a uniformly continuous as $\lim_{t\to 0} \|S(t) - I\| = 0$. This comes from the inequality $\|S(t) - I\| \leq t\|A\|e^{t\|A\|}$. It is also noteworthy that this example is a natural extension of the exponential of a square $n \times n$ matrix M when the system acts in an n-dimensional space. This motivates the following definition:

Definition 2.3.2 The *generator of the semigroup* is the operator $\mathscr{A} : \mathscr{D}(|A) \to \mathscr{H}$ defined by

$$\mathscr{D}(\mathscr{A}) = \left\{ x \in \mathscr{H};\ \lim_{t \to 0^+} \frac{S(t)x - x}{t} \text{ exists} \right\}$$

and

$$\mathscr{A}x = \lim_{t \to 0^+} \frac{S(t)x - x}{t}, \quad \text{for all } x \in \mathscr{D}(\mathscr{A}).$$

In this case, $\mathscr{D}(\mathscr{A})$ is called the domain of $\mathscr{A}$.

It is worth mentioning that according to the above definition, the generator of a semigroup of linear operators is also linear.

One particular but frequent type of semigroup is the contractive one:

Definition 2.3.3 If a C_0-semigroup satisfies $\|S(t)\|_{\mathscr{L}(\mathscr{H})} \leq 1$, for all $t \geq 0$, then it is called a C_0-semigroup of *contractions* .

Now, we can state the main properties of contractive C_0-semigroups:

Theorem 2.3.1 *Let $\mathscr{A}$ be a generator of a C_0-semigroup $(S(t))_{t \geq 0}$ of contractions on $\mathscr{H}$. It holds:*

(i) For all $x \in \mathscr{H}$, the mapping $t \to S(t)x$ is continuous from $[0, +\infty)$ into $\mathscr{H}$.
(ii) For all $x \in \mathscr{H}$ and all $t \geq 0$, we have

$$\lim_{\tau \to 0} \frac{1}{\tau} \int_t^{t+\tau} S(r)x\, dr = S(t)x. \tag{2.3.2}$$

(iii) For all $x \in \mathscr{H}$ and all $t > 0$, the element $\int_0^t S(r)x\, dr$ belongs to $\mathscr{D}(\mathscr{A})$ and

$$\mathscr{A}\left(\int_0^t S(r)x\, dr \right) = S(t)x - x. \tag{2.3.3}$$

(iv) For all $x \in \mathscr{D}(\mathscr{A})$, and all $t > 0$, we have that $S(t)x \in \mathscr{D}(\mathscr{A})$, and the mapping $t \to S(t)x$ is continuous differentiable from $(0, +\infty)$ into $\mathscr{H}$. Moreover, we have

$$\frac{d}{dt} S(t)x = \mathscr{A} S(t)x = S(t)\mathscr{A}x, \ \textit{for all } t > 0, \tag{2.3.4}$$

and

$$S(t)x - S(s)x = \int_s^t S(r)\mathscr{A}x\,dr = \int_s^t \mathscr{A}S(r)x\,dr, \text{ for all } t > s \geq 0.$$

(v) *The generator* $\mathscr{A}$ *is a closed linear operator and densely defined in* $\mathscr{H}$.

Proof

(i) Obviously, the property holds at $t = 0$ by (2.3.1). Now, let $t > 0$ and $x \in \mathscr{H}$. Then for $\tau \geq 0$, we have thanks to the contraction property

$$\|S(t+\tau)x - S(t)x\|_{\mathscr{H}} = \|S(t)(S(\tau)x - x)\|_{\mathscr{H}} \leq \|S(\tau)x - x\|_{\mathscr{H}},$$

and the continuity of $t \to S(t)x$ is verified. Now, picking up $\tau < 0$ such that $t + \tau > 0$, one can use once again the contraction property to write

$$\begin{aligned}\|S(t+\tau)x - S(t)x\|_{\mathscr{H}} &= \|S(t+\tau)(x - S(-\tau)x)\|_{\mathscr{H}} \leq \|S(-\tau)x - x\|_{\mathscr{H}} \\ &\leq \|S(-\tau)x - x\|_{\mathscr{H}}.\end{aligned}$$

This gives the desired property.

(ii) This is a simple consequence of (i). Indeed, $\frac{1}{\tau}\int_t^{t+\tau} S(r)x\,dr = \frac{1}{\tau}\int_0^\tau S(t+s)x\,ds = \frac{1}{\tau}\int_0^\tau S(s)(S(t)x)\,ds$, which is the average of the mapping $s \to S(s)(S(t)x)$ over $[0, \tau]$.

(iii) Let $x \in \mathscr{H}$ and $\tau > 0$. We have

$$\begin{aligned}\frac{S(\tau) - I}{\tau}\int_0^t S(r)x\,dr &= \frac{1}{\tau}\int_0^t (S(r+\tau)x - S(r)x)\,dr \\ &= \frac{1}{\tau}\int_t^{t+\tau} S(r)x\,dr - \frac{1}{\tau}\int_0^\tau S(r)x\,dr.\end{aligned}$$

In light of (2.3.3), the right-hand side goes to $S(t)x - x$ as τ approaches 0, and the desired results follow from Definition 2.3.2.

(iv) Taking $x \in \mathscr{D}(\mathscr{A})$, $t > 0$ and $\tau > 0$. Using the semigroup properties along with the definition of $\mathscr{A}$, it follows that

$$\lim_{\tau \to 0+} \frac{S(\tau) - I}{\tau} S(t)x = S(t) \lim_{\tau \to 0+} \left(\frac{S(\tau) - I}{\tau}\right) x = S(t)\mathscr{A}x.$$

Whereupon $S(t)x$ belongs to $\mathscr{D}(\mathscr{A})$ and $\mathscr{A}S(t)x = S(t)\mathscr{A}x$. Additionally, the right derivative of $S(t)x$ exists with

$$\frac{d^+}{dt}S(t)x = \mathscr{A}S(t)x = S(t)\mathscr{A}x.$$

For the left derivative, let $0 < \tau < t$, and write

$$\frac{S(t)x - S(t-\tau)x}{\tau} - S(t)\mathscr{A}x = S(t-\tau)\left(\frac{S(\tau)x - x}{\tau} - \mathscr{A}x\right) + (S(t-\tau)\mathscr{A}x - S(t)\mathscr{A}x).$$

Now, invoking the contraction property of the semigroup yields

$$\left\| \frac{S(t)x - S(t-\tau)x}{\tau} - S(t)\mathscr{A}x \right\|_{\mathscr{H}} \leq \left\| \frac{S(\tau)x - x}{\tau} - \mathscr{A}x \right\|_{\mathscr{H}} + \|S(t-\tau)\mathscr{A}x - S(t)\mathscr{A}x\|_{\mathscr{H}}.$$

Letting τ go to zero, the proof of (iv) is achieved after using (i) and the fact that $x \in \mathscr{D}(\mathscr{A})$. Finally, the last identity can be established by simply integrating (2.3.4) from s to t.

(v) As mentioned earlier, the operator $\mathscr{A}$ is linear and so let us show that $\mathscr{A}$ is closed. To this end, take a sequence $(x_n) \subset \mathscr{D}(\mathscr{A})$, for $n \in \mathbb{N}$ such that

$$x_n \to x \text{ and } \mathscr{A}x_n \to y \text{ in } \mathscr{H}, \quad \text{as } n \to \infty.$$

In view of (iv), it follows that for any $t > 0$:

$$S(t)x_n - x_n = \int_0^t S(r)\mathscr{A}x_n \, dr.$$

Thereafter, recalling the uniform convergence of $S(\cdot)\mathscr{A}x_n$ on $[0, t]$ for $n \to \infty$, the last identity yields

$$S(t)x - x = \int_0^t S(r)y \, dr.$$

Consequently,

$$\lim_{t\to 0} \frac{S(t)x - x}{t} = \lim_{t\to 0}\left(\frac{1}{t}\int_0^t S(u)y \, du\right) = y,$$

by means of (ii). This implies that $x \in \mathscr{D}(\mathscr{A})$, $\mathscr{A}x = y$, and so $\mathscr{A}$ is closed.

Concerning the density of $\mathscr{D}(\mathscr{A})$, let $x \in \mathscr{H}$ and $t > 0$. In view of (ii) and (iii), the element $\frac{1}{t}\int_0^t S(r)x \, dr$ belongs to $\mathscr{D}(\mathscr{A})$ and converges to x in $\mathscr{H}$. Therefore, $\mathscr{D}(\mathscr{A})$ is dense in $\mathscr{H}$. □

Subsequently, we introduce the so-called Yosida approximation, which is needed to "approach" unbounded operators by bounded ones. First, recall that $R(\lambda, \mathscr{A}) = (\lambda I - \mathscr{A})^{-1}$ is the resolvent operator (see Definition 2.1.20).

Definition 2.3.4 For every $\lambda > 0$, the Yosida approximation of $\mathscr{A}$ is defined by

$$\mathscr{A}_\lambda = \lambda \mathscr{A} R(\lambda, \mathscr{A}) = \lambda R(\lambda, \mathscr{A})\mathscr{A} = \lambda(\lambda R(\lambda, \mathscr{A}) - I). \tag{2.3.5}$$

Notice that the operator $\mathscr{A}_\lambda$ is a bounded linear operator from $\mathscr{H}$ into itself. Thereby, it generates a uniformly continuous semigroup denoted by $\left(e^{t\mathscr{A}_\lambda}\right)_{t\geq 0}$ on $\mathscr{H}$.

The next result reflects the meaning of approximation in the previous definition:

Theorem 2.3.2 *Suppose that $\mathscr{A}$ is a densely defined linear operator satisfying* (2.2.3). *Then, we have*

$$\lim_{\lambda\to\infty} \lambda R(\lambda, \mathscr{A})x = x, \forall x \in \mathscr{H}. \tag{2.3.6}$$

Moreover, it holds

$$\lim_{\lambda\to\infty} \mathscr{A}_\lambda x = \mathscr{A} x, \forall x \in \mathscr{D}(\mathscr{A}). \tag{2.3.7}$$

Finally, the uniformly continuous semigroup of contractions $\left(e^{t\mathscr{A}_\lambda}\right)_{t\geq 0}$ on $\mathscr{H}$ generated by the Yosida approximation $\mathscr{A}_\lambda$ satisfies

$$\left\| e^{t\mathscr{A}_\lambda} x - e^{t\mathscr{A}_\mu} x \right\|_{\mathscr{H}} \leq t \left\| \mathscr{A}_\lambda x - \mathscr{A}_\mu x \right\|_{\mathscr{H}}, \forall x \in \mathscr{H}, \lambda, \mu > 0. \tag{2.3.8}$$

Proof Thanks to a density argument together with $\|R(\lambda, \mathscr{A})\| \leq 1$ (see (2.2.3)), it suffices to show (2.3.6) for all x in $\mathscr{D}(\mathscr{A})$. In this case, we can write

$$\lambda R(\lambda, \mathscr{A})x - x = R(\lambda, \mathscr{A})\mathscr{A} x,$$

and hence,

$$\|\lambda R(\lambda, \mathscr{A})x - x\|_{\mathscr{H}} \leq \frac{1}{\lambda}\|\mathscr{A} x\|_{\mathscr{H}},$$

by means of (2.2.3). This leads to the desired result (2.3.6). In turn, combining the latter with (2.3.5) leads to (2.3.7).

By virtue of the discussion that follows Definition 2.3.4, we just need to show (2.3.8). First, (2.3.5) along with (2.2.3) gives

$$\left\| e^{t\mathscr{A}_\lambda} \right\|_{\mathscr{L}(\mathscr{H})} = e^{-\lambda t} \left\| e^{t\lambda^2 R(\lambda, \mathscr{A})} \right\|_{\mathscr{L}(\mathscr{H})} \leq e^{-\lambda t} e^{t\lambda \|\lambda R(\lambda, \mathscr{A})\|_{\mathscr{L}(\mathscr{H})}} \leq 1. \tag{2.3.9}$$

Thereby, $\left(e^{t\mathscr{A}_\lambda}\right)_{t\geq 0}$ is contractive.

Lastly, as the operators $e^{t\mathscr{A}_\lambda}$, $e^{t\mathscr{A}_\mu}$, $\mathscr{A}_\lambda$, and $\mathscr{A}_\mu$ commute, we can write

$$\begin{aligned}\left\| e^{t\mathscr{A}_\lambda}x - e^{t\mathscr{A}_\mu}x \right\| &= \left\| \int_0^1 \frac{d}{dr}\left(e^{tr\mathscr{A}_\lambda} e^{t(1-r)\mathscr{A}_\mu} x \right) dr \right\| \\ &= t \left\| \int_0^1 e^{tr\mathscr{A}_\lambda} e^{t(1-r)\mathscr{A}_\mu} \left(\mathscr{A}_\lambda - \mathscr{A}_\mu \right) x \, dr \right\|.\end{aligned}$$

Inequality (2.3.8) holds thanks to the contractions property of $\left(e^{t\mathscr{A}_\lambda}\right)_{t\geq 0}$. □

One of the most used results when dealing with the well-posedness of linear evolution PDEs is:

Theorem 2.3.3 (Hille-Yosida) *A linear operator $\mathscr{A} : \mathscr{D}(\mathscr{A}) \subset \mathscr{H} \to \mathscr{H}$ generates a C_0-semigroup of contractions $S(t)$on $\mathscr{H}$ if and only if:*

1. *$\mathscr{A}$ is a densely defined closed linear operator in $\mathscr{H}$.*
2. *The resolvent set $\rho(\mathscr{A})$ contains* $(0, \infty)$*, and* (2.2.3) *holds for all $\lambda > 0$, that is,*

$$\lambda \left\| R(\lambda, \mathscr{A}) \right\|_{\mathscr{L}(\mathscr{H})} \leq 1, \ \forall \lambda > 0.$$

Proof (Necessity) The first assertion is already shown (see part (v) of Theorem (2.3.1)). To prove the second, given any $x \in \mathscr{H}$ and $\lambda > 0$, we define the operator $R(\lambda) : \mathscr{H} \to \mathscr{H}$ by

$$R(\lambda)x = \int_0^\infty e^{-\lambda t} S(t)x \, dt.$$

Using the continuity and uniform boundedness of the mapping $t \to S(t)x$, one can claim that R is a well-defined (the integral converges) linear operator on $\mathscr{H}$ and also bounded since

$$\| R(\lambda)x \|_{\mathscr{H}} \leq \| x \|_{\mathscr{H}} \int_0^\infty e^{-\lambda t} \, dt = \frac{\| x \|_{\mathscr{H}}}{\lambda}. \tag{2.3.10}$$

Owing to the semigroup property, we have for any $\tau > 0$,

$$\frac{S(\tau)x - x}{\tau} R(\lambda)x = \frac{e^{\lambda\tau} - 1}{\tau} R(\lambda)x - \frac{e^{\lambda\tau}}{\tau} \int_0^\tau e^{-\lambda t} S(t)x \, dt.$$

Invoking (2.3.2), we deduce that

$$\lim_{\tau \to 0} \left(\frac{S(\tau)x - x}{\tau} R(\lambda)x \right) = \lambda R(\lambda)x - x.$$

This means that $R(\lambda)x \in \mathscr{D}(\mathscr{A})$ and

$$\mathscr{A} R(\lambda)x = \lambda R(\lambda)x - x,$$

that is,

$$(\lambda I - \mathscr{A})R(\lambda) = I.$$

In turn, in view of property (iv) of Theorem 2.3.1, we can claim that for any $x \in \mathscr{D}(\mathscr{A})$,

$$R(\lambda)\mathscr{A} x = \mathscr{A} R(\lambda)x,$$

and hence,

$$R(\lambda)(\lambda I - \mathscr{A})x = x.$$

Gathering these last identities together with (2.3.10), one can claim that the inverse of $\lambda I - \mathscr{A}$ is $R(\lambda)$, which satisfies the desired inequality.

(Sufficiency) Let $x \in \mathscr{D}(\mathscr{A})$ and $t > 0$. In view of (2.3.8) and the triangular inequality, we have, for any positive integers n and m,

$$\left\| e^{t\mathscr{A}_m} x - e^{t\mathscr{A}_n} x \right\|_{\mathscr{H}} \leq t \left(\|\mathscr{A}_m x - \mathscr{A} x\|_{\mathscr{H}} + \|\mathscr{A} x - \mathscr{A}_n x\|_{\mathscr{H}} \right).$$

This, together with (2.3.7), implies that $\left(e^{t\mathscr{A}_n} x\right)_{n\in\mathbb{N}}$ is a Cauchy sequence and hence has a limit called $S(t)x$, namely

$$S(t)x = \lim_{n\to\infty} e^{t\mathscr{A}_\lambda} x. \tag{2.3.11}$$

Using the density of $\mathscr{D}(\mathscr{A})$ in $\mathscr{H}$, this limit also exists for all $x \in \mathscr{H}$. Thereafter, owing to the semigroup properties of $\left(e^{t\mathscr{A}_n}\right)_{t\geq 0}$, we claim that $(S(t))_{t\geq 0}$ is also a semigroup. Moreover, $(S(t))_{t\geq 0}$ is contractive by means of (2.3.9).

In order to complete the proof, one has to check that $\mathscr{A}$ is the generator of $(S(t))_{t\geq 0}$. Suppose that $\mathscr{P}$ is the generator of $(S(t))_{t\geq 0}$, and let $x \in \mathscr{D}(\mathscr{A})$. Then, it follows from Theorems 2.3.1 and (2.3.11) that

$$S(t)x - x = \lim_{n\to\infty} \left(e^{t\mathscr{A}_n} x - x\right) = \lim_{n\to\infty} \int_0^t e^{r\mathscr{A}_n} \mathscr{A}_n x \, dr = \int_0^t S(r)\mathscr{A} x \, dr.$$

Whereupon, we have

$$\lim_{t\to 0} \frac{S(t)x - x}{t} = \lim_{t\to 0} \frac{1}{t} \int_0^t S(r)\mathscr{A} x \, dr = \mathscr{A} x.$$

Consequently, $x \in \mathscr{D}(\mathscr{P})$ and $\mathscr{P}x = \mathscr{A}x$, that is,

$$\mathscr{D}(\mathscr{P}) \supseteq \mathscr{D}(\mathscr{A}) \text{ and } \mathscr{P}x = \mathscr{A}x. \tag{2.3.12}$$

It remains to prove that $\mathscr{D}(\mathscr{P}) = \mathscr{D}(\mathscr{A})$. To proceed, recall the assumption $(0, \infty) \subset \rho(\mathscr{A})$, and hence, $1 \in \rho(\mathscr{A})$. On the other hand, since $\mathscr{P}$ is the generator of $(S(t))_{t\geq 0}$, one can write $1 \in \rho(\mathscr{P})$. Therefore, $1 \in \rho(\mathscr{A}) \cap \rho(\mathscr{P})$, which together with (2.3.12) yields

$$(I - \mathscr{P})\mathscr{D}(\mathscr{A}) = (I - \mathscr{A})\mathscr{D}(\mathscr{A}) = \mathscr{H}.$$

Thus

$$\mathscr{D}(\mathscr{A}) = (I - \mathscr{P})^{-1}\mathscr{H} = \mathscr{D}(\mathscr{P}).$$

The proof of the theorem is achieved. □

Recalling the Definition 2.2.2, we shall state and prove the second important theorem that is often used to solve the existence and uniqueness problem of solutions to evolution PDEs.

Theorem 2.3.4 (Lumer-Phillips) *A densely defined linear operator $\mathscr{A} : \mathscr{D}(\mathscr{A}) \subset \mathscr{H} \to \mathscr{H}$ generates a C_0-semigroup of contractions $(S(t))_{t\geq 0}$ on $\mathscr{H}$ if and only if $\mathscr{A}$ is m-dissipative.*

Proof (Necessity) As $\mathscr{A}$ generates a C_0-semigroup of contractions $(S(t))_{t\geq 0}$ on $\mathscr{H}$, Hille-Yosida theorem (see Theorem 2.3.3) permits to write that $\rho(\mathscr{A})$ contains $(0, \infty)$, and hence, $\lambda I - \mathscr{A}$ is onto $\mathscr{H}$ for all $\lambda > 0$. Next, if $x \in \mathscr{D}(\mathscr{A})$, then the contraction of $S(t)$ gives

$$|< S(t)x, x >_{\mathscr{H}}| \leq \|x\|^2_{\mathscr{H}},$$

and hence, for any $t > 0$, there holds

$$\frac{1}{t}\Re e < S(t)x - x, x >_{\mathscr{H}} = \frac{1}{t}\left(\Re e < S(t)x, x >_{\mathscr{H}} - \|x\|^2_{\mathscr{H}}\right) \leq 0.$$

Then, letting t go to zero and using the definition of $\mathscr{A}$, we reach that $\mathscr{A}$ is dissipative.

(Sufficiency) Suppose that $\mathscr{A}$ is m-dissipative operator. Thanks to Theorem 2.2.1, we claim that the conditions of Hille-Yosida Theorem (see Theorem 2.3.3) hold, and hence, $\mathscr{A}$ generates a C_0-semigroup of contractions. □

The following result is sometimes useful:

Theorem 2.3.5 *Let $\mathscr{A}$ be a densely defined and closed linear operator in a Hilbert space $\mathscr{H}$. If both $\mathscr{A}$ and $\mathscr{A}^*$ are dissipative, then $\mathscr{A}$ generates a C_0-semigroup of contractions on $\mathscr{H}$.*

2.3.2 Linear Evolution Equations

Consider the following linear evolution equation:

$$\begin{cases} \dfrac{d}{dt}u(t) = \mathscr{A}u(t), t > 0, \\ u(0) = u_0 \in \mathscr{H}. \end{cases} \tag{2.3.13}$$

The *abstract Cauchy problem* is to seek a "solution" $u(t)$ of (2.3.13), where the meaning of solution is given below:

Definition 2.3.5 A function $u : [0, \infty) \to \mathscr{H}$ is a *classical (strong) solution* of (2.3.13) if u is continuous for all $t \geq 0$, continuously differentiable for all $t > 0$, and satisfies (2.3.13).

The next result provides the relationship between the semigroups theory and the Cauchy problem [158]:

Theorem 2.3.6 *Let $\mathscr{A}$ be a densely defined linear operator on $\mathscr{H}$ with nonempty resolvent. Let $u_0 \in \mathscr{D}(\mathscr{A})$. Then the Cauchy problem* (2.3.13) *has a unique classical solution if and only if $\mathscr{A}$ generates a C_0-semigroup $S(t)$ on $\mathscr{H}$. In such an event, $u(t) = S(t)u_0$.*

Note that if $u_0 \notin \mathscr{D}(\mathscr{A})$, the problem (2.3.13) does not necessarily have a strong solution. The function $u(t) = S(t)u_0$ is then called a *mild solution* . More precisely, u is a mild solution of (2.3.13) if u is continuous on $[0, \infty)$, and there exists a sequence $(u_n) \subset \mathscr{D}(\mathscr{A})$ such that $u_n \to u_0$ as $n \to \infty$ and $S(t)u_n \to u(t)$ uniformly on bounded intervals.

2.3.3 Semigroups Stability

There are three main types of stability of semigroups.

Definition 2.3.6 Let $S(t)$ be a C_0-semigroup on a Hilbert space $\mathscr{H}$. This semigroup is:

(i) *Weakly stable* if $\lim_{t\to\infty} < S(t)x, y >= 0$, for every x and y in $\mathscr{H}$
(ii) *Strongly stable* if $\lim_{t\to\infty} S(t)x = 0$, for every x in $\mathscr{H}$

(iii) *Exponentially stable* if there exist two positive constants M and ω such that $\|S(t)\| \leq Me^{-\omega t}$

Note that the exponential stability, as defined above, is often called uniformly exponentially stable. Also, it is obvious that the exponential stability implies the strong one, which further gives the weak stability. Moreover, if the generator $\mathscr{A}$ of the semigroup $S(t)$ has a compact resolvent, then the weak and strong stabilities are equivalent [96].

It is well-known that the exponential stability is a very desirable property in systems control. This motivated many authors to seek the simplest characterization of such property. In fact, the definition requires that $S(t)$ is known, which is not the case in practice. There are basically two ways to overcome this difficulty. The first one is called the time-domain method (also called energy or Lyapunov method), whereas the second one is known as the frequency-domain (or resolvent) method. The basic idea behind the time-domain (energy or Lyapunov) method is to seek a suitable energy norm that decays exponentially. This approach is usually simple and works even for nonlinear problems, but requires certain skills to find such energy. In turn, the resolvent method often involves length computations and is limited to linear systems.

First, recall the following definition.

Definition 2.3.7 Let $\mathscr{A}$ be the generator of a C_0-semigroup $S(t)$ on a Hilbert space $\mathscr{H}$:

(i) The *growth bound* of $S(t)$ is the number

$$\omega_0 = \inf_{t>0} \frac{1}{t} \log \|S(t)\|. \tag{2.3.14}$$

Note that the growth bound belongs to $\mathbb{R} \cup \{-\infty\}$.

(ii) The *spectral bound of the operator* $\mathscr{A}$ is

$$s_0 = \{\Re e\lambda;\ \lambda \in \sigma(\mathscr{A})\}. \tag{2.3.15}$$

(iii) If $\omega_0 = s_0$, then we say that the spectrum-determined-growth condition holds.

Now, we can state the following result that characterizes the exponential stability of C_0-semigroup (see [78, 96, 158]):

Theorem 2.3.7 *Let $\mathscr{A}$ be the generator of a C_0-semigroup of contractions $S(t)$ on a Hilbert space $\mathscr{H}$. The following statements are equivalent:*

(i) The semigroup $S(t)$ is exponentially stable.

(ii) $\lim_{t\to\infty} \|S(t)\|_{\mathscr{L}(\mathscr{H})} = 0$.

(iii) There exists a $T > 0$ such that $\|S(T)\|_{\mathscr{L}(\mathscr{H})} < 1$.

(iv) There exists $p \geq 1$ such that $\int_0^\infty \|S(t)x\|^p\, dt < \infty$, for all $x \in \mathscr{H}$.

(v) There exists a positive definite operator $B \in \mathscr{L}(\mathscr{H})$ *such that* $< \mathscr{A}x, Bx > + < Bx, \mathscr{A}x >= -\|x\|^2$ *for all* $x \in \mathscr{D}(\mathscr{A})$.
(vi) The growth bound ω_0 *is negative.*

Second, we recall the resolvent method result for the exponential stability (see [118, 141, 163]):

Theorem 2.3.8 *Let* $\mathscr{A}$ *be the generator of a* C_0*-semigroup of contractions* $S(t)$ *on a Hilbert space* $\mathscr{H}$. *The semigroup* $S(t)$ *is exponentially stable if and only if the following two conditions hold:*

$$\rho(\mathscr{A}) \supset \{i\gamma;\ \gamma \in \mathbb{R}\} \equiv i\mathbb{R}, \tag{2.3.16}$$

and

$$\limsup_{|\gamma|\to\infty} \|(i\gamma I - \mathscr{A})^{-1}\|_{\mathscr{L}(\mathscr{H})} < \infty, \tag{2.3.17}$$

where $\rho(\mathscr{A})$ *denotes the resolvent set of the operator* $\mathscr{A}$.

Naturally, there are other types of stability in addition to those in Definition (2.3.6). In general, if there exist a positive function $f(t)$ with $\lim_{t\to\infty} f(t) = 0$ and a positive constant α such that

$$\|S(t)x_0\| \leq f(t)\|x_0\|_{\mathscr{D}(\mathscr{A})^\alpha},\ t > 0,$$

we say that the semigroup $S(t)$ is stable with a rate of $f(t)$ for all $x_0 \in \mathscr{D}(\mathscr{A})^\alpha$.

In particular, if $f(t) = Mt^{-\omega}$, for some positive constants M and ω, then $S(t)$ is said to be *polynomially stable* . In the case where $f(t) = M(\log t)^{-\beta}$, then $S(t)$ is said to be *logarithmically stable* .

We will also employ the following frequency-domain theorem for logarithmic stability from [34, Theorem A] (see also [42]) of a C_0-semigroup of contractions on a Hilbert space:

Theorem 2.3.9 *A* C_0*-semigroup* $S(t)$ *of contractions on a Hilbert space* $\mathscr{H}$ *satisfies, for all* $t > 0$,

$$||S(t)||_{\mathscr{L}(\mathscr{D}(\mathscr{A}),\mathscr{H})} \leq \frac{C}{\log(2+t)}$$

for some constant $C > 0$ *if* (2.3.16) *holds and*

$$\limsup_{|\gamma|\to\infty} \|e^{-c|\gamma|}(i\gamma I - \mathscr{A})^{-1}\|_{\mathscr{L}(\mathscr{H})} < \infty,\ \textit{for some}\ c > 0, \tag{2.3.18}$$

where $\rho(\mathscr{A})$ *denotes the resolvent set of the operator* $\mathscr{A}$.

Furthermore, the frequency-domain theorem for polynomial stability [38] (see also [34, 140] for weaker variants) of a C_0-semigroup of contractions on a Hilbert space is given below:

Theorem 2.3.10 *A C_0-semigroup $S(t)$ of contractions on a Hilbert space $\mathscr{H}$ satisfies*

$$||S(t)U_0||_{\mathscr{H}} \leq \frac{C}{t^{\frac{1}{\theta}}}||U_0||_{\mathscr{D}(\mathscr{A})},$$

for some constant $C > 0$ and for $\theta > 0$ if and only if (2.3.16) *holds and*

$$\limsup_{|\beta|\to\infty} \frac{1}{|\beta|^{\theta}} \|(i\beta I - \mathscr{A})^{-1}\|_{\mathscr{L}(\mathscr{H})} < \infty, \tag{2.3.19}$$

where $\rho(\mathscr{A})$ denotes the resolvent set of the operator $\mathscr{A}$.

Third and last, we note that there is a powerful tool to verify the strong stability of linear semigroups, known as LaSalle invariance principle. Some background is needed to discuss this result.

Definition 2.3.8 (ω-Limit Set) The set

$$\omega(x) = \{y \in \mathscr{H};\ \exists (t_n) \to +\infty \text{ such that } S(t_n)x \to y \text{ as } n \to +\infty\}$$

is called the ω-limit set of x.

Proposition 2.3.1 ([114]) *Let $x \in \mathscr{H}$. Then for all $t \geq 0$, we have $\omega(S(t)x) = \omega(x)$ and $S(t)(\omega(x)) \subset \omega(x)$. The set $\omega(x)$ is said to be invariant by $S(t)$.*

If the set $\cup_{t\geq 0}\{S(t)x\}$ is precompact in $\mathscr{H}$, then $S(t)(\omega(x)) = \omega(x) \neq \emptyset$. Moreover, $\omega(x)$ is a compact connected set of $\mathscr{H}$ and $S(t)x \to \omega(x)$ as $t \to \infty$.

Theorem 2.3.11 (LaSalle Invariance Principle [114]) *Suppose that there exists a positive functional V so that $\frac{dV}{dt} \leq 0$ along the trajectories of $S(t)$ and a positive real λ such that the resolvent $R(\lambda, \mathscr{A})$ is compact. Then for any $x \in \mathscr{H}$, we have:*

(1) The trajectories set $\{S(t)x\}_{t\geq 0}$ is precompact in $\mathscr{H}$.
(2) If $x \in \mathscr{D}(\mathscr{A})$, then the ω-limit set $\omega(x)$ is a subset of M, where M is the largest invariant subset of

$$\{u \in \mathscr{D}(\mathscr{A});\ \frac{dV}{dt}(u(t)) = 0,\ \forall t \geq 0\}.$$

Remark 2.3.1

(i) In practice, LaSalle invariance principle is used in such a way that we show that M is reduced to $\{0\}$, and hence, the strong stability holds.

(ii) It is also important to point out that the compactness of the resolvent plays a crucial role when using Lasalle principle. In turn, observe that a resolvent operator is compact for one $\lambda \in \rho(\mathscr{A})$ if and only if it is compact for all $\lambda \in \rho(\mathscr{A})$. This follows from the *resolvent identity*

$$R(\lambda, \mathscr{A}) - R(\mu, \mathscr{A}) = (\mu - \lambda) R(\lambda, \mathscr{A}) R(\mu, \mathscr{A}), \quad \text{for } \lambda, \mu \in \rho(\mathscr{A}). \tag{2.3.20}$$

The latter can be obtained from the identities

$$(\lambda R(\lambda, \mathscr{A}) - \mathscr{A} R(\lambda, \mathscr{A}))\, R(\mu, \mathscr{A}) = R(\mu, \mathscr{A}),$$
$$R(\lambda, \mathscr{A})\, (\mu R(\mu, \mathscr{A}) - \mathscr{A} R(\mu, \mathscr{A})) = R(\lambda, \mathscr{A}).$$

(iii) Assume that $\rho(\mathscr{A}) \neq \emptyset$. Then, one may assume that $0 \in \rho(\mathscr{A})$. Otherwise, we use the shifting operator

$$\widetilde{\mathscr{A}} := \mu I + c\mathscr{A}, \quad \text{for some } c > 0, \mu \in \mathbb{C},$$

and the identity $R(\lambda, \widetilde{\mathscr{A}}) = \dfrac{1}{c} R([\lambda - \mu]/c, \widetilde{\mathscr{A}})$, for $\lambda \in \rho(\widetilde{\mathscr{A}})$.

It turns out that the strong stability of a semigroup could be characterized via the spectrum of the infinitesimal generator. In fact, we have:

Theorem 2.3.12 ([118]) *Let $\mathscr{A}$ be the infinitesimal generator of a C_0-semigroup $S(t)$ in a Hilbert space $\mathscr{H}$ such that $\mathscr{A}$ has compact resolvent. Then, $S(t)$ is strongly stable if and only if it is uniformly bounded and $\Re e(\lambda) < 0$, for any λ in the spectrum of $\mathscr{A}$.*

Finally, we end this section by stating the following helpful result, as it characterizes the compactness of the resolvent:

Theorem 2.3.13 *Let $\mathscr{A}$ be a linear operator on a Hilbert space $\mathscr{H}$ such that $\rho(\mathscr{A}) \neq \emptyset$. Then $\mathscr{A}$ has compact resolvent if and only if the canonical injection $i : \mathscr{D}(\mathscr{A}) \to \mathscr{H}$ is compact, where $\mathscr{D}(\mathscr{A})$ is endowed with the graph norm.*

Proof Given $\lambda \in \rho(\mathscr{A})$, we introduce a new norm as follows:

$$\|x\|_* = \|(\lambda I - \mathscr{A})x\|,$$

which is clearly equivalent to the graph norm. Thereby, the operator $\widetilde{\mathscr{A}} := \lambda I - \mathscr{A} : (\mathscr{D}(\mathscr{A}), \|\ \|_*) \to \mathscr{H}$ is an isomorphism with continuous inverse $R(\lambda, \mathscr{A})$. Lastly, the desired result follows from the fact $i = \widetilde{\mathscr{A}} \circ R(\lambda, \mathscr{A})$. □

2.4 Semigroups of Nonlinear Operators

In this section, we assume that $\mathscr{A}$ is a nonlinear operator with domain $\mathscr{D}(\mathscr{A}) \subset \mathscr{H}$.

Definition 2.4.1 Let $\mathscr{A}$ be a nonlinear operator and $\lambda > 0$. $\mathscr{A}$ is maximal monotone if:

(i) (Monotonicity) For every $x_1,\ x_2 \in \mathscr{D}(\mathscr{A})$, we have $\langle \mathscr{A}x_1 - \mathscr{A}x_2, x_1 - x_2\rangle \geq 0$.
(ii) (Maximality) The operator $I + \mathscr{A}$ is onto $\mathscr{H}$.

Let us define $J_\lambda =: I + \lambda\mathscr{A} : \mathscr{D}(\mathscr{A}) \to \mathscr{H}$. Thereafter, it is shown that $\mathscr{A}$ is maximal monotone if and only if J_λ is bijective and the operator $J_\lambda^{-1} : \mathscr{H} \to \mathscr{H}$ is a contraction [187].

Definition 2.4.2 A family of nonlinear operators $\{S(t) : \mathscr{H} \to \mathscr{H},\ t \geq 0\}$ is called a nonlinear semigroup of contractions if:

i) $S(0) = I$ *and* $S(t)S(\tau) = S(t + \tau),\ \forall t, \tau \geq 0.$
ii) $\lim_{t\downarrow 0} S(t)x = x,\ \forall x \in \mathscr{H}.$
iii) $\|S(t)x - S(t)y\| \leq \|x - y\|,\ \forall x, y \in \mathscr{H}$ and $\forall t \geq 0$.

As for the linear case, if $S(t)$ is a semigroup of contractions, then for every $x \in \mathscr{H}$, the mapping $t \to S(t)x$ is continuous on $[0, \infty)$. Moreover, we define the generator $\mathscr{A}$ of $S(t)$ exactly as Definition 2.3.2.

Theorem 2.4.1 ([187]) *Let $S(t)$ be a semigroup of contractions on $\mathscr{H}$ and let $\mathscr{A}$ be its generator. Then, for all x in $\mathscr{D}(\mathscr{A})$, we have:*

(i) $S(t)x \in \mathscr{D}(\mathscr{A}),\ \forall t \geq 0$, *and* $t \to \mathscr{A}S(t)x$ *is continuous on* $[0, \infty)$.
(ii) $t \to S(t)x$ *is differentiable a. e. on* $[0, \infty)$ *and* $\dfrac{d}{dt}S(t)x = AS(t)x,\ \forall t \geq 0$ *a.e.*

Given a maximal monotone operator $\mathscr{A}$ on $\mathscr{H}$, consider the evolution equation:

$$\begin{cases} \dot{u}(t) + \mathscr{A}u(t) = 0, \\ u(0) = u_0. \end{cases} \tag{2.4.1}$$

We have the following theorem:

Theorem 2.4.2 ([187]) *For every $u_0 \in \mathscr{D}(\mathscr{A})$, there exists a unique function $u(t) \in \mathscr{H}$ such that:*

(1) $u(0) = u_0$ *and* $u(t) \in \mathscr{D}(\mathscr{A})$, *for all* $t > 0$.
(2) $\dfrac{du}{dt} \in L^\infty([0, \infty); \mathscr{H})$ *and* $\|\dfrac{du}{dt}\|_{L^\infty([0,\infty);\mathscr{H})} \leq \|\mathscr{A}u_0\|$.
(3) $u(t)$ *is a solution (strong) of* (2.4.1) *on* $(0, \infty)$. *Moreover,* $\dfrac{d^+u}{dt}(t) + \mathscr{A}u(t) = 0$, $t \geq 0$.
(4) The mapping $t \to Au(t)$ *is continuous and* $t \to \|Au(t)\|$ *is nonincreasing.*

Remark 2.4.1 In view of Theorem (2.4.2), $S(t) : \mathscr{D}(\mathscr{A}) \to \mathscr{H}$ so that $S(t)u_0 = u(t)$ for all $t \geq 0$. Then $S(t)$ is a contractive semigroup on $\mathscr{D}(\mathscr{A})$, which can be uniquely extended to a contractive semigroup on $\overline{\mathscr{D}(\mathscr{A})}$. Also, $S(t)$ is the semigroup generated by $-\mathscr{A}$. Finally, setting $u(t) = S(t)u_0$, for $u_0 \in \mathscr{H}$, then u is called the generalized solution of (2.4.1).

We closed this chapter by raising two points. The first one is related to the concept of strong and exponential stability of a semigroup of nonlinear operators. In fact, Definition 2.3.6 remains the same for nonlinear operators that generate semigroups. Needless to mention that for nonlinear semigroups, the energy method is the usual way to establish stability results. Also the following result (LaSalle invariance principle) is often used to show the strong stability of nonlinear semigroups.

Theorem 2.4.3 (Dafermos and Slemrod [80]) *Let $\mathscr{A}$ be a nonlinear maximal monotone operator that generates a C_0-semigroup of contractions $S(t)$ on $\mathscr{H}$. If $0 \in \Im m(\mathscr{A})$ and J_λ^{-1} is compact for some $\lambda > 0$, then for all $x_0 \in \mathscr{D}(\mathscr{A})$, $\omega(x_0)$ is a nonempty compact set and $x(t) = S(t)x_0 \to \omega(x_0)$ as $t \to \infty$.*

The second point concerns the stabilization issue of systems. This problem consists in designing an appropriate input control (either internal or boundary) so that the closed-loop system remains stable in the sense of Definition 2.3.6. Nevertheless, we point out that, in many instances, the system has a nonzero steady state. In such an event, the stabilization of a system is interpreted by finding a control so that the solutions of the system tend to the equilibrium state. This definition is often used throughout this book.

It turns out that the abstract mathematical formulation of the stabilization problem of a system is not simple in general as the input operator is often unbounded. For instance, if we consider two Hilbert spaces $\mathscr{H}$ and U and assume that $\mathscr{A}$ generates a C_0-semigroup $S(t)$ on $\mathscr{H}$. Then, let $\mathscr{H}_1 = \mathscr{D}(\mathscr{A})$ endowed with the norm $\|x\|_1 = \|(\alpha I - \mathscr{A})x\|$, where α is a fixed complex number of $\rho(\mathscr{A})$. In addition, we define $\mathscr{H}_{-1}$ as the completion of $\mathscr{H}_1$ with respect to the norm $\|x\|_{-1} = \|(\alpha I - \mathscr{A})^{-1}x\|$. Next, consider the abstract differential equation

$$\frac{d}{dt}x(t) = \mathscr{A}x(t) + Bu(t), \quad x(0) = x_0 \in \mathscr{H},$$

where $u \in L^2_{\text{loc}}([0, \infty); U)$ and $B \in \mathscr{L}(U, \mathscr{H}_1)$ so that all mild solutions of the equation are $\mathscr{H}$-valued continuous functions.

We say that the equation is strongly (resp. exponentially or polynomially or logarithmically) stabilizable if there exists an operator $F \in \mathscr{L}(\mathscr{H}_1, U)$ such that $u = Fx$ and $\mathscr{A} + BF$ generates a strongly (resp. exponentially or polynomially or logarithmically) stable semigroup (see [180] for more details). Notwithstanding, in order to make the book maximally lucid, we shall eschew as much as possible these kinds of abstract approaches.

Chapter 3
Stabilization of a Multidimensional Wave Equation

3.1 Motivation and State of the Art

This chapter deals with the stabilization problem of the multidimensional wave equation subject to a time delay or memory in either the boundary or the interior.

Unless stated otherwise, Ω denotes an open bounded connected set of $\mathbb{R}^n$, where n is a positive integer greater than 1. It is also assumed that Ω has a smooth boundary $\Gamma = \partial\Omega$ of class C^2, which is turn has a partition (Γ_0, Γ_1).

The simplest form of the wave equation is

$$y_{tt}(x,t) - \Delta y(x,t) = 0, \qquad \text{in } \Omega \times (0,\infty),$$

where $y(x,t)$ is the wave displacement at the spatial variable x and time t. It models a wide variety of physical waves such as elastic waves, sound waves, and pressure waves in a fluid. This equation can be derived by applying fundamental physical principles (conservation laws and Newton's second law). For sake of completeness, we shall discuss this point with some details.

3.1.1 Two-Dimensional Wave Equation: a Vibrating Membrane

Consider a highly stretched membrane such as for a vibrating drum. First, let $u(x,y,t)$ be the displacement of the membrane at the spatial variables x, y and time t. Then, we focus on a small piece of the membrane lying above the rectangle $\Delta x \Delta y$ (see Fig. 3.1) and denote byT^x (respectively, T^y), the tensile force on the constant-x (respectively, y) edge. For sake of simplicity, assume that there is a constant tension T such that $T^x \approx T\Delta y$ and $T^y \approx T\Delta x$. Thereby, the external force exerted on the

K. Ammari, B. Chentouf, *Stabilization of Wave-Type Equations with Delay or Memory*, Advances in Mechanics and Mathematics 57,
https://doi.org/10.1007/978-3-032-28496-9_3

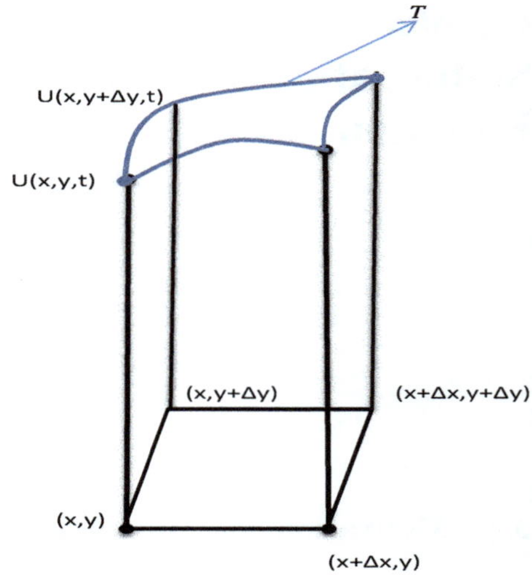

Fig. 3.1 Stretched membrane with constant tension T

membrane is

$$F = T^x u_{xx} \Delta x + T^y u_{yy} \Delta y \simeq T \Delta x \Delta y (u_{xx} + u_{yy}).$$

Now, denote by ρ the membrane mass per unit area, and hence, one can approximate the mass m of the small piece of the membrane by $m = \rho \Delta x \Delta y$. Finally, Newton's second law yields the two-dimensional wave equation:

$$\rho u_{tt} = T(u_{xx} + u_{yy}).$$

3.1.2 *Three-Dimensional D-Wave Equation: Fluid Pressure Wave*

Consider a fluid with density $\rho(x, y, z)$ and velocity $\mathbf{V}(x, y, z)$. As usual, consider a small rectangular "piece" of the fluid of size Δx, Δy and Δz. Then, the mass of the small "piece" is given by $m = \rho(x, y, z)\Delta x \Delta y \Delta z$ (see Fig. 3.2). Furthermore, the net force in the x direction due to the pressure $P(x, y, z)$ is

$$f^x = P(x, y, z)\Delta y \Delta z - P(x + \Delta x, y, z)\Delta y \Delta z \approx -\Delta x \Delta y \Delta z P_x.$$

Applying again Newton's second law, we get

$$\rho(x, y, z)\Delta x \Delta y \Delta z \frac{\mathrm{d}\mathbf{V}^x}{\mathrm{d}t} = -\Delta x \Delta y \Delta z P_x.$$

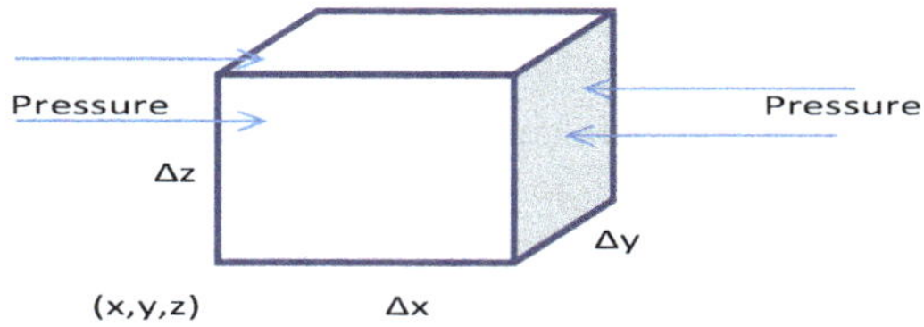

Fig. 3.2 Pressure on fluid element

Arguing as above for the y and z pressures, we have

$$\rho \frac{\mathrm{d}\mathbf{V}}{\mathrm{d}t} = -\nabla P.$$

In turn, the continuity equation gives

$$\frac{\partial \rho}{\partial t} = -\nabla \cdot \rho \mathbf{V}.$$

Assuming that the pressure P and the density are small perturbations of uniform quantities, one can write $P(x, y, z) = P_0 + p(x, y, z)$, $\rho(x, y, z) = \rho_0 + u(x, y, z)$. This, together with the assumption that $\mathbf{V}$ is also small, permits to write the above equations

$$\rho_0 \mathbf{V}_t = -\frac{\mathrm{d}p}{\mathrm{d}u} \nabla u,$$

and

$$u_t = -\rho_0 \nabla \cdot \mathbf{V}.$$

Whereupon, we have

$$u_{tt} = c^2 (u_{xx} + u_{yy} + u_{zz}),$$

where $c = \sqrt{dP/d\rho|_{\rho} - \rho_0}$.

When studying the controllability or stabilization or observability of the wave equation, a condition on the domain Ω often emerges. Such a condition may take several forms such as strictly convex or star shaped [46, 131, 144, 165, 168]. In turn, the most used geometric condition is the so-called Bardos-Lebeau-Rauch (BLR) geometric condition [32, 133]. Roughly speaking, it requires that the control region must meet, within a finite time, each ray at a nondiffractive point, which is a point where the ray's direction is not parallel to the boundary of the control region.

The wave equation has been actively and continuously analyzed either qualitatively or numerically or both. In particular, the study of long-time behavior of solutions to the wave equation has been the subject of an active research and mathematical endeavor. This has led to hundreds of research papers, and hence, it is quasi-impossible to cite all of them (see, for instance, such as [32, 42, 46–49, 129, 133–135, 144, 147, 165, 168, 171, 179]). Notwithstanding, in order to highlight the main contribution and the novelty of this section, we will merely point out the articles whose content is closely related to our problem and shall survey those bearing a resemblance to (3.2.1)–(3.2.2). The author in [128] has been mainly preoccupied with the asymptotic convergence of solutions of (3.2.1)–(3.2.2) in the case where there is no delay term ($\beta = 0$). More recently, a nonstandard energy norm has been provided in [57–61] in order to analyze the long-time behavior of solutions to the wave equation without any displacement term and where no delay arises.

Since the pioneer work of [81, 83, 84], where it has been shown that the presence of a delay in a one-dimensional wave equation could create an instability phenomenon, delayed wave equation has come under the spotlight. Indeed, one very active area of mathematical control theory has been the investigation of the delay effect in the stabilization of hyperbolic systems, and many authors have shown that delays can destabilize a system that is asymptotically stable in the absence of delays (see [5, 6, 84, 149, 151, 154] for more details). This research endeavor is mainly motivated by the fact that delay effects arise in many applications and practical problems. Furthermore, it is well-known that an arbitrarily small delay may destabilize a system that is uniformly asymptotically stable in the absence of delay (see, e.g., [81, 84, 128, 129, 191]). Specifically, it has been proved in [82, Example 3.5] that the system

$$\begin{cases} w_{tt} - w_{xx} - a w_{xxt} = 0, & x \in (0,1),\ t > 0, \\ w(0,t) = 0, \qquad w_x(1,t) = -k w_t(1, t-\tau), & t > 0, \end{cases} \tag{3.1.1}$$

where a, k and τ are positive constants, becomes unstable for arbitrarily small values of τ and any values of a and k. In (3.1.1) and even in the presence of the strong damping $-aw_{xxt}$, without any other damping, the overall structure can be unstable. Subsequently, Datko et al. [84] treated the following one -dimensional problem:

$$\begin{cases} u_{tt}(x,t) - u_{xx}(x,t) + 2au_t(x,t) + a^2 u(x,t) = 0, & 0 < x < 1,\ t > 0, \\ u(0,t) = 0, & t > 0, \\ u_x(1,t) = -ku_t(1, t-\tau), & t > 0, \end{cases} \tag{3.1.2}$$

which models the vibrations of a string clamped at one end and free at the other end, where $u(x,t)$ is the displacement of the string. Also, the string is controlled by a boundary control force (with a delay) at the free end. They showed that if the positive constants a and k satisfy

$$k\frac{e^{2a}+1}{e^{2a}-1} < 1,$$

then the delayed feedback system (3.1.2) is stable for all sufficiently small delays. On the other hand, if

$$k\frac{e^{2a}+1}{e^{2a}-1} > 1,$$

then there exists a dense open set D in $(0, \infty)$ such that for each $\tau \in D$, system (3.1.2) admits exponentially unstable solutions.

It is well-known that in the absence of delay in (3.1.2), that is for $\tau = 0$, system (3.1.2) is uniformly asymptotically stable under the condition $a^2 + k^2 > 0$, and the total energy of the solution satisfies for all $t > 0$,

$$E(t,u) = \int_0^1 (u_t^2 + u_x^2 + a^2u^2)dx \leq CE(0,u)\, e^{-\alpha t} \tag{3.1.3}$$

for some positive constant α (see [46] for more details).

Thereafter, Ammari et al. [20] have treated the wave equation

$$\begin{cases} u_{tt}(x,t) - \Delta u(x,t) + a u_t(x,t-\tau) = 0, & x \in \Omega,\ t > 0, \\ u(x,t) = 0, & x \in \Gamma_0,\ t > 0, \\ \dfrac{\partial u}{\partial \nu}(x,t) = -ku(x,t), & x \in \Gamma_1,\ t > 0, \\ u(x,0) = u_0(x),\ u_t(x,0) = u_1(x), & x \in \Omega, \\ u_t(x,t-\tau) = f_0(x,t-\tau), & x \in \Gamma_1,\ t \in (0,\tau), \end{cases} \tag{3.1.4}$$

where the boundary $\partial\Omega = \Gamma_0 \cup \Gamma_1$ and $\Gamma_0 \cap \Gamma_1 = \emptyset$. Under the usual geometric condition on the domain Ω, they showed an exponential stability result, provided that the delay coefficient a is sufficiently small.

In [149], the authors examined a system of wave equations with a linear boundary damping term with a delay. Namely, they looked to the following system:

$$\begin{cases} u_{tt} - \Delta u = 0, & x \in \Omega,\ t > 0, \\ u(x,t) = 0, & x \in \Gamma_0,\ t > 0, \\ \dfrac{\partial u}{\partial \nu}(x,t) = \mu_1 u_t(x,t) + \mu_2 u_t(x,t-\tau) & x \in \Gamma_1,\ t > 0, \\ u(x,0) = u_0(x), & x \in \Omega, \\ u_t(x,0) = u_1(x) & x \in \Omega, \\ u_t(x,t-\tau) = g_0(x,t-\tau) & x \in \Omega, \tau \in (0,1), \end{cases} \tag{3.1.5}$$

and proved under the assumption

$$\mu_2 < \mu_1 \tag{3.1.6}$$

(which means that the weight of the feedback with delay is smaller than the one without delay) that null stationary state is exponentially stable. On the contrary, if (3.1.6) does not hold, they found a sequence of delays for which the corresponding solution of (3.1.5) will be unstable. The main approach used in [149] is an observability inequality obtained with a Carleman estimate.

The case of time-varying delay (i.e., $\tau = \tau(t)$ is a function depending on t) has been studied by Nicaise, Valein and Fridman [155] in one space dimension. In their work, an exponential stability result was given under the condition:

$$\mu_2 < \sqrt{1-d}\,\mu_1, \tag{3.1.7}$$

where d is a constant such that

$$\tau'(t) \leq d < 1, \text{ for all } t > 0. \tag{3.1.8}$$

Let us also mention the work of Ammari and Nicaise, [19], in which the authors performed a complete study of the stabilization of elastic systems by collocated feedback with or without delay. Lastly, we point out that the well-posedness in evolution equations with delay is not always guaranteed. In fact, Dreher, Quintilla and Racke have shown some ill-posedness results for a wide range of evolution equations with a delay term [89]. It is also noteworthy that numerous papers reveal that particular choices of the delay may restitute exponential stability property (see [21, 110] for linear systems and [35, 56] for nonlinear problems).

The main goal of this chapter is to study the long-time behavior of the delayed wave equation under various situations of the delay term and the boundary conditions. In the first section, we assume that the boundary conditions are static and a delay occurs at one part of the boundary of the domain. Moreover, we suppose that there is no displacement term in the problem. Then, we show a logarithmic convergence result without the BLR condition. Furthermore, we establish an exponential convergence result if the BLR condition holds. In Sect. 3.3, we treat the dynamic delayed boundary conditions. The principal outcome of this section is to prove the exponential stability of the system with the Kelvin-Voigt damping. In Sect. 3.4, we consider a special case of the wave equation in a three-dimensional domain containing a trapped ray, and hence, the BLR condition is violated. A localized interior damping acts as a compensator of a distributed delay. Despite the geometry of the domain, a convergence rate is obtained. Section 3.5 is devoted to the case of space-dependent wave equation with memory but without any displacement term. Using the history approach, we manage to show an exponential convergence result. The same outcome is obtained for the case of a distributed delay and several systems. In Sect. 3.6, we deal with the boundary feedback stabilization problem of a wave equation with an interior delay. A boundary delay control is proposed to prove an exponential stability result under Lions geometric condition. Finally, this chapter ends with a study of an abstract-wave system with a delay and Kelvin-Voigt damping. Using appropriate damping coefficients, the exponential stability of the system is established. Note that the rate of convergence of solutions to each problem studied in this chapter is obtained via the resolvent method, except in Sect. 3.6, where the energy method is used.

3.2 Static Boundary Delay: Logarithmic and Exponential Stabilization

In the first section, we suppose that the delay occurs in the boundary of the domain. To be more specific, we consider the case of a boundary control and then introduce the problem in Sect. 3.2.1. Next, in Sect. 3.2.2, preliminaries are given, and the problem is set up. Section 3.2.3 is dedicated to the proof of existence and uniqueness of solutions of our system. Section 3.2.4 deals with the asymptotic convergence of solutions to a steady state. In Sect. 3.2.5, it is proved that the convergence is in fact logarithmic without the BLR geometric condition [32, 133]. Finally, we establish in Sect. 3.2.6 an exponential convergence result under the BLR condition.

3.2.1 Introduction

We will be concerned with the wave equation

$$y_{tt}(x,t) - \Delta y(x,t) = 0, \qquad \text{in } \Omega \times (0,\infty), \tag{3.2.1}$$

combined with the following boundary and initial conditions:

$$\begin{cases} \frac{\partial y}{\partial \nu}(x,t) = 0, & \text{on } \Gamma_0 \times (0,\infty), \\ \frac{\partial y}{\partial \nu}(x,t) = -\alpha y_t(x,t) - \beta y_t(x,t-\tau), & \text{on } \Gamma_1 \times (0,\infty), \\ y(x,0) = y_0(x), \;\; y_t(x,0) = z_0(x), & x \in \Omega, \\ y_t(x,t) = f(x,t), & (x,t) \in \Gamma_1 \times (-\tau,0), \end{cases} \tag{3.2.2}$$

in which $\alpha > 0$, $\beta \in \mathbb{R}$, and ν is the unit normal of Γ pointing toward the exterior of Ω. Additionally, Γ_1 is supposed to be non-empty, while Γ_0 may be empty.

The present section places primary emphasis on the study of the asymptotic behavior of solutions of the system (3.2.1)–(3.2.2) without any additional condition on the domain Ω. The main concern is to highlight the impact of a boundary delay term (in contrast to [128] and [57–61]) on the solutions to the wave equation when no displacement term occurs (contrary to [149]). This extends the outcome in [128] and [57–61] in the sense that the presence of a delay phenomenon is taken into account. Additionally, in contrast to [149], we shall be able to provide a convergence result of the solutions to the system (3.2.1)–(3.2.2) despite the absence of any displacement term. Last but not least, we show that the convergence is logarithmic. Thereby, this finding may be considered as optimal since no extra geometric condition is required. In the event when the BLR geometric condition [32, 133] on the domain is imposed, the exponential convergence of solutions to their equilibrium state is proved.

3.2.2 Preliminaries and Problem Statement

In this subsection, we set our problem in an appropriate functional space. To proceed, we assume, without loss of generality, that $\beta > 0$. Thereafter, consider the standard change of state variable [184] (see also [149])

$$u(x, \rho, t) = y_t(x, t - \tau\rho), \qquad x \in \Gamma_1,\ \rho \in (0, 1),\ t > 0,$$

and the state space

$$\mathcal{H} = H^1(\Omega) \times L^2(\Omega) \times L^2(\Gamma_1 \times (0, 1)),$$

equipped with the inner product

$$\begin{aligned}\langle (y, z, u), (\tilde{y}, \tilde{z}, \tilde{u})\rangle_{\mathcal{H}} &= \int_\Omega (\nabla y \cdot \nabla \tilde{y} + z\tilde{z})\, dx + \xi \int_0^1 \int_{\Gamma_1} u\tilde{u}\, d\sigma\, d\rho \\ &+\varpi \left[\int_\Omega z\, dx + (\alpha+\beta)\int_{\Gamma_1} y\, d\sigma - \beta\tau \int_0^1\int_{\Gamma_1} u\, d\sigma\, d\rho\right] \\ &\times \left[\int_\Omega \tilde{z}\, dx + (\alpha+\beta)\int_{\Gamma_1} \tilde{y}\, d\sigma - \beta\tau \int_0^1\int_{\Gamma_1} \tilde{u}\, d\sigma\, d\rho\right].\end{aligned} \tag{3.2.1}$$

Here $\varpi > 0$ is a positive constant to be determined, while α, β and ξ satisfy

$$\begin{aligned} &0 < \beta < \alpha, \\ &\tau\beta < \xi < \tau(2\alpha - \beta). \end{aligned} \tag{3.2.2}$$

The result below addresses the issue of equivalence of the usual norm of our state space

$$\mathcal{H} = H^1(\Omega) \times L^2(\Omega) \times L^2(\Gamma_1 \times (0, 1))$$

and that induced by the inner product defined by (3.2.1).

Proposition 3.2.1 *The state space $\mathcal{H}$ equipped with the inner product (3.2.1) is a Hilbert space, provided that ϖ is sufficiently small.*

Proof The immediate task is to show the existence of two positive constants L_1 and L_2 such that

$$\begin{aligned} &L_1\|(y,z,u)\|_{H^1(\Omega)\times L^2(\Omega)\times L^2(\Gamma_1\times(0,1))} \\ &\quad\le \|(y,z,u)\|_{\mathscr{H}} \le L_2\|(y,z,u)\|_{H^1(\Omega)\times L^2(\Omega)\times L^2(\Gamma_1\times(0,1))}. \end{aligned} \tag{3.2.3}$$

Invoking Young's and Hölder's inequalities, (3.2.1) yields

$$\begin{aligned} \|(y,z,u)\|_{\mathscr{H}}^2 \le &\int_\Omega |\nabla y|^2 dx + (1+3\varpi\, mes(\Omega))\int_\Omega z^2\,dx + (\xi+3\beta^2\tau^2) \\ &\times\int_0^1\int_{\Gamma_1} u^2\,d\sigma\,d\rho \\ &+3\varpi(\alpha+\beta)^2 mes(\Gamma_1)\int_{\Gamma_1} y^2\,d\sigma. \end{aligned}$$

Subsequently, the above inequality becomes after using the trace Theorem [4]:

$$\begin{aligned} \|(y,z,u)\|_{\mathscr{H}}^2 \le &(1+3K\varpi(\alpha+\beta)^2 mes(\Gamma_1))\int_\Omega |\nabla y|^2 dx \\ &+3K\varpi(\alpha+\beta)^2 mes(\Gamma_1)\int_\Omega y^2 dx \\ &+(1+3\varpi\, mes(\Omega)))\int_\Omega z^2\,dx + (\xi+3\beta^2\tau^2)\int_0^1\int_{\Gamma_1} u^2\,d\sigma\,d\rho, \end{aligned}$$

where K is a positive constant depending on Ω [4]. Whereupon, the right inequality of (3.2.3) immediately follows. With regard to the left one, we use (3.2.1) and apply Young's inequality to get

$$\begin{aligned} \|(y,z,u)\|_{\mathscr{H}}^2 = &\int_\Omega \left(|\nabla y|^2+z^2\right)dx + \varpi\left[\int_\Omega z\,dx - \beta\tau\int_0^1\int_{\Gamma_1} u\,d\sigma\,d\rho\right]^2 \\ &+\varpi(\alpha+\beta)^2\left[\int_{\Gamma_1} y\,d\sigma\right]^2 \\ &+\xi\int_0^1\int_{\Gamma_1} u^2\,d\sigma\,d\rho + 2\varpi(\alpha+\beta)\left[\int_{\Gamma_1} y\,d\sigma\right] \\ &\times\left[\int_\Omega z\,dx - \beta\tau\int_0^1\int_{\Gamma_1} u\,d\sigma\,d\rho\right] \\ \ge &\int_\Omega \left(|\nabla y|^2+z^2\right)dx + \xi\int_0^1\int_{\Gamma_1} u^2\,d\sigma\,d\rho \end{aligned}$$

$$
+\varpi(\alpha+\beta)\,[\alpha+\beta-\delta]\left[\int_{\Gamma_1} y\,d\sigma\right]^2
$$

$$
+\varpi\left[1-(\alpha+\beta)\delta^{-1}\right]\left[\int_{\Omega} z\,dx-\beta\tau\int_0^1\int_{\Gamma_1} u\,d\sigma\,d\rho\right]^2, \quad (3.2.4)
$$

for any positive constant δ. Now, we apply the generalized Poincaré inequality (see Sect. 2.1.1)

$$
\int_{\Omega} y^2 dx \leq C\left\{\int_{\Omega} |\nabla y|^2 dx+\left(\int_{\Gamma_1} y\,d\sigma\right)^2\right\},
$$

and $C>0$ is the Poincaré constant that depends solely on Ω (see Sect. 2.1.1). The latter implies that

$$
\begin{aligned}
\|(y,z,u)\|_{\mathscr{H}}^2 \geq\ & \varpi(\alpha+\beta)(\alpha+\beta-\delta)C_2^{-1}\int_{\Omega} y^2\,dx \\
& +[1-\varpi(\alpha+\beta)(\alpha+\beta-\delta)]\int_{\Omega}|\nabla y|^2\,dx \\
& +\int_{\Omega} z^2\,dx+\varpi\left[1-(\alpha+\beta)\delta^{-1}\right] \\
& \times\left[\int_{\Omega} z\,dx-\beta\tau\int_0^1\int_{\Gamma_1} u\,d\sigma\,d\rho\right]^2 \\
& +\xi\int_0^1\int_{\Gamma_1} u^2\,d\sigma\,d\rho, \qquad (3.2.5)
\end{aligned}
$$

provided that δ satisfies $\delta<\alpha+\beta$. Applying again Young's and Hölder's inequalities, it follows from (3.2.5) that for any $\delta<\alpha+\beta$

$$
\begin{aligned}
\|(y,z,u)\|_{\mathscr{H}}^2 \geq\ & \varpi(\alpha+\beta)(\alpha+\beta-\delta)C^{-1}\int_{\Omega} y^2\,dx \\
& +[1-\varpi(\alpha+\beta)(\alpha+\beta-\delta)]\int_{\Omega}|\nabla y|^2\,dx \\
& +\left[1+2\varpi(1-(\alpha+\beta)\delta^{-1})mes(\Omega)\right]\int_{\Omega} z^2\,dx \\
& +\left[\xi+2\varpi(1-(\alpha+\beta)\delta^{-1})\beta^2\tau^2 mes(\Gamma_1)\right] \\
& \times\int_0^1\int_{\Gamma_1} u^2\,d\sigma\,d\rho. \qquad (3.2.6)
\end{aligned}
$$

Lastly, we pick up ϖ such that

$$\varpi < \min\left\{\frac{1}{(\alpha+\beta)(\alpha+\beta-\delta)}, \frac{\delta}{2(\alpha+\beta-\delta)mes(\Omega)}, \frac{\delta\xi}{2(\alpha+\beta-\delta)mes(\Gamma_1)}\right\},$$

which gives rise to the second inequality of (3.2.3). □

Let us formulate the system (3.2.1)–(3.2.2) in an abstract differential equation in the Hilbert state space $\mathscr{H}$ equipped with the inner product (3.2.1). To do so, recall that $u(x,\rho,t) = y_t(x, t-\tau\rho)$, $x \in \Gamma_1$, $\rho \in (0,1)$, $t > 0$, and let $z = y_t$, $\Phi = (y, z, u)$. Whereupon, the system can be written as follows:

$$\begin{cases} \Phi_t(t) = \mathscr{A}\Phi(t), \\ \Phi(0) = \Phi_0 = (y_0, z_0, f), \end{cases} \tag{3.2.7}$$

where $\mathscr{A}$ is an unbounded linear operator defined by

$$\mathscr{D}(\mathscr{A}) = \Big\{(y,z,u) \in H^1(\Omega) \times H^1(\Omega) \times L^2(\Gamma_1 \times (0,1));\ \Delta y \in L^2(\Omega);\ \frac{\partial y}{\partial \nu} = 0 \text{ on } \Gamma_0; \\ \frac{\partial y}{\partial \nu} + \alpha z + \beta u(\cdot,1) = 0, \text{ and } z = u(\cdot,0) \text{ on } \Gamma_1\Big\}, \tag{3.2.8}$$

and

$$\mathscr{A}(y,z,u) = (z, \Delta y, -\tau^{-1}u_\rho), \quad \forall (y,z,u) \in \mathscr{D}(\mathscr{A}). \tag{3.2.9}$$

3.2.3 Well-Posedness of the Problem

This subsection addresses the problem of existence and uniqueness of solutions of the system (3.2.7) in $\mathscr{H}$. To this end, we shall invoke semigroups theory (see Sect. 2.3). We have:

Proposition 3.2.2 *Assume that condition (3.2.2) holds. Then, the linear operator $\mathscr{A}$ generates a C_0-semigroup of contractions $S(t)$ on $\mathscr{H} = \overline{\mathscr{D}(\mathscr{A})}$. Additionally, for any initial data $\Phi_0 \in \mathscr{D}(\mathscr{A})$, the system (3.2.7) possesses a unique strong solution $\Phi(t){=}S(t)\Phi_0 \in \mathscr{D}(\mathscr{A})$ for all $t \geq 0$ such that $\Phi(\cdot) \in C^1(\mathbb{R}^+;\mathscr{H}) \cap C(\mathbb{R}^+;\mathscr{D}(\mathscr{A}))$. In turn, if $\Phi_0 \in \mathscr{H}$, then the system (3.2.7) has a unique weak solution $\Phi(t) = S(t)\Phi_0 \in \mathscr{H}$ such that $\Phi(\cdot) \in C^0(\mathbb{R}^+;\mathscr{H})$.*

Proof The ultimate outcome is to prove that the operator $\mathscr{A}$ is m-dissipative.

First, let $\Phi = (y, z, u) \in \mathscr{D}(\mathscr{A})$. In view of (3.2.1) and (3.2.9), we have

$$\langle \mathscr{A}\Phi, \Phi\rangle_{\mathscr{H}} = \int_{\Omega} (\nabla y \cdot \nabla z + \Delta y z)\, dx - \xi\tau^{-1}\int_0^1 \int_{\Gamma_1} u_\rho u \, d\sigma d\rho +$$
$$\varpi \left[\int_{\Omega} \Delta \, dx + (\alpha+\beta)\int_{\Gamma_1} z\, d\sigma + \beta \int_0^1 \int_{\Gamma_1} u_\rho \, d\sigma d\rho\right]$$
$$\left[\int_{\Omega} z\, dx + (\alpha+\beta)\int_{\Gamma_1} y\, d\sigma - \beta\tau \int_0^1 \int_{\Gamma_1} u \, d\sigma d\rho\right].$$

Thanks to Green's formula (2.1.1) and (3.2.8), the latter gives

$$\langle \mathscr{A}\Phi, \Phi\rangle_{\mathscr{H}} = -\alpha \int_{\Gamma_1} z^2\, d\sigma - \beta \int_{\Gamma_1} zu(x,1)\, d\sigma - \xi\tau^{-1}\int_0^1 \int_{\Gamma_1} u_\rho u\, d\sigma d\rho. \tag{3.2.10}$$

Subsequently, we apply Young's inequality and use the identity

$$2\int_0^1 \int_{\Gamma_1} u_\rho u\, d\sigma d\rho = \int_{\Gamma_1} (u^2(x,1) - u^2(x,0))\, d\sigma = \int_{\Gamma_1} (u^2(x,1) - z^2)\, d\sigma,$$

to deduce from (3.2.10) that

$$\langle \mathscr{A}\Phi, \Phi\rangle_{\mathscr{H}} \le \frac{1}{2}\left(\beta - 2\alpha + \xi\tau^{-1}\right)\int_{\Gamma_1} z^2\, d\sigma + \frac{1}{2}\left(\beta - \xi\tau^{-1}\right)\int_{\Gamma_1} u^2(x,1)\, d\sigma. \tag{3.2.11}$$

Now, exploring (3.2.2), one can claim that $\mathscr{A}$ is dissipative.

It remains to show that $(I - \mathscr{A})$ is onto, which amounts to proving that given $(f, g, v) \in \mathscr{H}$, we seek $(y, z, u) \in \mathscr{D}(\mathscr{A})$ such that $(I - \mathscr{A})(y, z, u) = (f, g, v)$, that is,

$$\begin{cases} y - z = f, & \text{in } \Omega \\ z - \Delta y = g, & \text{in } \Omega \\ u_\rho = -\tau u + \tau v, & \text{on } \Gamma_1 \times (0,1), \\ \dfrac{\partial y}{\partial \nu} = 0, & \text{on } \Gamma_0, \\ u(\cdot, 0) = z, & \text{on } \Gamma_1, \\ \dfrac{\partial y}{\partial \nu} + \alpha z + \beta u(\cdot, 1) = 0, & \text{on } \Gamma_1. \end{cases} \tag{3.2.12}$$

Whereupon, $z = y - f$, and then solving the equation of u in (3.2.12) and recalling that $u(\cdot, 0) = z$ on Γ_1, we get

$$u(x,\rho) = e^{-\tau\rho} y(x) - e^{-\tau\rho} f(x) + \tau \int_0^\rho e^{\tau(\eta - \rho)} v(x,\eta)\, d\eta. \tag{3.2.13}$$

Hence,

$$u(x,1) = e^{-\tau} y(x) + v_f(x),$$

where

$$v_f(x) = -e^{-\tau} f(x) + \tau \int_0^1 e^{\tau(\eta-1)} v(x,\eta)\, d\eta.$$

By virtue of the above arguments, we only seek $y \in H^2(\Omega)$ satisfying

$$\begin{cases} y^2 - \Delta y = f + g, & \text{in } \Omega \\ \dfrac{\partial y}{\partial \nu} = 0, & \text{on } \Gamma_0, \\ \dfrac{\partial y}{\partial \nu} + \left(\alpha + \beta e^{-\tau}\right) y - \alpha f + \beta v_f = 0, & \text{on } \Gamma_1. \end{cases} \tag{3.2.14}$$

Invoking once again Green's formula (2.1.1), the problem (3.2.14) has the following variational form:

$$\begin{aligned} &\int_\Omega \left(y\phi + \nabla y \cdot \nabla \phi \right) dx + \int_{\Gamma_1} \left(\alpha + \beta e^{-\tau}\right) y\phi \, d\sigma \\ &= \int_\Omega (f+g)\, \phi \, dx + \int_{\Gamma_1} (\alpha f - \beta v_f)\phi \, d\sigma, \end{aligned} \tag{3.2.15}$$

for any $\phi \in H^1(\Omega)$. Invoking Lax-Milgram theorem (see Theorem 2.1.6), one can prove that (3.2.15) admits a unique solution $y \in H^1(\Omega)$. Next, in light of standard arguments used for solving elliptic linear equations, one can recover the boundary conditions in (3.2.14). Herewith, we reach the desired result $I - \mathscr{A}$ is onto. Finally, $\mathscr{A}$ is densely defined closed in $\mathscr{H}$, and as an immediate consequence of Lumer-Phillips Theorem 2.3.4, the operator $\mathscr{A}$ generates a C_0-semigroup of contractions $S(t)$ on $\mathscr{H}$. The rest of the claims in Proposition 3.2.2 can be derived from semigroups theory [158]. □

3.2.4 Asymptotic Behavior

In this subsection, we establish an asymptotic behavior result for the unique solution of (3.2.7) in $\mathscr{H}$.

The first main result is as follows:

Theorem 3.2.1 *Assume that condition (3.2.2) holds. Then, for any initial data* $\Phi_0 = (y_0, z_0, f) \in \mathcal{H}$*, the solution* $\Phi(t) = (y(\cdot,t), y_t(\cdot,t)), y_t(\cdot, t-\tau\rho))$ *of (3.2.7) tends in* $\mathcal{H}$ *to* $(\chi, 0, 0)$ *as* $t \longrightarrow +\infty$*, where*

$$\chi = ((\alpha+\beta)mes(\Gamma_1))^{-1}\left(\int_\Omega z_0\, dx + (\alpha+\beta)\int_{\Gamma_1} y_0\, d\sigma - \beta\tau\int_0^1\int_{\Gamma_1} f\, d\sigma d\rho\right).$$

Proof Using a standard density argument and the contraction of the semigroup $S(t)$, it suffices to prove Theorem 3.2.1 for smooth initial data $\Phi_0 \in \mathcal{D}(\mathcal{A}^2)$. Thereby, let $\Phi(t) = S(t)\Phi_0$ be the solution of (3.2.7). It follows from Proposition 3.2.2 that the solutions trajectory $\{\Phi(t)\}_{t\ge 0}$ is a bounded set for the graph norm and thus precompact. Applying LaSalle's principle (see Theorem 2.3.11), we can claim that $\omega(\Phi_0)$ is non-empty, compact, invariant under the semigroup $S(t)$, and in addition $S(t)\Phi_0 \longrightarrow \omega(\Phi_0)$ as $t \to +\infty$ (see Theorem 2.3.11). Thenceforth, it suffices to show that $\omega(\Phi_0)$ reduces to $(\chi, 0, 0)$. To this end, let $\tilde{\Phi}_0 = (\tilde{y}_0, \tilde{z}_0, \tilde{u}_0) \in \omega(\Phi_0) \subset \mathcal{D}(\mathcal{A})$ and consider the unique strong solution of (3.2.1)–(3.2.2) $\tilde{\Phi}(t) = (\tilde{y}(t), \tilde{y}_t(t), \tilde{y}_t(\cdot, t-\tau\rho)) = S(t)\tilde{\Phi}_0 \in \mathcal{D}(\mathcal{A})$. In view of the fact that $\|\tilde{\Phi}(t)\|_{\mathcal{H}}$ is constant [114], we have $< \mathcal{A}\tilde{\Phi}, \tilde{\Phi} >_{\mathcal{H}} = 0$. Putting the above deductions together with (3.2.11) yields $\tilde{z} = \tilde{y}_t = 0$ and $\tilde{u}(x,1) = \tilde{y}_t(x, t-\tau) = 0$ on Γ_1. Whence $\tilde{y}$ satisfies the following:

$$\begin{cases} \tilde{y}_{tt} - \Delta\tilde{y} = 0, & \text{in } \Omega \\ \dfrac{\partial\tilde{y}}{\partial\nu} = 0, & \text{on } \Gamma_0, \\ \tilde{y}_t = \dfrac{\partial\tilde{y}}{\partial\nu} = 0, & \text{on } \Gamma_1, \\ \tilde{y}(0) = \tilde{y}_0;\ \tilde{y}_t(0) = \tilde{z}_0, & \text{in } \Omega, \\ \tilde{y} \in H^2(\Omega), \end{cases} \tag{3.2.16}$$

which in turn implies that $\tilde{z} = \tilde{y}_t$ is a solution of

$$\begin{cases} \tilde{z}_{tt} - \Delta\tilde{z} = 0, & \text{in } \Omega \\ \tilde{z} = \dfrac{\partial\tilde{z}}{\partial\nu} = 0, & \text{on } \Gamma_0, \\ \tilde{z} = \dfrac{\partial\tilde{z}}{\partial\nu} = 0, & \text{on } \Gamma_1. \end{cases} \tag{3.2.17}$$

Thanks to Holmgren's uniqueness theorem for the system (3.2.17), it follows that $\tilde{z} = 0$, and thus the system (3.2.16) becomes

$$\begin{cases} \Delta\tilde{y} = 0, & \text{in } \Omega \\ \dfrac{\partial\tilde{y}}{\partial\nu} = 0, & \text{on } \Gamma_0, \\ \dfrac{\partial\tilde{y}}{\partial\nu} = 0, & \text{on } \Gamma_1. \end{cases}$$

Consequently, $\tilde{y}$ is constant. In summary, we have proved that given $\tilde{\Phi}_0 = (\tilde{y}_0, \tilde{z}_0, \tilde{u}_0) \in \omega(\Phi_0) \subset \mathscr{D}(\mathscr{A})$, the solution $\tilde{\Phi}(t) = (\tilde{y}(t), \tilde{y}_t(t), \tilde{y}_t(\cdot, t-\rho\tau)) = S(t)\tilde{\Phi}_0 \in \mathscr{D}(\mathscr{A})$ satisfies $(\tilde{y}(t), \tilde{y}_t(t), \tilde{y}_t(\cdot, t-\rho\tau)) = (\chi, 0, 0)$, for any $t \geq 0$, where χ is a real constant. Henceforth, the ω-limit set $\omega(\Phi_0)$ consists of constants $(\chi, 0)$. It remains now to provide an explicit form of χ. To proceed, assume that $(\chi, 0, 0) \in \omega(\Phi_0)$. This implies that there exists $\{t_n\} \to \infty$, as $n \to \infty$ such that

$$\Phi(t_n) = (y(t_n), y_t(t_n), y_t(\cdot, t_n - \rho\tau)) = S(t_n)\Phi_0 \longrightarrow (\chi, 0, 0), \quad \text{as } n \to \infty, \tag{3.2.18}$$

in $\mathscr{H}$. On the other hand, we claim that any solution of (3.2.1)–(3.2.2) that obeys the following property

$$\mathscr{E}(t) = \int_\Omega y_t \, dx + (\alpha+\beta)\int_{\Gamma_1} y \, d\sigma - \beta\tau \int_0^1 \int_{\Gamma_1} y_t(x, t-\rho\tau) \, d\sigma \, d\rho$$

is time invariant. To ascertain the correctness of this claim, let us differentiate the above expression with respect to t and then use (3.2.1)–(3.2.2) together with Green's formula (2.1.1). A straightforward computation gives

$$\begin{aligned}
\mathscr{E}_t(t) &= -\int_{\Gamma_1} (\alpha y_t + \beta y_t(x, t-\tau)) \, d\sigma + (\alpha+\beta)\int_{\Gamma_1} y_t \, d\sigma \\
&\quad -\beta\tau \int_0^1 \int_{\Gamma_1} y_{tt}(x, t-\rho\tau) \, d\sigma \, d\rho \\
&= -\beta \int_{\Gamma_1} y_t(x, t-\tau) \, d\sigma + \beta \int_{\Gamma_1} y_t \, d\sigma \\
&\quad -\beta\tau \int_0^1 \int_{\Gamma_1} \tau^{-2} y_{\rho\rho}(x, t-\rho\tau) \, d\sigma \, d\rho \\
&= -\beta \int_{\Gamma_1} y_t(x, t-\tau) \, d\sigma + \beta \int_{\Gamma_1} y_t \, d\sigma \\
&\quad -\beta\tau^{-1} \int_{\Gamma_1} (y_\rho(x, t-\tau) - y_\rho(x, t)) \, d\sigma \\
&= -\beta \int_{\Gamma_1} y_t(x, t-\tau) \, d\sigma + \beta \int_{\Gamma_1} y_t \, d\sigma \\
&\quad -\beta\tau^{-1} \int_{\Gamma_1} (-\tau)(y_t(x, t-\tau) - y_t(x, t)) \, d\sigma = 0,
\end{aligned}$$

which confirms our claim. This yields

$$\int_\Omega y_t \, dx + (\alpha+\beta)\int_{\Gamma_1} y \, d\sigma - \beta\tau \int_0^1 \int_{\Gamma_1} y_t(x, t-\rho\tau) \, d\sigma \, d\rho$$

$$
\begin{aligned}
&= \int_{\Omega} z_0\, dx + (\alpha+\beta)\int_{\Gamma_1} y_0\, d\sigma \\
&\quad -\beta\tau \int_0^1 \int_{\Gamma_1} f\, d\sigma d\rho. \qquad (3.2.19)
\end{aligned}
$$

It suffices now to pick up $t = t_n$ in (3.2.19), and let $n \to \infty$. This, together with (3.2.18), implies that

$$
0 + (\alpha+\beta)\int_{\Gamma_1} \chi\, d\sigma - 0 = \int_{\Omega} z_0\, dx + (\alpha+\beta)\int_{\Gamma_1} y_0\, d\sigma - \beta\tau \int_0^1 \int_{\Gamma_1} f\, d\sigma d\rho,
$$

and hence,

$$
\chi = ((\alpha+\beta) mes(\Gamma_1))^{-1} \left(\int_{\Omega} z_0\, dx + (\alpha+\beta)\int_{\Gamma_1} y_0\, d\sigma - \beta\tau \int_0^1 \int_{\Gamma_1} f\, d\sigma d\rho \right).
$$

□

3.2.5 *Logarithmic Convergence*

In the previous subsection, the solutions of (3.2.1)–(3.2.2) (see also (3.2.7)) are shown to be asymptotically converging to a steady state. It is then legitimate to wonder what type of convergence we can expect. In this subsection, we will mainly be concerned with the answer of such a question by showing, without geometric condition (BLR) on the control zone Γ_1, that the convergence is actually logarithmic.

Let $\dot{\mathscr{H}}$ the closed subspace of $\mathscr{H}$ and of codimension 1 given by

$$
\begin{aligned}
\dot{\mathscr{H}} = \Big\{ &(y, z, u) \in \mathscr{H};\ \int_{\Omega} z(x)\, dx - \beta\tau \\
&\times \int_0^1 \int_{\Gamma_1} u(x,\rho)\, d\sigma\, d\rho + (\alpha+\beta)\int_{\Gamma_1} y\, d\sigma = 0 \Big\},
\end{aligned}
$$

and denote by $\dot{\mathscr{A}}$ a new operator defined as follows:

$$
\begin{aligned}
&\dot{\mathscr{A}} : \mathscr{D}(\dot{\mathscr{A}}) := \mathscr{D}(\mathscr{A}) \cap \dot{\mathscr{H}} \subset \dot{\mathscr{H}} \to \dot{\mathscr{H}}, \\
&\dot{\mathscr{A}}(y, z, u) = \mathscr{A}(y, z, u),\ \forall (y, z, u) \in \mathscr{D}(\dot{\mathscr{A}}). \qquad (3.2.20)
\end{aligned}
$$

Assume that condition (3.2.2) holds. Then, we see, thanks to the results of previous subsections, that the new operator $\dot{\mathscr{A}}$ defined by (3.2.20) generates a C_0-semigroup of contractions $e^{t\dot{\mathscr{A}}}$ on $\dot{\mathscr{H}}$. Furthermore, $\sigma(\dot{\mathscr{A}})$, the spectrum of $\dot{\mathscr{A}}$, consists of isolated eigenvalues of finite algebraic multiplicity only (see Remark 2.1.2).

More importantly, the semigroup operator $e^{t\dot{\mathscr{A}}}$ is logarithmically stable on $\dot{\mathscr{H}}$. Indeed, our second main result is:

Theorem 3.2.2 *Under the conditions (3.2.2), there exists $C > 0$ such that for all $t > 0$, we have*

$$\left\| e^{t\dot{\mathscr{A}}} \right\|_{\mathscr{L}(\mathscr{D}(\dot{\mathscr{A}}),\dot{\mathscr{H}})} \leq \frac{C}{\log(2+t)}.$$

Proof We will use the frequency-domain theorem for logarithmic stability (see Theorem 2.3.9). The proof of Theorem 3.2.2 is based on the following lemmas.

We first look at the point spectrum. □

Lemma 3.2.1 *If γ is a real number, then $i\gamma$ is not an eigenvalue of $\dot{\mathscr{A}}$.*

Proof We will show that the equation

$$\dot{\mathscr{A}} Z = i\gamma Z \tag{3.2.21}$$

with $Z = (y, z, u) \in \mathscr{D}(\dot{\mathscr{A}})$ and $\gamma \in \mathbb{R}$ has only one trivial solution. Clearly, the system (3.2.21) writes

$$z = i\gamma y \tag{3.2.22}$$

$$\Delta y = i\gamma z, \tag{3.2.23}$$

$$-\frac{u_\rho}{\tau} = i\gamma u, \tag{3.2.24}$$

$$\int_\Omega z(x)\,dx - \beta\tau \int_0^1 \int_{\Gamma_1} u(x,\rho)\,d\sigma\,d\rho + (\alpha+\beta)\int_{\Gamma_1} y\,d\sigma = 0, \tag{3.2.25}$$

$$\frac{\partial y}{\partial \nu} = 0 \text{ on } \Gamma_0, \tag{3.2.26}$$

$$\frac{\partial y}{\partial \nu} + \alpha z + \beta u(\cdot,1) = 0, \text{ and } z = u(\cdot,0) \text{ on } \Gamma_1. \tag{3.2.27}$$

Let us first deal with the case where $\gamma = 0$. In such an event, it is clear, thanks to (3.2.22)–(3.2.27), that the only solution of (3.2.21) is the trivial one.

Let us now suppose that $\gamma \neq 0$. Taking the inner product of (3.2.21) with Z and using inequality (3.2.11), we get

$$\begin{aligned} \Re e\left(< \dot{\mathscr{A}} Z, Z >_{\dot{\mathscr{H}}}\right) \leq \frac{1}{2}\Big((\beta - 2\alpha + \xi\tau^{-1}) \\ \times \int_{\Gamma_1} |z|^2 \sigma + (\beta - \xi\tau^{-1}) \int_{\Gamma_1} |u(x,1)|^2\,d\sigma\Big)\ (\leq 0). \end{aligned} \tag{3.2.28}$$

Whereupon, $z = u(\cdot, 1) = 0$ on Γ_1, and consequently, the only solution of (3.2.21) is the trivial one. □

The second lemma is:

Lemma 3.2.2 *The resolvent operator of* $\dot{\mathscr{A}}$ *satisfies condition* (2.3.18).

Proof Suppose that condition (2.3.18) is false. By virtue of Banach-Steinhaus theorem (see Theorem 2.1.7), there exist a sequence of real numbers $\gamma_n \to \infty$ and a sequence of vectors $Z_n = (y_n, z_n, u_n) \in \mathscr{D}(\dot{\mathscr{A}})$ with $\|Z_n\|_{\dot{\mathscr{H}}} = 1$ such that

$$\|e^{c|\gamma_n|}(i\gamma_n I - \dot{\mathscr{A}})Z_n\|_{\dot{\mathscr{H}}} \to 0 \quad \text{as} \quad n \to \infty, \tag{3.2.29}$$

i.e.,

$$e^{c|\gamma_n|}(i\gamma_n y_n - z_n) \equiv e^{c|\gamma_n|} f_n \to 0 \quad \text{in} \quad H^1(\Omega), \tag{3.2.30}$$

$$e^{c|\gamma_n|}(i\gamma_n z_n - \Delta y_n) \equiv e^{c|\gamma_n|} g_n \to 0 \quad \text{in} \quad L^2(\Omega), \tag{3.2.31}$$

$$e^{c|\gamma_n|}\left(i\gamma_n u_n + \frac{(u_n)_\rho}{\tau}\right) \equiv e^{c|\gamma_n|} v_n \to 0 \quad \text{in} \quad L^2(\Gamma_1; (0,1)). \tag{3.2.32}$$

The task ahead is to derive from (3.2.29) that $\|Z_n\|_{\dot{\mathscr{H}}}$ converges to zero. Obviously, such a result contradicts the fact that for all $n \in \mathbb{N}$ $\|Z_n\|_{\dot{\mathscr{H}}} = 1$. For sake of clarity, we shall prove our result by proceeding by steps:

- **First step.**
  We first notice that we have

$$||e^{c|\gamma_n|}(i\gamma_n I - \dot{\mathscr{A}})Z_n||_{\dot{\mathscr{H}}} \geq \left|\Re e\left(\langle e^{c|\gamma_n|}(i\gamma_n I - \dot{\mathscr{A}})Z_n, Z_n\rangle_{\dot{\mathscr{H}}}\right)\right|. \tag{3.2.33}$$

Thus, (3.2.28) and (3.2.29) yield

$$e^{\frac{c|\gamma_n|}{2}} z_n \to 0 \text{ in } L^2(\Gamma_1),\; e^{\frac{c|\gamma_n|}{2}} u_n(\cdot, 1) \to 0 \text{ in } L^2(\Gamma_1), \tag{3.2.34}$$

and hence,

$$u_n(\cdot, 0) \to 0 \text{ in } L^2(\Gamma_1). \tag{3.2.35}$$

Also, we have according to (3.2.30) and (3.2.34)

$$i\gamma_n e^{\frac{c|\gamma_n|}{2}} y_n = e^{\frac{c|\gamma_n|}{2}} z_n + e^{\frac{c|\gamma_n|}{2}} f_n \to 0 \text{ in } L^2(\Gamma_1). \tag{3.2.36}$$

Therefore,

$$e^{\frac{c|\gamma_n|}{2}} y_n \to 0 \text{ in } L^2(\Gamma_1). \tag{3.2.37}$$

In light of (3.2.32), we have

$$u_n(x,\rho) = u_n(x,0)\, e^{-i\tau\gamma_n x} + \tau \int_0^\rho e^{-i\tau\gamma_n(\rho-s)}\, v_n(s)\, ds.$$

This, together with (3.2.32) and (3.2.35), gives

$$u_n \to 0 \ \text{ in } \ L^2(\Gamma_1, (0,1))\,. \tag{3.2.38}$$

- **Second step.**
  We take the inner product of (3.2.31) with $z_n = i\gamma_n y_n - f_n$ in $L^2(\Omega)$ to obtain

$$\int_\Omega |z_n|^2\, dx - \int_\Omega |\nabla y_n|^2\, dx =$$

$$- \int_{\Gamma_1} \frac{\partial y_n}{\partial \nu}\, \overline{y}\, d\sigma + \frac{1}{i\gamma_n} \int_\Omega \nabla y_n \cdot \nabla \overline{f}_n\, dx - \frac{1}{i\gamma_n} \int_{\Gamma_1} \frac{\partial y_n}{\partial \nu}\, \overline{f}_n\, d\sigma$$
$$+ \frac{1}{i\gamma_n} \int_\Omega g_n\, \overline{z}_n\, dx = \text{o}(1). \tag{3.2.39}$$

Moreover, according to the interpolation inequality given in [133, Theorem 3], there exists a constant $C > 0$ such that for sufficiently large n we have

$$\|y_n\|_{H^1(\Omega)} \le C\, e^{\frac{c|\gamma_n|}{2}} \left(\|f_n\|_{H^1(\Omega)} + \|g_n\|_{L^2(\Omega)} + \|y_n\|_{L^2(\Gamma_1)} \right).$$

Thereafter, it follows from (3.2.30), (3.2.31), and (3.2.37) that

$$y_n \to 0 \text{ in } H^1(\Omega). \tag{3.2.40}$$

Combining (3.2.39) and (3.2.40), we get

$$z_n \to 0 \text{ in } L^2(\Omega). \tag{3.2.41}$$

Lastly, amalgamating (3.2.38), (3.2.40), and (3.2.41), we clearly reach the contradiction with $\|Z_n\|_{\mathcal{H}} = 1$, for all $n \in \mathbb{N}$.

□

The two hypotheses of Lemma 2.3.9 are proved, and hence, the proof of Theorem 3.2.2 is achieved.

Remark 3.2.1 It is worth mentioning that one may consider dynamical boundary conditions, that is, instead of (3.2.2), one can treat

$$\begin{cases} m(x)y_{tt}(x,t) + \dfrac{\partial y}{\partial \nu}(x,t) = 0, & (x,t) \in \Gamma_0 \times (0,\infty) \\ M(x)y_{tt}(x,t) + \dfrac{\partial y}{\partial \nu} = -\alpha y_t(x,t) - \beta y_t(x,t-\tau), & (x,t) \in \Gamma_1 \times (0,\infty) \\ y(x,0) = y_0(x) \in H^1(\Omega),\ y_t(x,0) = z_0(x) \in L^2(\Omega), & \\ y_t|_{\Gamma_0}(x,0) = w_0^0(x) \in L^2(\Gamma_0),\ y_t|_{\Gamma_1}(x,0) & \\ = w_1^0(x) \in L^2(\Gamma_1), & \\ y_t(x,t) = f(x,t), & (x,t) \in \Gamma_1 \times (-\tau,0), \end{cases} \tag{3.2.42}$$

in which

$$\begin{cases} m \in L^\infty(\Gamma_0);\ m(x) \geq m_0 > 0, \text{ for all } x \in \Gamma_0; \\ M \in L^\infty(\Gamma_1);\ M(x) \geq M_1 > 0, \text{ for all } x \in \Gamma_1. \end{cases} \tag{3.2.43}$$

Note that in this case, most of the arguments used run on much the same lines as we did previously but, of course, with a number of changes born out of necessity. For instance, the new state space is

$$\mathscr{H}_d = H^1(\Omega) \times L^2(\Omega) \times L^2(\Gamma_1 \times (0,1)) \times L^2(\Gamma_0) \times L^2(\Gamma_1),$$

equipped with the inner product

$$\begin{aligned} &\Big\langle (y,z,u,w_0,w_1), (\tilde{y},\tilde{z},\tilde{u},\tilde{w}_0,\tilde{w}_1) \Big\rangle_{\mathscr{H}_d} \\ &= \int_\Omega (y\tilde{y} + z\tilde{z})\,dx + \xi \int_0^1 \int_{\Gamma_1} u\tilde{u}\,d\sigma\,d\rho + \int_{\Gamma_0} m w_0 \tilde{w}_0\,d\sigma + \\ &\int_{\Gamma_1} M w_1 \tilde{w}_1\,d\sigma + \varpi \left(\int_\Omega z\,dx + \int_{\Gamma_0} m w_0\,d\sigma + \int_{\Gamma_1} (M w_1 + (\alpha+\beta) y)\,d\sigma \right. \\ &\left. -\beta\tau \textstyle\int_0^1 \int_{\Gamma_1} \tilde{u}\,d\sigma\,d\rho \right) \\ &\times \left(\int_\Omega \tilde{z}\,dx + \int_{\Gamma_0} m\tilde{w}_0\,d\sigma + \int_{\Gamma_1} (M\tilde{w}_1 + (\alpha+\beta)\tilde{y})\,d\sigma - \beta\tau \int_0^1 \int_{\Gamma_1} \tilde{u}\,d\sigma\,d\rho \right). \end{aligned} \tag{3.2.44}$$

Thereafter, arguing as in the proof of Proposition 3.2.1, one can show that the above inner product generates an equivalent norm to the standard one provided that $\varpi > 0$ is small enough. Additionally, the novel system operator has just two additional components w_0 and w_1 compared to $\mathscr{A}$ defined in (3.2.8)–(3.2.9). Lastly, it is a simple task to check that all the outcomes of this section such as well-posedness, asymptotic convergence, and logarithmic decay remain valid with dynamical boundary conditions (3.2.42).

3.2.6 Exponential Convergence

Assuming in the first lieu that Γ_1 is non-empty, while Γ_0 may be empty, we shall invoke the BLR geometric condition [32, 133] in the "control zone" Γ_1 in order to show that the solutions to the system (3.2.1)–(3.2.2) (see also (3.2.7)) converge exponentially to an equilibrium state. The outcomes of this subsection will improve those previously obtained, where only the logarithmic convergence has been established.

First, recalling the notations of Sect. 3.2.5 (particularly (3.2.20)), we have:

Theorem 3.2.3 *Assume that assumption (3.2.2) holds. If Γ_1 satisfies a BLR geometrical control condition, then there exist two positive constants C and ω such that*

$$\left\| e^{t\dot{\mathscr{A}}} \right\|_{\mathscr{L}(\dot{\mathscr{H}})} \leq C\, e^{-\omega t}, \text{ for all } t > 0.$$

Therefore, the solutions of system (3.2.1)–(3.2.2) (see also (3.2.7)*) exponentially approach the steady equilibrium state χ.*

Proof Recalling the resolvent method (see Theorem 2.3.8) for the exponential stability of a C_0-semigroup of contractions on a Hilbert space, it suffices to show that

$$\rho(\dot{\mathscr{A}}) \supset \{i\gamma;\ \gamma \in \mathbb{R}\} \equiv i\mathbb{R}, \tag{3.2.45}$$

$$\limsup_{|\gamma|\to+\infty} \|(i\gamma I - \dot{\mathscr{A}})^{-1}\|_{\mathscr{L}(\dot{\mathscr{H}})} < \infty, \tag{3.2.46}$$

where $\rho(\dot{\mathscr{A}})$ denotes the resolvent set of the operator $\dot{\mathscr{A}}$.

The first concern is to prove that $i\gamma$ is not an eigenvalue of $\dot{\mathscr{A}}$ for any real number γ, which clearly gives (3.2.45). To proceed, one has to check that the only solution to the equation $\dot{\mathscr{A}}Z = i\gamma Z,\ Z = (y, z, u) \in \mathscr{D}(\dot{\mathscr{A}}),\ \gamma \in \mathbb{R}$, is the trivial solution. This desired result is already obtained in Lemma 3.2.1.

Next, assume that (3.2.46) does not hold. This gives rise, thanks to Banach-Steinhaus theorem (see Theorem 2.1.7), to the existence of a sequence of real numbers $\gamma_n \to \infty$ and a sequence of vectors $Z_n = (y_n, z_n, u_n) \in \mathscr{D}(\dot{\mathscr{A}})$ with $\|Z_n\|_{\dot{\mathscr{H}}} = 1$ such that

$$\|(i\gamma_n I - \dot{\mathscr{A}})Z_n\|_{\dot{\mathscr{H}}} \to 0, \quad \text{as} \quad n \to \infty, \tag{3.2.47}$$

that is,

$$i\gamma_n y_n - z_n \equiv f_n \to 0 \quad \text{in} \quad H^1(\Omega), \tag{3.2.48}$$

$$i\gamma_n z_n - \Delta y_n \equiv g_n \to 0 \quad \text{in} \quad L^2(\Omega), \tag{3.2.49}$$

$$i\gamma_n u_n + \frac{(u_n)_\rho}{\tau} \equiv v_n \to 0 \quad \text{in} \quad L^2(\Gamma_1 \times (0,1)). \tag{3.2.50}$$

The ultimate outcome is to show the convergence of $\|Z_n\|_{\dot{\mathscr{H}}}$ to zero as $n \to \infty$, and hence a contradiction with $\|Z_n\|_{\dot{\mathscr{H}}} = 1$, for all $n \in \mathbb{N}$. First, note that

$$\begin{aligned}&\left|\Re e\left(\langle (i\gamma_n I - \dot{\mathscr{A}})Z_n, Z_n\rangle_{\dot{\mathscr{H}}}\right)\right| \\ &\quad \geq \frac{1}{2}\left[(2\alpha - \beta - \xi\tau^{-1})\int_{\Gamma_1} |z|^2\, d\sigma + (\xi\tau^{-1} - \beta)\int_{\Gamma_1} |u(x,1)|^2\, d\sigma\right],\end{aligned}$$

and hence, (3.2.29) yields

$$z_n \to 0, \quad \text{and} \quad u_n(\cdot, 1) \to 0 \text{ in } L^2(\Gamma_1). \tag{3.2.51}$$

Therewith

$$u_n(\cdot, 0) \to 0 \text{ in } L^2(\Gamma_1). \tag{3.2.52}$$

Amalgamating (3.2.48) and (3.2.51), we get $i\gamma_n\, y_n = z_n + f_n \to 0$ in $L^2(\Gamma_1)$. Thus, we have

$$y_n \to 0 \text{ in } L^2(\Gamma_1). \tag{3.2.53}$$

Furthermore, (3.2.50) yields

$$u_n(x,\rho) = u_n(x,0)\, e^{-i\tau\gamma_n x} + \tau \int_0^\rho e^{-i\tau\gamma_n(\rho - s)}\, v_n(s)\, ds.$$

Combining the above deduction with (3.2.50) and (3.2.52), we deduce

$$u_n \to 0 \quad \text{in} \quad L^2(\Gamma_1 \times (0,1))\,. \tag{3.2.54}$$

Now, let us take the inner product of (3.2.49) with $z_n = i\gamma_n y_n - f_n$ in $L^2(\Omega)$. A straightforward calculation gives

$$\begin{aligned}\int_\Omega |z_n|^2\, dx - \int_\Omega |\nabla y_n|^2\, dx &= -\int_{\Gamma_1} \frac{\partial y_n}{\partial \nu}\, \overline{y}_n\, d\sigma + \frac{1}{i\gamma_n}\int_\Omega \nabla y_n \cdot \nabla \overline{f}_n\, dx \\ &\quad - \frac{1}{i\gamma_n}\int_{\Gamma_1} \frac{\partial y_n}{\partial \nu}\, \overline{f}_n\, d\sigma + \frac{1}{i\gamma_n}\int_\Omega g_n\, \overline{z}_n\, dx \\ &= \circ(1).\end{aligned} \tag{3.2.55}$$

The intention now is to evoke the BLR condition to our situation. To do so, consider a new space $\dot{\mathscr{H}_0}$ defined as follows:

$$\dot{\mathscr{H}_0} := \left\{(y, z) \in \mathscr{H}_0;\ \int_\Omega z(x)\,dx + \alpha \int_{\Gamma_1} y\,d\sigma = 0\right\}.$$

Note that $\dot{\mathscr{H}_0}$ is a closed subspace of $\mathscr{H}_0 := H^1(\Omega) \times L^2(\Omega)$ and of codimension 1.

Thereupon, consider the operator $\dot{\mathscr{A}_0}$ defined by

$$\begin{aligned}
&\mathscr{D}(\dot{\mathscr{A}_0}) = \left\{(y, z) \in \mathscr{H}_0;\ (z, \Delta y) \in \mathscr{H}_0,\ \tfrac{\partial y}{\partial \nu} = 0 \text{ on } \Gamma_0,\ \tfrac{\partial y}{\partial \nu} + \alpha\, z = 0 \text{ on } \Gamma_1\right\} \\
&\cap \dot{\mathscr{H}_0}, \\
&\dot{\mathscr{A}_0}(y, z) = (z, \Delta y),\ \forall\, (y, z) \in \mathscr{D}(\dot{\mathscr{A}_0}).
\end{aligned} \tag{3.2.56}$$

In view of the BLR geometric control condition [32], we have

$$\limsup_{|\gamma| \to +\infty} \|(i\gamma I - \dot{\mathscr{A}_0})^{-1}\|_{\mathscr{L}(\dot{\mathscr{H}_0})} < \infty. \tag{3.2.57}$$

Next, exploring the fact that y_n satisfies the following problem:

$$\begin{cases}
-\gamma_n^2 y_n - \Delta y_n = i\gamma_n f_n + g_n, \text{ on } \Omega, \\
\frac{\partial y_n}{\partial \nu} = 0, \text{ on } \Gamma_0, \quad \frac{\partial y_n}{\partial \nu} + i\,\alpha\,\gamma_n\, y_n = \alpha\, f_n - \beta\, u_n(\cdot, 1), \text{ on } \Gamma_1,
\end{cases}$$

we claim due to (3.2.57) that there exists a positive constant C such that for sufficiently large n, we have

$$\|(y_n, z_n)\|_{\mathscr{H}_0} \le C \left(\|(f_n, g_n)\|_{\mathscr{H}_0} + \|u_n(., 1)\|_{L^2(\Gamma_1)}\right).$$

This, together with (3.2.48)–(3.2.49) and (3.2.53), yields

$$y_n \to 0 \text{ in } H^1(\Omega). \tag{3.2.58}$$

Merging (3.2.55) and (3.2.58), we get

$$z_n \to 0 \text{ in } L^2(\Omega). \tag{3.2.59}$$

By virtue of (3.2.54) and (3.2.58)–(3.2.59), we reach that $\|Z_n\|_{\dot{\mathscr{H}}} \to 0$, which is our target. The proof of Theorem 3.2.3 is completed. ◇ □

3.2.7 Conclusion

Two convergence results (logarithmic and exponential) are established in this section for a boundary delayed wave equation. The type of convergence depends on the availability of the BLR condition.

One promising research avenue is to study the long-time behavior of solutions to the wave equation (3.2.1)–(3.2.2) but when the delay is time dependent.

Another interesting problem to tackle is the wave equation (3.2.1) with nonlinear boundary delay, that is,

$$\frac{\partial y}{\partial \nu}(x,t) = -f(y_t(x,t)) - g(y_t(x,t-\tau)), \quad \text{on } \Gamma_1 \times (0,\infty),$$

or the time-dependent case

$$\frac{\partial y}{\partial \nu}(x,t) = -f(y_t(x,t)) - g(y_t(x,t-\tau(t))), \quad \text{on } \Gamma_1 \times (0,\infty),$$

where f and g are nonlinear functions. This problem is definitely more challenging.

3.3 Dynamic Boundary Delay: Exponential Stabilization

In this second section, we deal with an interior stabilization problem for the wave equation with dynamic boundary delay. Assuming that $meas(\Gamma_0) > 0$, we shall prove stabilization results under the choice of an interior damping control. The proof of the main result is based on the resolvent method together with a contradiction argument.

3.3.1 Introduction

The system under consideration is

$$\begin{cases} u_{tt} - \Delta u + a\,u_t = 0, & x \in \Omega,\ t > 0, \\ u(x,t) = 0, & x \in \Gamma_0,\ t > 0, \\ u_{tt}(x,t) = -\dfrac{\partial u}{\partial \nu}(x,t) - \mu u_t(x,t-\tau), & x \in \Gamma_1,\ t > 0, \\ u(x,0) = u_0(x),\ u_t(x,0) = u_1(x), & x \in \Omega, \\ u_t(x,t-\tau) = f_0(x,t-\tau), & x \in \Gamma_1,\ t \in (0,\tau), \end{cases} \tag{3.3.1}$$

where the constant $\tau > 0$ is the time delay, a and μ are positive numbers, and the initial data u_0, u_1, f_0 are given functions belonging to suitable spaces that will be specified later.

Let us first review some results for particular cases that seem to us interesting. Before doing so, note that the problem (3.3.1) without delay ($\tau = 0$) is

$$\begin{cases} u_{tt} - \Delta u + a\, u_t = 0, & x \in \Omega,\ t > 0, \\ u(x,t) = 0, & x \in \Gamma_0,\ t > 0, \\ u_{tt}(x,t) = -\dfrac{\partial u}{\partial \nu}(x,t) - \mu u_t(x,t), & x \in \Gamma_1,\ t > 0, \\ u(x,0) = u_0(x),\ u_t(x,0) = u_1(x), & x \in \Omega. \end{cases} \tag{3.3.2}$$

This kind of system arises in modeling longitudinal vibrations in a homogeneous bar in which there are viscous effects. The damping au_t indicates that the stress is proportional not only to the strain, but also to the displacement rate (see, for instance, [43]). From the mathematical point of view, this problem does not neglect acceleration terms on the boundary, and hence, such types of boundary conditions are often called *dynamic boundary conditions*. Such boundary conditions are not only important from the theoretical point of view but also arise in several physical applications. For example, in the case of one-dimensional space variable, the system (3.3.2) is a good model for the dynamic evolution of a viscoelastic rod that is fixed at one end and has a tip mass attached to its free end. The dynamic boundary conditions reflect Newton's law for the attached mass (see [27, 41, 75] for more details). If the space variable has two components, it is shown in [169] and in the references therein that these boundary conditions arise when we consider the transverse motion of a flexible membrane whose boundary may be affected by the vibrations only in a region. Also some dynamic boundary conditions, as in the problem (3.3.2), appear when we assume that Ω is an exterior domain of $\mathbb{R}^3$ in which a homogeneous fluid is at rest except for sound waves. Each point of the boundary is subject to small normal displacements into the obstacle (see [36] for more details). These types of dynamic boundary conditions are known as acoustic boundary conditions.

In recent years, many authors have investigated the well-posedness and longtime behavior of miscellaneous equations that resemble (3.3.1), but without delay on bounded domains (see, e.g., [101, 102, 159, 160]). Among the early results dealing with this type of boundary conditions are those of Grobbelaar-Van Dalsen [101–103], in which the author has made contributions to this field. In fact, the author in [101] introduced a model that describes the damped longitudinal vibrations of a homogeneous flexible horizontal rod of length L, when the end $x = 0$ is rigidly fixed, while the other end $x = L$ is free to move with an attached load. This yields to a system of two second-order equations of the form

$$\begin{cases} u_{tt} - u_{xx} - u_{txx} = 0, & x \in (0, L),\ t > 0, \\ u(0,t) = u_t(0,t) = 0, & t > 0, \\ u_{tt}(L,t) = -\left[u_x + u_{tx}\right](L,t), & t > 0, \\ u\,(x,0) = u_0\,(x)\,,\ u_t\,(x,0) = v_0\,(x)\,, & x \in (0, L), \\ u\,(L,0) = \eta, \qquad u_t\,(L,0) = \mu\,. \end{cases} \tag{3.3.3}$$

By rewriting (3.3.3) within the framework of the abstract theory of the so-called B-evolution theory, the existence of a unique solution in the strong sense has been shown. An exponential decay result was also proved in [103] for a problem related to (3.3.3), which describes weakly damped vibrations of an extensible beam (see [103] for more details).

Later, the authors in [188] considered the problem

$$\begin{cases} u_{tt} - p\,(u_x)_{xt} - q\,(u_x)_x = 0, & x \in (0,1)\,,\ t > 0, \\ u\,(0,t) = 0, & t \geq 0\,, \\ p\,(u_x)_t + q\,(u_x)\,(1,t) + k u_{tt}\,(1,t) = 0, & t \geq 0\,, \\ u\,(x,0) = u_0\,(x)\,, \qquad u_t\,(x,0) = u_1\,(x)\,, & x \in (0,1)\,. \end{cases} \tag{3.3.4}$$

Using Nakao inequality, and under appropriate conditions on p and q, both exponential and polynomial decay rates for the energy are established depending on the form of the terms p and q.

Similarly, and still in the absence of the delay term, Pellicer and Solà-Morales [160] considered the one-dimensional problem of (3.3.1) as an alternative model for the classical spring-mass-damper system. Then using the dominant eigenvalues method, they proved that for small values of the parameter a, the partial differential equation in (3.3.2) has the classical second-order differential equation

$$m_1 u''(t) + d_1 u'(t) + k_1 u(t) = 0,$$

as a limit, where the parameters m_1 , d_1 and k_1 are determined from the values of the spring-mass-damper system. Thus, the asymptotic stability of the model has been determined as a consequence of this limit. However, no convergence rate is provided. This result was followed by more recent works [159, 161]. In particular in [161], the authors considered a one-dimensional nonlocal nonlinear strongly damped wave equation with dynamical boundary conditions. Specifically, the following problem is treated:

$$\begin{cases} u_{tt} - u_{xx} - \alpha u_{txx} + \varepsilon f\left(u(1,t), \frac{u_t(1,t)}{\sqrt{\varepsilon}}\right) = 0, \\ u(0,t) = 0, \\ u_{tt}(1,t) = -\varepsilon\left[u_x + \alpha u_{tx} + r u_t\right](1,t) - \varepsilon f\left(u(1,t), \frac{u_t(1,t)}{\sqrt{\varepsilon}}\right), \end{cases} \tag{3.3.5}$$

with $x \in (0,1)$, $t > 0$, $r, \alpha > 0$ and $\varepsilon \geq 0$. The above system models a spring-mass-damper system, where the term $\varepsilon f\left(u(1,t), \frac{u_t(1,t)}{\sqrt{\varepsilon}}\right)$ represents a control acceleration at $x = 1$. Using invariant manifold theory, the authors proved that for small values of the parameter ε, the solutions of (3.3.5) are attracted to a two-dimensional invariant manifold (see [161] for further details).

The main difficulty of the problem considered is related to the nonstandard boundary conditions defined on Γ_1. To the authors' best knowledge, very little attention has been paid to this kind of boundary conditions. We can mention only a few particular results in the one-dimensional space [88, 105, 124, 160].

The stability of (3.3.1) with $\tau = 0, a = 0$ has been studied in [1], where it has been shown that the system is stable under some geometric condition on Γ_1 (as in [32]). Moreover, if $\mu = 0$, that is, in the absence of delay, the solutions of (3.3.1), with $a > 0$, are exponentially stable. In turn, in the presence of a delay term, there are instability phenomena [149].

Another problem related to (3.3.2) is the following:

$$
\begin{aligned}
u_{tt} - \Delta u + g(u_t) &= f, && \text{in } \Omega \times (0,T) \\
\frac{\partial u}{\partial \nu} + K(u)u_{tt} + h(u_t), &= 0, && \text{on } \partial\Omega \times (0,T) \\
u(x,0) &= u_0(x), && \text{in } \Omega \\
u_t(x,0) &= u_1(x), && \text{in } \Omega,
\end{aligned}
$$

where the boundary term $h(u_t) = |u_t|^\rho u_t$ arises when studying the flows of gas in a channel with porous walls. The term u_{tt} on the boundary appears from the internal forces, and the nonlinearity $K(u)u_{tt}$ on the boundary represents the internal forces when the density of the medium depends on the displacement. This problem has been examined in [87, 88]. In fact, using the Fadeo-Galerkin approximations and a compactness argument, the authors proved the global existence and the exponential decay of the solution of the problem.

We also recall some results related to the interaction of an elastic medium with a rigid mass. Using semigroups theory, Littman and Markus [136] established a uniqueness result for a particular Euler-Bernoulli beam rigid body structure. They also proved, using feedback boundary damping, the asymptotic stability of the structure. In [137], the authors considered the Euler-Bernoulli beam equation that describes the dynamics of the clamped elastic beam in which one segment of the beam is made of viscoelastic material, while the other consists of elastic material. By combining the resolvent method with the multiplier technique, they proved the exponential decay for the transversal motion but not for the longitudinal motion of the model, when the Kelvin-Voigt damping is distributed only on a subinterval of the domain. One can also consult the work of Chen et *al.* [50] concerning the Euler-Bernoulli beam equation with the global or local Kelvin-Voigt damping. Note also that numerous models of vibrating strings with local viscoelasticity and Boltzmann damping, instead of the Kelvin-Voigt one, were considered in [139], and an exponential energy decay rate was established. Afterward, Grobbelaar-Van Dalsen [104] considered an extensible thermo-elastic beam that is suspended at one end with a rigid body attached to its free end, i.e., a one-dimensional hybrid thermo-elastic structure, and showed that the method used in [157] is still valid to establish a uniform stabilization of the system. Concerning the controllability of the hybrid system, we refer to the work by Castro and Zuazua [44], in which they considered flexible beams connected by point mass and the model takes account of rotational inertia.

Motivated by the work [82], where (3.1.1) is treated, the main objective of this section is to study problem (3.3.1), in which a delay term acts in the dynamic boundary conditions. Obviously, the two systems are significantly different due to the dimension of the spatial variable and also due to the nature of the boundary conditions in each problem.

The principal idea is to contrast the effect of the time delay by using the dissipative feedback (i.e., by giving the control in the feedback form $a\, u_t(x,t)$ or $-a\Delta u_t(x,t), \quad x \in \Omega, \quad t > 0$).

In the next subsection, we will show the global existence of solutions to (3.3.1) by transforming the delay term and using semigroups theory. Next, the natural problem we study is the stability of the solutions of (3.3.1). This is the goal of Sect. 3.3.3. We will show that a "shifted" problem is asymptotically stable with a polynomial decay rate, and we cannot answer the question of the stability of problem (3.3.1). In fact, in the last subsection, numerical experiments in the one-dimensional vase show that under certain conditions, the system (3.3.1) is unstable.

To stabilize the problem (3.3.1), we will demonstrate that a Kelvin-Voigt damping is efficient. This is done in Sect. 3.3.4.

Lastly, we will conduct some numerical examples in 1D to illustrate these stability or instability results.

3.3.2 *Well-Posedness of Problem (3.3.1)*

In this subsection, we first transform the delay boundary conditions by introducing, as in the previous section, an auxiliary variable. Then we will use Lumer-Phillips' theorem (see Theorem 2.3.4) to prove the existence and uniqueness of the solutions to the problem (3.3.1).

3.3.2.1 Setup and Notations

We present here some materials that we shall use in order to prove the existence of solutions of (3.3.1).

First, we denote

$$H^1_{\Gamma_0}(\Omega) = \left\{u \in H^1(\Omega);\ u_{\Gamma_0} = 0\right\}.$$

We set γ_1 the trace operator from $H^1_{\Gamma_0}(\Omega)$ on $L^2(\Gamma_1)$ and $H^{1/2}(\Gamma_1) = \gamma_1\big(H^1_{\Gamma_0}(\Omega)\big)$. We denote by $(.,.)$ the standard scalar product in $L^2(\Omega)$, i.e., $(u, v) = \int_\Omega u(x)v(x)\,dx$. Also, $\|.\|_q$ denotes the $L^q(\Omega)$ norm, for $1 \le q \le \infty$, while $\|.\|_{q,\Gamma_1}$ denotes the $L^q(\Gamma_1)$ norm.

Let $T > 0$ be a real number and X a Banach space endowed with norm $\|.\|_X$. We denote by $L^p(0, T; X)$, $1 \le p < \infty$ the space of functions f that are L^p over $(0, T)$ with values in X, which are measurable and $\|f\|_X \in L^p(0, T)$. This space is a Banach space endowed with the norm

$$\|f\|_{L^p(0,T;X)} = \left(\int_0^T \|f\|_X^p dt\right)^{1/p}.$$

$L^\infty(0, T; X)$ denotes the space of functions $f :]0, T[\to X$, which are measurable, and $\|f\|_X \in L^\infty(0, T)$. This space is a Banach space endowed with the norm:

$$\|f\|_{L^\infty(0,T;X)} = \operatorname*{ess\ sup}_{0<t<T} \|f\|_X .$$

Also, we recall that if X and Y are two Banach spaces such that $X \hookrightarrow Y$ (continuous embedding), then

$$L^p(0, T; X) \hookrightarrow L^p(0, T; Y),\ 1 \le p \le \infty .$$

3.3.2.2 Semigroup Formulation of the Problem

In this subsection, we prove the global existence and the uniqueness of the solution of (3.3.1). We first transform as usual the problem (3.3.1) into (3.3.8) by introducing an auxiliary variable and then use the semigroups approach to prove the existence of the unique solution of problem (3.3.8) (see Sect. 2.3).

Let us consider the new variable [149, 184]:

$$z(x, \rho, t) = u_t(x, t - \tau\rho),\ x \in \Gamma_1,\ \rho \in (0, 1),\ t > 0. \tag{3.3.6}$$

Then, we have

$$\tau z_t(x, \rho, t) + z_\rho(x, \rho, t) = 0,\ \text{in } \Gamma_1 \times (0, 1) \times (0, +\infty). \tag{3.3.7}$$

Then (3.3.1) takes the following form:

$$\begin{cases} u_{tt} - \Delta u + a\, u_t = 0, & x \in \Omega,\ t > 0\,, \\ u(x,t) = 0, & x \in \Gamma_0,\ t > 0\,, \\ u_{tt}(x,t) = -\dfrac{\partial u}{\partial \nu}(x,t) - \mu z(x,1,t), & x \in \Gamma_1,\ t > 0\,, \\ \tau z_t(x,\rho,t) + z_\rho(x,\rho,t) = 0, & x \in \Gamma_1, \rho \in (0,1)\,,\ t > 0\,, \\ z(x,0,t) = u_t(x,t), & x \in \Gamma_1,\ t > 0\,, \\ u(x,0) = u_0(x),\ u_t(x,0) = u_1(x), & x \in \Omega\,, \\ z(x,\rho,0) = f_0(x,-\tau\rho) & x \in \Gamma_1,\ \rho \in (0,1)\,. \end{cases} \tag{3.3.8}$$

In this subsection, we will provide a sufficient condition that guaranties that (3.3.8) is well-posed. For this purpose, we will use a semigroups formulation. Let $V := (u, u_t, \gamma_1(u_t), z)$ and define the energy space:

$$\mathcal{H} = H^1_{\Gamma_0}(\Omega) \times L^2\,(\Omega) \times L^2(\Gamma_1) \times L^2(\Gamma_1 \times (0,1)).$$

Clearly, $\mathcal{H}$ is a Hilbert space with respect to the inner product

$$\begin{aligned} \langle V_1, V_2\rangle_{\mathcal{H}} = &\int_\Omega \nabla u_1 \cdot \nabla u_2 dx + \int_\Omega v_1 v_2 dx + \int_{\Gamma_1} w_1 w_2 d\sigma \\ &+ \xi \int_{\Gamma_1}\int_0^1 z_1 z_2 d\rho d\sigma \end{aligned} \tag{3.3.9}$$

for $V_1 = (u_1, v_1, w_1, z_1)$, $V_2 = (u_2, v_2, w_2, z_2)$, and $\xi > 0$ a nonnegative real number defined later.

Therefore, if $V_0 \in \mathcal{H}$ and $V \in \mathcal{H}$, problem (3.3.8) is formally equivalent to the following abstract evolution equation in the Hilbert space $\mathcal{H}$:

$$\begin{cases} V'(t) = \mathcal{A}V(t),\ t > 0, \\ V\,(0) = V_0, \end{cases}, \tag{3.3.10}$$

where $'$ denotes the derivative with respect to time t, $V_0 := (u_0, u_1, \gamma_1(u_1), f_0(.,-.\tau))$, and the operator $\mathcal{A}$ is defined by

$$\mathcal{A}\,(u,v,w,z) = \left(v,\ \Delta u - a\,v,\ -\frac{\partial u}{\partial \nu} - \mu z\,(.,1)\,,\ -\frac{1}{\tau} z_\rho\right).$$

The domain of $\mathcal{A}$ is the set of $V = (u, v, w, z)$ such that

$$(u,v,w,z) \in \left(H^1_{\Gamma_0}(\Omega) \cap H^2(\Omega)\right) \times H^1_{\Gamma_0}(\Omega) \times L^2(\Gamma_1)$$

$$\times L^2\left(\Gamma_1; H^1(0,1)\right), \tag{3.3.11}$$

$$w = \gamma_1(v) = z(.,0) \text{ on } \Gamma_1. \tag{3.3.12}$$

Let us also define

$$\xi^\star := \mu\tau.$$

Then, for all $\xi > \xi^\star$, we set

$$\mu_1 = \frac{\xi}{2\tau} + \frac{\mu}{2}.$$

Finally, we define the *shifted operator*

$$\mathcal{A}_d = \mathcal{A} - \mu_1 I.$$

The well-posedness of (3.3.8) is stated below:

Theorem 3.3.1 *Let $V_0 \in \mathcal{H}$; then there exists a unique solution $V \in C(\mathbb{R}_+; \mathcal{H})$ of problem (3.3.10). Moreover, if $V_0 \in \mathcal{D}(\mathcal{A})$, then*

$$V \in C(\mathbb{R}_+; \mathcal{D}(\mathcal{A})) \cap C^1(\mathbb{R}_+; \mathcal{H}).$$

Proof First, we prove that there exists a unique solution $V \in C(\mathbb{R}_+; \mathcal{H})$ to the shifted problem:

$$\begin{cases} V'(t) = \mathcal{A}_d V(t), \; t > 0, \\ V(0) = V_0, \end{cases} \tag{3.3.13}$$

Then, since $\mathcal{A} = \mathcal{A}_d + \mu_1 I$, we can deduce that there exists a solution $V \in C(\mathbb{R}_+; \mathcal{H})$ to the original problem (3.3.10).

As usual, in order to prove the existence and uniqueness of the solution of problem (3.3.13), we use the semigroup approach along with Lumer-Phillips' theorem (see Theorem 2.3.4).

Let $V = (u, v, w, z) \in \mathcal{D}(\mathcal{A})$. Thanks to the definition of $\mathcal{A}$ and the scalar product of $\mathcal{H}$, we have

$$\begin{aligned} \langle \mathcal{A}V, V\rangle_{\mathcal{H}} = & \int_\Omega \nabla u \cdot \nabla v dx + \int_\Omega v \Delta u dx - \int_\Omega a|v(x)|^2 dx \\ & + \int_{\Gamma_1} w\left(-\frac{\partial u}{\partial \nu} - \mu z(x,1)\right) d\sigma - \frac{\xi}{\tau}\int_{\Gamma_1}\int_0^1 z z_\rho d\rho d\sigma. \end{aligned}$$

By Green's formula (see (2.1.1)), the latter becomes

$$\langle \mathcal{A}V, V\rangle_{\mathcal{H}} = -\int_{\Omega} a|v(x)|^2 dx - \mu \int_{\Gamma_1} z(x,1)\, w d\sigma - \frac{\xi}{\tau}\int_{\Gamma_1}\int_0^1 z_\rho z d\rho dx. \tag{3.3.14}$$

We also have the identity

$$\begin{aligned}\frac{\xi}{\tau}\int_{\Gamma_1}\int_0^1 z_\rho z\, d\rho\, d\sigma &= \frac{\xi}{2\tau}\int_{\Gamma_1}\int_0^1 \frac{\partial}{\partial\rho} z^2\, d\rho\, d\sigma \\ &= \frac{\xi}{2\tau}\int_{\Gamma_1}\Big(z^2(x,1) - z^2(x,0)\Big)\, d\sigma\,. \end{aligned} \tag{3.3.15}$$

In view of the compatibility condition in (3.3.12), we have

$$-\frac{\xi}{\tau}\int_{\Gamma_1}\int_0^1 z_\rho z\, d\rho\, d\sigma = \frac{\xi}{2\tau}\int_{\Gamma_1}\Big(v^2 - z^2(x,1)\Big)\, d\sigma\,.$$

Therefore, (3.3.14) becomes

$$\begin{aligned}\langle \mathcal{A}V, V\rangle_{\mathcal{H}} = &-\int_{\Omega} a|v(x)|^2 dx - \frac{\xi}{2\tau}\int_{\Gamma_1}\int_0^1 z^2(x,1) d\sigma + \frac{\xi}{2\tau}\int_{\Gamma_1}|v|^2 d\sigma \\ &-\mu\int_{\Gamma_1} z(x,1)\, w d\sigma\,. \end{aligned} \tag{3.3.16}$$

In turn, the compatibility condition and Young's inequality yield

$$-\int_{\Gamma_1} wz(x,1)\, d\sigma \le \frac{1}{2}\int_{\Gamma_1} z^2(x,1)\, d\sigma + \frac{1}{2}\int_{\Gamma_1} v^2 d\sigma\,. \tag{3.3.17}$$

Hence, we obtain

$$\begin{aligned}\langle \mathcal{A}V, V\rangle_{\mathcal{H}} &+ \int_{\Omega} a|v(x)|^2 dx - \left(\frac{\xi}{2\tau} + \frac{\mu}{2}\right)\int_{\Gamma_1}|v|^2 d\sigma \\ &+ \left(\frac{\xi}{2\tau} - \frac{\mu}{2}\right)\int_{\Gamma_1} z^2(x,1) d\sigma \le 0\,, \end{aligned} \tag{3.3.18}$$

that is,

$$\langle \mathcal{A}V, V\rangle_{\mathcal{H}} + \int_{\Omega} a|v(x)|^2 dx - \mu_1\int_{\Gamma_1}|v|^2 d\sigma + \left(\frac{\xi}{2\tau} - \frac{\mu}{2}\right)\int_{\Gamma_1} z^2(x,1) d\sigma \le 0\,.$$

From the preceding inequality, we get

$$\left\langle \left(\mathcal{A} - \mu_1 I\right) V, V \right\rangle_{\mathcal{H}} \leq - \int_{\Omega} a|v(x)|^2 dx - \left(\frac{\xi}{2\tau} - \frac{\mu}{2}\right) \int_{\Gamma_1} z^2(x, 1) d\sigma \,. \tag{3.3.19}$$

Now, since $\forall \xi > \xi^\star$, $\left(\frac{\xi}{2\tau} - \frac{\mu}{2}\right) > 0$, we finally get

$$\left\langle \left(\mathcal{A} - \mu_1 I\right) V, V \right\rangle_{\mathcal{H}} \leq 0 \,. \tag{3.3.20}$$

Thus, the operator $\mathcal{A}_d = \mathcal{A} - \mu_1 I$ is dissipative.

The next step is to show that for any $\lambda > 0$ and any $\xi > \xi^\star$, the operator $\lambda I - \mathcal{A}_d$ is surjective. Clearly, it suffices to show that $\lambda I - \mathcal{A}$ is surjective for all $\lambda > 0$. To reach this result, consider an element $F = (f_1, f_2, f_3, f_4) \in \mathcal{H}$, and let us seek a solution $V = (u, v, w, z) \in \mathcal{D}(\mathcal{A})$ of the equation

$$(\lambda I - \mathcal{A}) V = F,$$

which can be written as

$$\lambda u - v = f_1, \tag{3.3.21}$$

$$\lambda v - \Delta u + a v = f_2, \tag{3.3.22}$$

$$\lambda w + \frac{\partial u}{\partial \nu} + \mu z(., 1) = f_3, \tag{3.3.23}$$

$$\lambda z + \frac{1}{\tau} z_\rho = f_4. \tag{3.3.24}$$

In order to find $V = (u, v, w, z) \in \mathcal{D}(\mathcal{A})$ solution of the system (3.3.21), (3.3.22), (3.3.23), and (3.3.24), we suppose that u is determined with the appropriate regularity. Then (3.3.21) gives

$$v = \lambda u - f_1 \,. \tag{3.3.25}$$

Next, the compatibility condition on Γ_1, together with (3.3.12), permits to determine $z(., 0)$

$$z(x, 0) = v(x) = \lambda u(x) - f_1(x), \quad \text{for } x \in \Gamma_1. \tag{3.3.26}$$

Thereby, it follows from (3.3.24) that z satisfies

$$\begin{cases} z_\rho = \tau\Big(f_4(x) - \lambda z(x, \rho)\Big), & \text{for } x \in \Gamma_1\,,\ \rho \in (0, 1), \\ z(x, 0) = \lambda u(x) - f_1(x), & \end{cases} \tag{3.3.27}$$

which means

$$z(x,\rho) = \lambda u(x)e^{-\lambda\rho\tau} - f_1 e^{-\lambda\rho\tau} + \tau e^{-\lambda\rho\tau}\int_0^\rho f_4(x,\sigma)e^{\lambda\sigma\tau}d\sigma$$

$$\text{for } x \in \Gamma_1\,,\ \rho \in (0,1). \tag{3.3.28}$$

This implies that

$$z(x,1) = \lambda u(x)e^{-\lambda\tau} + z_1(x), \qquad \text{for } x \in \Gamma_1 \tag{3.3.29}$$

with

$$z_1(x) = -f_1 e^{-\lambda\tau} + \tau e^{-\lambda\tau}\int_0^1 f_4(x,\sigma)e^{\lambda\sigma\tau}d\sigma, \qquad \text{for } x \in \Gamma_1.$$

Since $f_1 \in H^1_{\Gamma_0}(\Omega)$ and $f_4 \in L^2(\Gamma_1) \times L^2(0,1)$, we have $z_1 \in L^2(\Gamma_1)$.

To sum up, if u is known, then we get v from (3.3.25), and z from (3.3.28), while using (3.3.29), we deduce w from (3.3.23). Additionally, recall that since $V = (u, v, w, z) \in \mathcal{D}(\mathcal{A})$, we get $w = \gamma_1(v)$.

Note that in view of (3.3.22), (3.3.23), and (3.3.25), u must satisfy:

$$\lambda(\lambda + a)u - \Delta u = f_2 + (\lambda + a)f_1, \quad \text{in } \Omega \tag{3.3.30}$$

with the boundary conditions

$$u = 0, \quad \text{on } \Gamma_0 \tag{3.3.31}$$

$$\frac{\partial u}{\partial \nu} = f_3 - \lambda z(.,0) - \mu z(.,1), \quad \text{on } \Gamma_1. \tag{3.3.32}$$

Using z1-dyn and (3.3.25), we have

$$\frac{\partial u}{\partial \nu} = -\left(\lambda^2 + \mu\lambda e^{-\lambda\tau}\right)u + f(x), \qquad \text{for } x \in \Gamma_1 \tag{3.3.33}$$

with

$$f(x) = f_3(x) + \lambda f_1(x) - \mu z_1(x), \qquad \text{for } x \in \Gamma_1\,.$$

Due to the regularity of f_3, f_2, z_1, we have $f \in L^2(\Gamma_1)$.

In turn, the variational form of (3.3.30), (3.3.31), and (3.3.33) can be formulated as follows: Find $u \in H^1_{\Gamma_0}(\Omega)$ such that

$$\begin{aligned}\int_\Omega \lambda(\lambda+a)u\omega + \nabla u \cdot \nabla \omega dx + \int_{\Gamma_1} \left(\lambda^2 + \mu\lambda e^{-\lambda\tau}\right) u\omega d\sigma, \qquad (3.3.34)\\ = \int_\Omega (f_2 + (\lambda+a) f_1)\, \omega dx + \int_{\Gamma_1} f\omega d\sigma,\end{aligned}$$

for any $\omega \in H^1_{\Gamma_0}(\Omega)$.

Since $\lambda > 0$, $\mu > 0$, the left-hand side of (3.3.34) defines a coercive bilinear form on $H^1_{\Gamma_0}(\Omega)$, while the right-hand side defines a continuous linear form. Thus, Lax-Milgram theorem (see Theorem 2.1.6) allows us to claim that there exists a unique $u \in H^1_{\Gamma_0}(\Omega)$ solution of (3.3.34). Now, choosing $\omega \in \mathcal{C}_c^\infty(\Omega)$, we deduce that u is a solution of (3.3.30) in the sense of distribution, and hence, $u \in H^2(\Omega) \cap H^1_{\Gamma_0}(\Omega)$. Moreover, using the Green's formula and exploiting (3.3.30) on Ω, we obtain

$$\int_{\Gamma_1} \left(\lambda^2 + \mu\lambda e^{-\lambda\tau}\right) u\omega d\sigma + \left\langle \frac{\partial u}{\partial \nu}; \omega \right\rangle_{\Gamma_1} = \int_{\Gamma_1} f\omega d\sigma \ \forall \omega \in H^1_{\Gamma_0}(\Omega)\,.$$

Whereupon, $u \in H^2(\Omega) \cap H^1_{\Gamma_0}(\Omega)$ verifies (3.3.33), and consequently, we are able to recover u and v, and then by (3.3.28), we obtain z. Finally, setting $w = \gamma_1(v)$, we have found $V = (u, v, w, z) \in \mathcal{D}(\mathcal{A})$ solution of $(I - \mathcal{A})\, V = F$.

In view of Lumer-Phillips' theorem, there exists a unique solution $V \in C(\mathbb{R}_+; \mathcal{H})$ of the shifted problem (3.3.13). This completes the proof of Theorem 3.3.1. □

Remark 3.3.1 According to the above result, the operator $\mathcal{A}_d$ generates a C_0-semigroup of contractions $e^{t\mathcal{A}_d}$ on $\mathcal{H}$.

3.3.3 Asymptotic Behavior

In this subsection, we show that if $\xi > \xi^*$, the semigroup $e^{t\mathcal{A}_d}$ decays to the null steady state with a polynomial decay rate for regular initial data. To obtain this, our technique is based on the resolvent domain method and combines a contradiction argument with the multiplier technique to carry out a special analysis for the resolvent.

Theorem 3.3.2 *Let $\xi > \xi^*$. Denote $\mu_1 = \dfrac{\xi}{2\tau} + \dfrac{\mu}{2}$ and $\mathcal{A}_d = \mathcal{A} - \mu_1 I$. Then there exists a constant $C > 0$ such that, for all $V_0 \in \mathcal{D}(\mathcal{A}_d)$, the semigroup $e^{t\mathcal{A}_d}$ satisfies the following estimate:*

$$\left\| e^{t\mathcal{A}_d} V_0 \right\|_{\mathcal{H}} \leq \frac{C}{\sqrt{t}} \, \| V_0 \|_{\mathcal{D}(\mathcal{A}_d)} \, , \forall \, t > 0. \tag{3.3.35}$$

Remark 3.3.2 Notice that although the semigroup $e^{t\mathcal{A}_d}$ generates a polynomial stability, we are unable to conclude any result on the stability of the semigroup $e^{t\mathcal{A}}$.

Indeed, let us consider $V_0 \in \mathcal{D}(\mathcal{A})$ and $\widetilde{V}_0 = V_0 - \mu_1 V_0$, and the following two problems:

$$\begin{cases} V' = \mathcal{A} V, \\ V_{t=0} = V_0; \end{cases} \quad \text{and} \quad \begin{cases} \widetilde{V}' = \mathcal{A}_d \widetilde{V}, \\ \widetilde{V}_{t=0} = \widetilde{V}_0 \, . \end{cases}$$

Given $V_0 = (u_0, u_1, \gamma_1(u_1), z_0) \in \mathcal{D}(\mathcal{A})$, the second problem can be written in terms of $V = (u, u_t, \gamma_1(u_t), z)$:

$$\begin{cases} u_{tt} - \dfrac{1}{1+\mu_1} \Delta u + \dfrac{a+\mu_1}{1+\mu_1} u_t = 0, & x \in \Omega, \ t > 0\,, \\ u(x,t) = 0, & x \in \Gamma_0, \ t > 0\,, \\ u_{tt}(x,t) = -\dfrac{\partial u}{\partial \nu}(x,t) - \mu z(x,1,t) - \mu u_t(x,t), & x \in \Gamma_1, \ t > 0\,, \\ \tau z_t(x,\rho,t) + z_\rho(x,\rho,t) + \mu_1 \tau z(x,\rho,t) = 0, & x \in \Gamma_1, \rho \in (0,1)\,, \ t > 0\,, \\ z(x,0,t) = u_t(x,t), & x \in \Gamma_1, \ t > 0\,, \\ u(x,0) = (1-\mu_1) u_0(x), & x \in \Omega\,, \\ u_t(x,0) = (1-\mu_1) u_1(x), & x \in \Omega\,, \\ z(x,\rho,0) = (1-\mu_1) f_0(x,-\tau\rho), & x \in \Gamma_1, \ \rho \in (0,1)\,. \end{cases} \tag{3.3.36}$$

We call this problem the “shifted” problem.

By Duhamel’s formula, we get

$$\forall t > 0\,, \; e^{-\mu_1 t} V(t) = \frac{-e^{-\mu_1 t}}{1+\mu_1} V_0 + \frac{\widetilde{V}(t)}{1+\mu_1} - \frac{\mu_1}{1+\mu_1} \int_0^t e^{-\mu_1 (t-s)} \, \widetilde{V}(s) ds\,. \tag{3.3.37}$$

The first two terms of the right-hand side of Eq. (3.3.37) tend to zero as t tends to infinity. Hence, we obtain

$$V(t) \simeq \int_0^t e^{\mu_1 s} \, \widetilde{V}(s) ds\,.$$

We only know at this stage that $\|\widetilde{V}(s)\|_{\mathcal{D}(\mathcal{A}_d)}$ tends to zero at least as $s^{-1/2}$, and thus $\|V(t)\|_{\mathcal{D}(\mathcal{A})}$ may tend to zero or blow-up in infinite time. We will investigate this behavior by considering numerical examples in 1D in the last Sect. 3.3.5.

Proof of Theorem 3.3.2 We will use the frequency-domain method for polynomial stability (see Theorem (2.3.10)). □

Remark 3.3.3 In view of this theorem, we need to identify the spectrum of $\mathcal{A}_d$ lying on the imaginary axis. Unfortunately, since the embedding of $L^2\left(\Gamma_1, H^1(0,1)\right)$ into $L^2(\Gamma_1 \times (0,1))$ is not compact, the operator $\mathcal{A}_d$ does not have a compact resolvent. Therefore, its spectrum $\sigma(\mathcal{A}_d)$ does not only consist of eigenvalues of $\mathcal{A}_d$. Hence, we have to show that:

1. If β is a real number, then $i\beta I - \mathcal{A}_d$ is injective.
2. If β is a real number, then $i\beta I - \mathcal{A}_d$ is surjective.

This will be achieved in the following two lemmas.

First, we look at the point spectrum of $\mathcal{A}_d$.

Lemma 3.3.1 *If β is a real number, then $i\beta$ is not an eigenvalue of $\mathcal{A}_d$.*

Proof We will show that the equation

$$\mathcal{A}_d Z = i\beta Z \tag{3.3.38}$$

with $Z = (u, v, w, z) \in \mathcal{D}(\mathcal{A}_d)$ and $\beta \in \mathbb{R}$ has only one trivial solution.

Clearly, (3.3.38) writes

$$(i\beta + \mu_1)u - v = 0, \tag{3.3.39}$$

$$(i\beta + \mu_1)v - \Delta u + av = 0, \tag{3.3.40}$$

$$(i\beta + \mu_1)w + \frac{\partial u}{\partial \nu} + \mu z(., 1) = 0, \tag{3.3.41}$$

$$(i\beta + \mu_1)z + \frac{1}{\tau} z_\rho = 0\,. \tag{3.3.42}$$

Taking the inner product of (3.3.38) with Z and using (3.3.19), we get

$$\Re e\,(< \mathcal{A}_d Z, Z >_{\mathcal{H}}) \leq -\int_\Omega a\,|v(x)|^2\,dx - \left(\frac{\xi}{2\tau} - \frac{\mu}{2}\right)\int_{\Gamma_1} |z(x,1)|^2\,d\sigma. \tag{3.3.43}$$

Thus

$$v = 0 \text{ and } z(., 1) = 0,$$

and hence, (3.3.39) gives $u = 0$. In turn, since $w = \gamma_1(v) = z(.,0)$, we also obtain $w = 0$ and $z(.,0) = 0$. Finally, (3.3.42) yields

$$z(.,\rho) = z(.,0)\, e^{-\tau(i\beta+\mu_1)\rho}.$$

But $z(.,0) = 0$, and hence, $z = 0$. Thus, the only solution of (3.3.38) is the trivial one. □

Next, we show that $\mathcal{A}_d$ has no continuous spectrum on the imaginary axis.

Lemma 3.3.2 *Let $\xi > \xi^*$. If β is a real number, then $i\beta$ belongs to the resolvent set $\rho(\mathcal{A}_d)$ of $\mathcal{A}_d$.*

Proof In view of Lemma 3.3.1, it suffices to show that $i\beta I - \mathcal{A}_d$ is surjective.

Given $F = (f_1, f_2, f_3, f_4) \in \mathcal{H}$, let $V = (u, v, w, z) \in \mathcal{D}(\mathcal{A}_d)$ be a solution of

$$(i\beta I - \mathcal{A}_d)\, V = F,$$

which gives

$$(i\beta + \mu_1)u - v = f_1, \tag{3.3.44}$$

$$(i\beta + \mu_1)v - \Delta u + av = f_2, \tag{3.3.45}$$

$$(i\beta + \mu_1)w + \frac{\partial u}{\partial \nu} + \mu z(.,1) = f_3, \tag{3.3.46}$$

$$(i\beta + \mu_1)z + \frac{1}{\tau} z_\rho = f_4. \tag{3.3.47}$$

Arguing as before, in order to have $V = (u, v, w, z) \in \mathcal{D}(\mathcal{A}_d)$ solution of (3.3.44), (3.3.45), (3.3.46), and (3.3.47), we suppose that u is determined with the appropriate regularity. Then from (3.3.44), we get

$$v = (i\beta + \mu_1)u - f_1\,. \tag{3.3.48}$$

Therefore, from the compatibility condition on Γ_1, Eq. (3.3.12), we determine $z(.,0)$ by

$$z(x,0) = v(x) = (i\beta + \mu_1)u(x) - f_1(x), \quad \text{for } x \in \Gamma_1. \tag{3.3.49}$$

Thus, from (3.3.47), z is the solution of the linear Cauchy problem:

$$\begin{cases} z_\rho = \tau\Big(f_4(x) - (i\beta + \mu_1)z(x,\rho)\Big), & \text{for } x \in \Gamma_1\,,\ \rho \in (0,1), \\ z(x,0) = (i\beta + \mu_1)u(x) - f_1(x). \end{cases} \tag{3.3.50}$$

The solution of the Cauchy problem (3.3.50) is given by for $x \in \Gamma_1$, $\rho \in (0, 1)$

$$z(x, \rho) = (i\beta + \mu_1)u(x)e^{-(i\beta+\mu_1)\rho\tau} - f_1 e^{-(i\beta+\mu_1)\rho\tau} + \tau e^{-(i\beta+\mu_1)\rho\tau} \int_0^\rho f_4(x, \sigma)e^{(i\beta+\mu_1)\sigma\tau} d\sigma\ . \tag{3.3.51}$$

Hence,

$$z(x, 1) = (i\beta + \mu_1)u(x)e^{-(i\beta+\mu_1)\tau} + z_1(x), \qquad \text{for } x \in \Gamma_1 \tag{3.3.52}$$

with

$$z_1(x) = -f_1 e^{-(i\beta+\mu_1)\tau} + \tau e^{-(i\beta+\mu_1)\tau} \int_0^1 f_4(x, \sigma)e^{(i\beta+\mu_1)\sigma\tau} d\sigma, \qquad \text{for } x \in \Gamma_1.$$

Since $f_1 \in H^1_{\Gamma_0}(\Omega)$ and $f_4 \in L^2(\Gamma_1) \times L^2(0, 1)$, then $z_1 \in L^2(\Gamma_1)$.

Consequently, knowing u, we can deduce v by (3.3.48), z by (3.3.51), and using (3.3.52), we deduce w by (3.3.46). Also, recall that $w = \gamma_1(v)$.

Using (3.3.45) and (3.3.46), u must solve the following:

$$(i\beta + \mu_1)(i\beta + \mu_1 + a)u - \Delta u = f_2 + (i\beta + \mu_1 + a) f_1, \quad \text{in } \Omega \tag{3.3.53}$$

with the boundary conditions

$$u = 0, \quad \text{on } \Gamma_0 \tag{3.3.54}$$

$$\frac{\partial u}{\partial \nu} = f_3 - (i\beta + \mu_1)z(., 0) - \mu z(., 1), \quad \text{on } \Gamma_1\ . \tag{3.3.55}$$

Using the preceding expression of $z(., 1)$ and the expression of v given by (3.3.48), we have

$$\frac{\partial u}{\partial \nu} = - \left((i\beta + \mu_1)^2 + \mu(i\beta + \mu_1)e^{-(i\beta+\mu_1)\tau}\right) u + f(x), \qquad \text{for } x \in \Gamma_1 \tag{3.3.56}$$

with

$$f(x) = f_3(x) + (i\beta + \mu_1) f_1(x) - \mu z_1(x), \qquad \text{for } x \in \Gamma_1\ .$$

From the regularity of f_3 , f_2 , z_1, we get $f \in L^2(\Gamma_1)$.

The variational formulation of (3.3.53), (3.3.54), (3.3.56) consists in finding $u \in H^1_{\Gamma_0}(\Omega)$ such that for any $\omega \in H^1_{\Gamma_0}(\Omega)$:

$$\begin{aligned}
&\int_\Omega \Big((i\beta + \mu_1)(i\beta + \mu_1 + a)u\bar{\omega} + \nabla u \cdot \nabla \bar{\omega}\Big)dx \\
&\quad + \int_{\Gamma_1} \Big((i\beta + \mu_1)^2 + \mu(i\beta + \mu_1)e^{-(i\beta+\mu_1)\tau}\Big)\, u\bar{\omega} d\sigma \\
&= \int_\Omega (f_2 + (i\beta + \mu_1 + a) f_1)\, \bar{\omega} dx + \int_{\Gamma_1} f \bar{\omega} d\sigma \,.
\end{aligned} \tag{3.3.57}$$

Multiplying the preceding equation by $-i\beta + \mu_1$ leads to, for any $\omega \in H^1_{\Gamma_0}(\Omega)$:

$$\begin{aligned}
&\int_\Omega \Big(|i\beta + \mu_1|^2(i\beta + \mu_1 + a)u\bar{\omega} + (-i\beta + \mu_1)\nabla u \cdot \nabla \bar{\omega}\Big)dx \\
&\quad + \int_{\Gamma_1} |i\beta + \mu_1|^2 \Big(i\beta + \mu_1 + \mu e^{-(i\beta+\mu_1)\tau}\Big)\, u\bar{\omega} d\sigma \\
&= \int_\Omega (-i\beta + \mu_1)\, (f_2 + (i\beta + \mu_1 + a) f_1)\, \bar{\omega} dx \\
&\quad + \int_{\Gamma_1} (-i\beta + \mu_1)\, f \bar{\omega} d\sigma \,.
\end{aligned} \tag{3.3.58}$$

Since $\mu_1 > \mu > 0$, the left-hand side of (3.3.58) defines a coercive sesquilinear form on $H^1_{\Gamma_0}(\Omega)$. Thus, applying the Lax-Milgram theorem (see Theorem 2.1.6), there exists a unique solution $u \in H^1_{\Gamma_0}(\Omega)$ of (3.3.57). Now, choosing $\omega \in \mathcal{C}_c^\infty(\Omega)$, u is a solution of (3.3.53) in the sense of distribution. Using the regularity of f_1 and f_2, we finally have $u \in H^2(\Omega) \cap H^1_{\Gamma_0}(\Omega)$. Subsequently, using Green's formula and exploiting the Eq. (3.3.53), it follows that $\forall \omega \in H^1_{\Gamma_0}(\Omega)$

$$\int_{\Gamma_1} \Big((i\beta + \mu_1)^2 + \mu(i\beta + \mu_1)e^{-(i\beta+\mu_1)\tau}\Big)\, u\omega d\sigma + \left\langle \frac{\partial u}{\partial \nu}; \omega \right\rangle_{\Gamma_1} = \int_{\Gamma_1} f \omega d\sigma \,.$$

Thus, $u \in H^2(\Omega) \cap H^1_{\Gamma_0}(\Omega)$ verifies (3.3.56). Then, we recover v from (3.3.44), while (3.3.51) gives z. Finally, setting $w = \gamma_1(v)$, we have found $V = (u, v, w, z) \in \mathcal{D}(\mathcal{A}_d)$ solution of $(i\beta I - \mathcal{A}_d)\, V = F$. □

The following result shows that (2.3.19) holds with $\mathscr{L} = \mathcal{A}_d$ and $\theta = 2$.

Lemma 3.3.3 *The resolvent operator of $\mathcal{A}_d$ satisfies condition* (2.3.19) *for $\theta = 2$.*

Proof Suppose that condition (2.3.19) is false with $\theta = 2$. In view of Banach-Steinhaus theorem (see Theorem 2.1.7), there exist a sequence of real numbers

$\beta_n \to +\infty$ and a sequence of vectors $Z_n = (u_n, v_n, w_n, z_n) \in \mathcal{D}(\mathcal{A}_d)$ with $\|Z_n\|_{\mathcal{H}} = 1$ such that

$$||\beta_n^2(i\beta_n I - \mathcal{A}_d)Z_n||_{\mathcal{H}} \to 0 \quad \text{as} \quad n \to \infty, \tag{3.3.59}$$

i.e.,

$$\beta_n \left((i\beta_n + \mu_1)u_n - v_n\right) \equiv f_n \to 0 \quad \text{in} \quad H^1_{\Gamma_0}(\Omega),$$

$$\beta_n \left(i\beta_n v_n - \Delta u_n + (\mu_1 + a)v_n\right) \equiv g_n \to 0 \quad \text{in} \quad L^2(\Omega),$$

$$\beta_n \left((i\beta_n + \mu_1)w_n + \frac{\partial u_n}{\partial \nu} + \mu z_n(.,1)\right) \equiv h_n \to 0 \quad \text{in} \quad L^2(\Gamma_1), \tag{3.3.60}$$

$$\beta_n \left((i\beta_n + \mu_1)z_n + \frac{1}{\tau}\partial_\rho z_n\right) \equiv k_n \to 0, \quad \text{in} \quad L^2(\Gamma_1 \times (0,1)) \tag{3.3.61}$$

since $\beta_n \leq \beta_n^2$.

Our goal is to derive from (3.3.59) that $||Z_n||_{\mathcal{H}}$ converges to zero, and thus there is a contradiction.

We first notice that we have

$$||\beta_n^2(i\beta_n I - \mathcal{A}_d)Z_n||_{\mathcal{H}} \geq |\Re e\left(\langle\beta_n^2(i\beta_n I - \mathcal{A}_d)Z_n, Z_n\rangle_{\mathcal{H}}\right)|.$$

Then, by (3.3.43) and (3.3.59),

$$\beta_n\, v_n \to 0, \quad \text{in} \quad L^2(\Omega), \quad \beta_n\, z_n(.,1) \to 0, \quad \text{in} \quad L^2(\Gamma_1), \tag{3.3.62}$$

and

$$u_n \to 0,\ \Delta u_n \to 0 \quad \text{in} \quad L^2(\Omega) \Rightarrow u_n \to 0 \quad \text{in} \quad H^1_{\Gamma_0}(\Omega)\,. \tag{3.3.63}$$

This further leads, by (3.3.60) and the trace theorem, to

$$w_n \to 0 \quad \text{in} \quad L^2(\Gamma_1). \tag{3.3.64}$$

Moreover, since $Z_n \in \mathscr{D}(\mathcal{A}_d)$, we have, by (3.3.64),

$$z_n(.,0) \to 0 \quad \text{in} \quad L^2(\Gamma_1). \tag{3.3.65}$$

We have

$$z_n(.,\rho) = z_n(.,0)\, e^{-(i\beta_n+\mu_1)\tau\rho} + \int_0^\rho e^{-(i\beta_n+\mu_1)\tau(\rho-s)}\, \frac{\tau k_n(s)}{\beta_n}\, ds. \tag{3.3.66}$$

This implies, according to (3.3.66), (3.3.65), and (3.3.61), that

$$z_n \to 0 \ \text{ in } \ L^2(\Gamma_1 \times (0,1))$$

and clearly contradicts $\|Z_n\|_{\mathcal{H}} = 1$. □

The two hypotheses of Lemma 2.3.10 are proved by Lemmas 3.3.3 and 3.3.2. Then (3.3.35) is satisfied. The proof of Theorem 3.3.2 is finished.

3.3.4 Changing the Damping Law

Let us consider now the system (3.3.1) but with a stronger damping of type Kelvin-Voigt, that is,

$$\begin{cases} u_{tt} - \Delta u - a\,\Delta u_t = 0, & x \in \Omega,\ t > 0, \\ u(x,t) = 0, & x \in \Gamma_0,\ t > 0, \\ u_{tt}(x,t) = -\dfrac{\partial u}{\partial \nu}(x,t) - a\,\dfrac{\partial u_t}{\partial \nu} - \mu u_t(x,t-\tau) & x \in \Gamma_1,\ t > 0, \\ u(x,0) = u_0(x),\ u_t(x,0) = u_1(x), & x \in \Omega, \\ u_t(x,t-\tau) = f_0(x,t-\tau), & x \in \Gamma_1,\ t \in (0,\tau). \end{cases} \tag{3.3.67}$$

The latter is equivalent to

$$\begin{cases} u_{tt} - \Delta u - a\,\Delta u_t = 0, & x \in \Omega,\ t > 0, \\ u(x,t) = 0, & x \in \Gamma_0,\ t > 0, \\ u_{tt}(x,t) = -\dfrac{\partial u}{\partial \nu}(x,t) - a\dfrac{\partial u_t}{\partial \nu}(x,t) - \mu z(x,1,t), & x \in \Gamma_1,\ t > 0, \\ \tau z_t(x,\rho,t) + z_\rho(x,\rho,t) = 0, & x \in \Gamma_1, \rho \in (0,1),\ t > 0, \\ z(x,0,t) = u_t(x,t), & x \in \Gamma_1,\ t > 0, \\ u(x,0) = u_0(x),\ u_t(x,0) = u_1(x), & x \in \Omega, \\ z(x,\rho,0) = f_0(x,-\tau\rho), & x \in \Gamma_1,\ \rho \in (0,1). \end{cases} \tag{3.3.68}$$

Let us define the operator $\mathcal{A}_{kv}$ by

$$\mathcal{A}_{kv}(u,v,w,z) = \left(v,\ \Delta u + a\,\Delta v,\ -\frac{\partial u}{\partial \nu} - a\frac{\partial v}{\partial \nu} - \mu z\,(.,1)\,,\ -\frac{1}{\tau} z_\rho\right).$$

The domain of $\mathcal{A}_{kv}$ is the set of $V = (u, v, w, z)$ such that:

$$(u, v, w, z) \in \left(H^1_{\Gamma_0}(\Omega) \cap H^2(\Omega)\right) \times H^1_{\Gamma_0}(\Omega) \times L^2(\Gamma_1) \times L^2\left(\Gamma_1; H^1(0, 1)\right), \tag{3.3.69}$$

$$\frac{\partial v}{\partial \nu} \in L^2(\Gamma_1), \tag{3.3.70}$$

$$w = \gamma_1(v) = z(., 0) \text{ on } \Gamma_1. \tag{3.3.71}$$

In the sequel, for any real constant c, we define

$$C_\Omega(c) := \inf_{u \in H^1_{\Gamma_0}(\Omega)} \frac{\|\nabla u\|_2^2 + c\|u\|_{2,\Gamma_1}^2}{\|u\|_2^2}. \tag{3.3.72}$$

$C_\Omega(c)$ is the first eigenvalue of the operator $-\Delta$ under the Dirichlet-Robin boundary conditions:

$$\begin{cases} u(x) = 0, & x \in \Gamma_0 \\ \dfrac{\partial u}{\partial \nu}(x) + cu(x) = 0 & x \in \Gamma_1 . \end{cases} \tag{3.3.73}$$

From Kato's perturbation theory [121] (see also [123, Theorem 1.3.1]), $C_\Omega(c)$ is a continuous increasing function. Moreover, in view of Poincaré's inequality (see Sect. 2.1.1) and the continuity of the trace operator γ_1, we have $C_\Omega(0) > 0$ and $C_\Omega(c) \to -\infty$ as $c \to -\infty$. Thus, there exists a unique $c^\star < 0$ such that

$$C_\Omega(c^\star) = 0 . \tag{3.3.74}$$

From now on, we fix $\xi = \mu\tau$ in the norm (3.3.9). This choice will be justified later.

Theorem 3.3.3 *Suppose that a and μ satisfy the following assumption:*

$$\mu < |c^\star| a . \tag{3.3.75}$$

Then, the operator $\mathcal{A}_{kv}$ generates a C_0-semigroup of contractions on $\mathcal{H}$. We have, in particular, if $V_0 \in \mathcal{H}$, then there exists a unique solution $V \in C(\mathbb{R}_+; \mathcal{H})$ of (3.3.67). Moreover, if $V_0 \in \mathcal{D}(\mathcal{A}_{kv})$, then

$$V \in C(\mathbb{R}_+; \mathcal{D}(\mathcal{A}_{kv})) \cap C^1(\mathbb{R}_+; \mathcal{H}) .$$

Proof We again use the semigroups theory and Theorem 2.3.4.

We first show that $\mathcal{A}_{kv}$ is dissipative. Indeed, let $V = (u, v, w, z) \in \mathcal{D}(\mathcal{A}_{kv})$. In light of the definition of $\mathcal{A}_{kv}$ and the scalar product of $\mathcal{H}$, we have

$$\begin{aligned}\langle \mathcal{A}_{kv} V, V\rangle_{\mathcal{H}} &= \int_\Omega \nabla u \cdot \nabla v dx + \int_\Omega v\left(\Delta u + a\Delta v\right) dx \\ &+ \int_{\Gamma_1} w\left(-\frac{\partial u}{\partial \nu} - a\frac{\partial v}{\partial \nu} - \mu z(x,1)\right) d\sigma - \frac{\xi}{\tau}\int_{\Gamma_1}\int_0^1 z z_\rho d\rho d\sigma.\end{aligned}$$

Applying Green's formula and the compatibility condition $w = \gamma_1(v)$, we obtain

$$\langle \mathcal{A}_{kv} V, V\rangle_{\mathcal{H}} = -\mu\int_{\Gamma_1} z(x,1)\, w d\sigma - a\int_\Omega |\nabla v|^2 dx - \frac{\xi}{\tau}\int_{\Gamma_1}\int_0^1 z_\rho z d\rho dx. \tag{3.3.76}$$

In turn, we have

$$\begin{aligned}\frac{\xi}{\tau}\int_{\Gamma_1}\int_0^1 z_\rho z\, d\rho\, d\sigma &= \frac{\xi}{2\tau}\int_{\Gamma_1}\int_0^1 \frac{\partial z^2}{\partial \rho} d\rho\, d\sigma \\ &= \frac{\xi}{2\tau}\int_{\Gamma_1}\left(z^2(x,1) - z^2(x,0)\right) d\sigma\,. \end{aligned}\tag{3.3.77}$$

Thus, from the compatibility condition (3.3.71), we get

$$-\frac{\xi}{\tau}\int_{\Gamma_1}\int_0^1 z_\rho z\, d\rho\, d\sigma = \frac{\xi}{2\tau}\int_{\Gamma_1}\left(v^2 - z^2(x,1)\right) d\sigma\,.$$

Therefore, (3.3.76) becomes

$$\begin{aligned}\langle \mathcal{A}_{kv} V, V\rangle_{\mathcal{H}} &= -a\int_\Omega |\nabla v|^2 dx + \frac{\xi}{2\tau}\int_{\Gamma_1} v^2 d\sigma - \frac{\xi}{2\tau}\int_{\Gamma_1}\int_0^1 z^2(x,1) d\sigma \\ &-\mu\int_{\Gamma_1} w z(x,1)\, d\sigma\,. \end{aligned}\tag{3.3.78}$$

Once again, using the compatibility conditions and applying Young's inequality, we have

$$-\int_{\Gamma_1} w z(x,1)\, d\sigma \le \frac{1}{2}\int_{\Gamma_1} z^2(x,1)\, d\sigma + \frac{1}{2}\int_{\Gamma_1} v^2 d\sigma\,. \tag{3.3.79}$$

Therefore, we get

$$\langle \mathcal{A}_{kv} V, V\rangle_{\mathcal{H}} + a\int_\Omega |\nabla v|^2 dx - \left(\frac{\xi}{2\tau} + \frac{\mu}{2}\right)\int_{\Gamma_1} v^2 d\sigma$$

$$+\left(\frac{\xi}{2\tau}-\frac{\mu}{2}\right)\int_{\Gamma_1} z^2(x,1)d\sigma \leq 0\,.$$

At this point, as $\xi = \mu\tau$, the previous inequality becomes

$$\langle \mathcal{A}_{kv} V, V\rangle_{\mathcal{H}} + a\int_{\Omega} |\nabla v|^2\,dx - \mu\int_{\Gamma_1} v^2 d\sigma \leq 0\,.$$

Denoting now $c = -\dfrac{\mu}{a}$, we get

$$\langle \mathcal{A}_{kv} V, V\rangle_{\mathcal{H}} + a\left(\int_{\Omega} |\nabla v|^2\,dx + c\int_{\Gamma_1} v^2 d\sigma\right) \leq 0\,.$$

By definition (3.3.72), we thus get

$$\langle \mathcal{A}_{kv} V, V\rangle_{\mathcal{H}} + a\,C_{\Omega}(c)\|v\|_2^2 \leq 0\,. \tag{3.3.80}$$

It follows from (3.3.75) that $C_{\Omega}(c) > 0$. This inequality proves that the operator $\mathcal{A}_{kv}$ is dissipative.

To show that $\lambda I - \mathcal{A}_{kv}$ is surjective for all $\lambda > 0$, we can easily adapt the proof of Theorem 3.3.1.

The proof of Theorem 3.3.3 follows from the Lumer-Phillips' theorem. □

More importantly, the semigroup operator $e^{t\mathcal{A}_{kv}}$ is exponentially stable on $\mathcal{H}$:

Theorem 3.3.4 *Suppose that the assumption* (3.3.75) *is satisfied. Then, there exists* $C, \omega > 0$ *such that for all* $t > 0$*, we have*

$$\left\| e^{t\mathcal{A}_{kv}} \right\|_{\mathcal{L}(\mathcal{H})} \leq C\,e^{-\omega t}\,.$$

Remark 3.3.4 We note here that, in the case where the damping operator is sufficiently unbounded for controlling the delay one, we obtain the exponential stability result.

Note that without internal damping (i.e., if $a = 0$), the previous model is destabilized for arbitrarily small delays for every value $\mu > 0$, see [83]. Thus, the internal damping $-a\Delta u_t$ makes the system robust with respect to time delays in the boundary condition if the coefficient a is sufficiently large with respect to μ.

Remark 3.3.5 In [152], the authors studied the existence and stability of a problem closely related to problem (3.3.67). They obtain a slightly different condition to ensure the existence of solutions and their stability. Namely, let us define C_P, a kind of Poincaré constant, by

$$C_P = \sup_{u\in H^1_{\Gamma_0}(\Omega)} \frac{\|u\|^2_{2,\Gamma_1}}{\|\nabla u\|_2^2}\,.$$

Using the semigroup approach and the Lumer-Phillips' theorem, the global existence of the solution is proved. Furthermore, via an exponential stability of the solutions is obtained using an observability inequality, but under the condition

$$\mu C_P < a\,.$$

A simple argument shows that $\frac{1}{C_P} \geq |c^\star|$. Thus, the bound for the damping coefficient a obtained in this section is of the same type as that obtained by Nicaise and Pignotti.

Proof of Theorem 3.3.4 We will once again employ Theorem 2.3.8.

The proof of Theorem 3.3.4 is based on the following lemmas.

For the same reason as before (see Remark 3.3.3), we have to show that there is no eigenvalue lying on the imaginary axis and that $\mathcal{A}_{kv}$ has no continuous spectrum on the imaginary axis.

We first look at the point spectrum. □

Lemma 3.3.4 *If β is a real number, then $i\beta$ is not an eigenvalue of $\mathcal{A}_{kv}$.*

Proof We will show that the equation

$$\mathcal{A}_{kv} Z = i\beta Z \tag{3.3.81}$$

with $Z = (u, v, w, z) \in \mathcal{D}(\mathcal{A}_{kv})$ and $\beta \in \mathbb{R}$ has only a trivial solution. Equation (3.3.81) writes

$$v = i\beta u, \tag{3.3.82}$$

$$\Delta u + a\,\Delta v = i\beta v, \tag{3.3.83}$$

$$-\frac{\partial u}{\partial \nu} - a\frac{\partial v}{\partial \nu} - \mu z\,(., 1) = i\beta w, \tag{3.3.84}$$

$$-\frac{1}{\tau} z_\rho = i\beta z\,. \tag{3.3.85}$$

Let us first consider the case $\beta = 0$. From (3.3.82), we deduce that $v = 0$, and from the compatibility condition (3.3.71), we deduce that $w = 0$ and $z(., 0) = 0$. From (3.3.85), we obtain $z(., \rho) = 0$. This, together with (3.3.83)–(3.3.84) and $u \in H^1_{\Gamma_0}(\Omega) \cap H^2(\Omega)$, implies that $u = 0$. Thus, if $\beta = 0$, then the only solution of (3.3.81) is the trivial one.

Now suppose that $\beta \neq 0$. Denoting now $c = -\dfrac{\mu}{a}$, by taking the inner product of (3.3.81) with Z, using (3.3.80), we get

$$\Re e\,(< \mathcal{A}_{kv} Z, Z >_{\mathcal{H}}) \leq -a\,C_\Omega(c)\|v\|_2^2\,. \tag{3.3.86}$$

It follows from (3.3.75) that $C_\Omega(c) > 0$, and thus we obtain $v = 0$. Thus, we deduce, from (3.3.82), that $u = 0$.

Next, since we have $w = \gamma_1(v)$, we also have $w = 0$. Moreover, since we have $w = z(.,0)$, we get $z(.,0) = 0$.

From (3.3.84), we get $z(.,1) = 0$.

As z satisfies (3.3.85), we obtain the following identity:

$$z(.,\rho) = e^{-i\beta\tau\rho} z(.,0) \; .$$

Thus, the only solution of (3.3.81) is the trivial one. □

By the same way, as in Lemma 3.3.2, we show that $\mathcal{A}_{kv}$ has no continuous spectrum on the imaginary axis.

Lemma 3.3.5 *If λ is a real number, then $i\lambda$ belongs to the resolvent set $\rho(\mathcal{A}_{kv})$ of $\mathcal{A}_{kv}$.*

Lemma 3.3.6 *The resolvent operator of $\mathcal{A}_{kv}$ satisfies condition* (2.3.17).

Proof Suppose that (2.3.17) is false. Then, it follows from Banach-Steinhaus theorem (see Theorem 2.1.7), there exist a sequence of real numbers $\beta_n \to +\infty$ and a sequence of vectors $Z_n = (u_n, v_n, w_n, z_n) \in \mathcal{D}(\mathcal{A}_{kv})$ with $\|Z_n\|_{\mathcal{H}} = 1$ such that

$$||(i\beta_n I - \mathcal{A}_{kv})Z_n||_{\mathcal{H}} \to 0 \quad \text{as} \quad n \to \infty, \tag{3.3.87}$$

i.e.,

$$(i\beta_n u_n - v_n) \equiv f_n \to 0 \;\; \text{in} \;\; H^1_{\Gamma_0}(\Omega), \tag{3.3.88}$$

$$(i\beta_n v_n - \Delta u_n - a\Delta v_n) \equiv g_n \to 0 \;\; \text{in} \;\; L^2(\Omega), \tag{3.3.89}$$

$$\left(i\beta_n w_n + \frac{\partial u_n}{\partial \nu} + a\frac{\partial v_n}{\partial \nu} + \mu z_n(.,1)\right) \equiv h_n \to 0 \;\; \text{in} \;\; L^2(\Gamma_1), \tag{3.3.90}$$

$$\left(i\beta_n z_n + \frac{1}{\tau}\partial_\rho z_n\right) \equiv k_n \to 0 \;\; \text{in} \;\; L^2(\Gamma_1 \times (0,1)) \; . \tag{3.3.91}$$

Our goal is to derive from (3.3.87) that $||Z_n||_{\mathcal{H}}$ converges to zero; thus, we reached a contradiction.

We first notice that we have

$$||(i\beta_n I - \mathcal{A}_{kv})Z_n||_{\mathcal{H}} \geq |\Re e\, (\langle (i\beta_n I - \mathcal{A}_{kv})Z_n, Z_n\rangle_{\mathcal{H}})| \; . \tag{3.3.92}$$

Consequently, by (3.3.86) and (3.3.87), we have

$$v_n \to 0 \text{ in } L^2(\Omega) \; .$$

From (3.3.88),

$$u_n \to 0 \text{ in } L^2(\Omega)\,.$$

But we also have

$$\Delta u_n \to 0 \text{ in } L^2(\Omega) \text{ and } \Delta v_n \to 0 \text{ in } L^2(\Omega)\,.$$

Thus, we obtain

$$u_n \to 0 \text{ in } H^1_{\Gamma_0}(\Omega) \text{ and } v_n \to 0 \text{ in } H^1_{\Gamma_0}(\Omega)\,. \tag{3.3.93}$$

In light of the trace theorem, we have

$$w_n = \gamma(v_n) \to 0 \text{ in } L^2(\Gamma_1)\,. \tag{3.3.94}$$

Moreover, since $Z_n \in \mathscr{D}(\mathcal{A}_{kv})$, $w_n = z_n(.,0)$. Hence, we get

$$z_n(.,0) \to 0 \text{ in } \quad L^2(\Gamma_1)\,. \tag{3.3.95}$$

Now, from (3.3.90), we also have

$$z_n(.,1) \to 0 \text{ in } \quad L^2(\Gamma_1)\,. \tag{3.3.96}$$

In turn, it is easy to verify that

$$z_n(.,\rho) = z_n(.,0)\, e^{-i\tau\beta_n\rho} + \tau \int_0^\rho e^{-i\tau\beta_n(\rho-s)} k_n(.,s)\, ds\,.$$

Thereby, it follows, from (3.3.91) and (3.3.95), that

$$z_n \to 0 \;\text{ in }\; L^2(\Gamma_1 \times (0,1))\,. \tag{3.3.97}$$

Identities (3.3.93), (3.3.94), and (3.3.97) clearly contradict

$$\forall\, n \in \mathbb{N},\; \|Z_n\|_{\mathcal{H}} = 1\,.$$

□

The two conditions of Theorem 2.3.8 are shown, and hence, the proof of Theorem 3.3.4 is finished. □

3.3.5 Comments and Numerical Illustrations

To illustrate the results presented in this section, we present numerical simulations of the problem (3.3.1) with the Kelvin-Voigt damping, namely (3.3.67), in 1D. Consider $\Omega = (0, 1)$, $\Gamma_0 = \{0\}, \Gamma_1 = \{1\}$.

To solve numerically problem (3.3.1) (resp., problem (3.3.67)), we have to consider its equivalent formulation, namely problem (3.3.8) (resp. problem (3.3.68)), which writes in the present case:

$$\begin{cases} u_{tt} - u_{xx} + a\, u_t = 0, & x \in (0, 1),\ t > 0\,, \\ u(0, t) = 0, & t > 0\,, \\ u_{tt}(1, t) = -u_x(1, t) - \mu z(x, 1, t), & t > 0\,, \\ \tau z_t(1, \rho, t) + z_\rho(1, \rho, t) = 0, & \rho \in (0, 1)\,,\ t > 0\,, \\ z(1, 0, t) = u_t(1, t), & t > 0\,, \\ u(x, 0) = u_0(x),\ u_t(x, 0) = u_1(x), & x \in (0, 1)\,, \\ z(1, \rho, 0) = f_0(1, -\tau\rho), & \rho \in (0, 1)\,. \end{cases} \tag{3.3.98}$$

As the stability result presented in Theorem 3.3.2 is for the "shifted" problem (3.3.36), we have to perform numerical simulations for both problems: the original one, namely (3.3.1), and the "shifted" problem (3.3.36).

To this end, in order to avoid a CFL condition between the mesh size and the time step, we decide to discretize the different problems by implicit first order in time, and finite difference method in space. For every simulation, the numerical parameters are the following:

$$\begin{aligned} &\tau = 2\,,\ \xi = 2\xi^\star\,,\ \Delta x = \tfrac{1}{20}\,,\ \Delta\rho = \tfrac{1}{20}\,,\ \Delta t = 0.1 \\ &u_0(x) = u_1(x) = x e^{10x}\,,\ f_0(1, \rho) = e^\rho e^{10}\,. \end{aligned}$$

For every time $t > 0$, we denote $E(t) = \big\|\big(u(.,t), u_t(.,t), u_t(1,t), z(1,.,t)\big)\big\|_{\mathcal{H}}$. The choice of u_0 , u_1, f_0 ensures a large initial energy.

In Figs. 3.3 and 3.4, we present the resulting simulation for the original problem and the "shifted" one. Let us first notice that the convergence rate for the shifted problem is better than the one expected: We have proved a polynomial decay rate, whereas numerically, we observe an exponential decay rate. This is most likely due to two facts:

1. The particular case of dimension 1 as proved in [184].
2. The numerical diffusion participates to the exponential stability as for the Kelvin-Voigt damping.

Moreover, as conjectured in Remark 3.3.2, the "shifted" problem converges for a large set of parameters a and μ, whereas the original problem does not and even worse, and it exhibits an exponential growth.

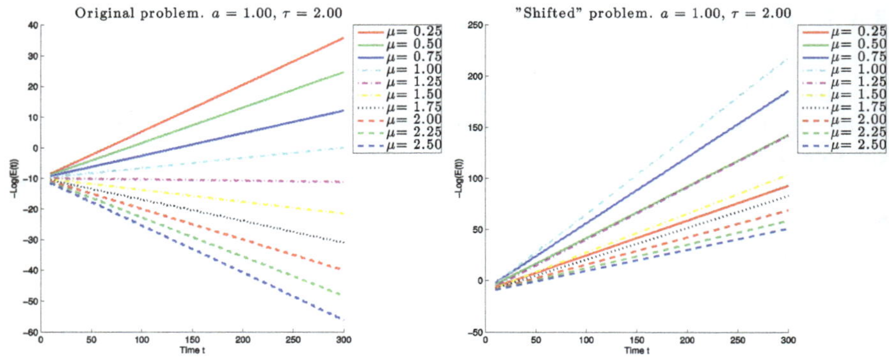

(a) (b)

Fig. 3.3 Energy (in -log scale) versus time: influence of μ. (**a**) Original problem. (**b**) Shifted problem

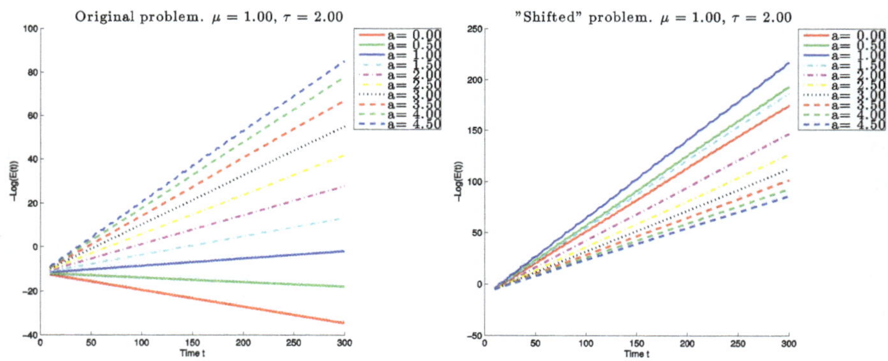

(a) (b)

Fig. 3.4 Energy (in -log scale) versus time: influence of a. (**a**) Original problem. (**b**) Shifted problem

In Fig. 3.5, we present the simulations for the case of the Kelvin-Voigt damping for which we have proved that under the condition $\mu < |c^\star|a$, we have an exponential decay rate.

From Eqs. (3.3.73) and (3.3.74), the constant $c^\star$ must satisfy

$$\begin{cases} u_{xx} = 0, & x \in (0,1), \\ u(0) = 0, & u_x(1) + c^\star u(1) = 0\,. \end{cases}$$

Thus, we obtain $c^\star = -1$. Let us first notice that, even though the condition between a and μ is not fulfilled, we have an exponential decay of the solution. This is also probably due to the particular case of one-dimensional space variable. Secondly, it seems that numerically the convergence rate does not depend on the parameter μ.

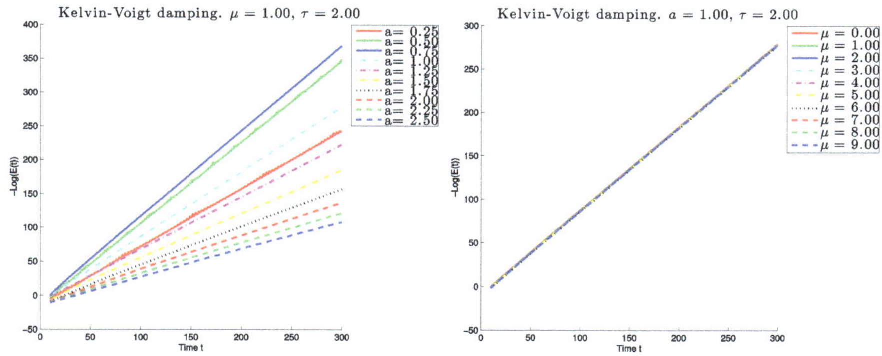

(a) (b)

Fig. 3.5 Energy (in -log scale) versus time. (**a**) Influence of a. (**b**) Influence of μ

3.3.6 Concluding Remarks

In this section, we proved an exponential stability result for the wave equation with dynamic delayed boundary conditions via the action of a strong Kelvin-Voigt damping.

As mentioned earlier, neither a decay nor a blow-up result could be established for the solutions to the original system (3.3.1), where the interior damping is "weak" ay_t. This problem remains unsolved, and it would be interesting to provide an answer to it.

It is also worth mentioning that the exponential stability result obtained for the Kelvin-Voigt damping problem (3.3.68) is valid with a global damping $a\ \Delta u_t$, where $a > 0$. It would be interesting to weaken the globality condition and consider a localized damping $\operatorname{div} a(x)\nabla u_t$, where $a(x) \geq 0$, for all $x \in \Omega$ and $a(x) > a_0$ only an open subdomain $\omega \subset \Omega$.

Lastly, the condition (3.3.75) played a crucial role in the stability result. Therefore, it would be good to investigate the case where this requirement is not fulfilled, in which we expect the system to be unstable. This should be investigated.

3.4 3D Wave Equation with Trapped Ray and Interior Delay: Polynomial Stabilization

In the present section, we study the long-time behavior of a wave equation in a *three-dimensional domain* Ω with a trapped ray, as in [162]. Thenceforth, the geometric control condition is violated. Additionally, an interior time delay occurs, and the task ahead is to neutralize any negative impact of the delay on the convergence of solutions to the problem under consideration. To do so, a localized interior damping is proposed and a polynomial convergence outcome is established.

3.4.1 Preliminaries

Consider the following delayed wave equation with an internal (distributed) localized damping:

$$\begin{cases} y_{tt}(x,t) - \Delta y(x,t) + a(x) y_t(x,t) & \\ +b(x) y_t(x, t-\tau) = 0, & (x,t) \in \Omega \times (0,\infty), \\ \dfrac{\partial y}{\partial \nu}(x,t) = 0, & \text{on } \Gamma \times (0,\infty), \\ y(x,0) = y_0(x),\ y_t(x,0) = z_0(x), & x \in \Omega, \\ y_t(x,t) = g(x,t), & (x,t) \in \Omega \times (-\tau, 0). \end{cases} \tag{3.4.1}$$

Here, a is a nonnegative function in $L^\infty(\Omega)$ and depends on a non-empty proper subset ω of Ω on which $1/a \in L^\infty(\omega)$, and, in particular, $\{x \in \Omega; a(x) > 0\}$ is a non-empty (or of positive measure) open set of Ω, and $b \in L^\infty(\Omega)$ is a nonnegative function.

Let

$$L_0^2(\Omega) = \left\{ u \in L^2(\Omega);\ \int_\Omega u(x)\,dx = 0 \right\}$$

and

$$H^{1,0}(\Omega) = H^1(\Omega) \cap L_0^2(\Omega).$$

Then, consider the state space

$$\mathscr{X} = H^{1,0}(\Omega) \times L_0^2(\Omega) \times L^2(\Omega \times (0,1)),$$

endowed with the inner product

$$\langle (y_1, z_1, u_1), (y_2, z_2, u_2) \rangle_{\mathscr{X}} = \int_\Omega (\nabla y_1 \cdot \nabla y_2 + z_1 z_2)\ dx + \xi \int_0^1 \int_\Omega u_1 u_2\, dx\, d\rho. \tag{3.4.2}$$

Additionally, we assume that a, b, and ξ verify the following conditions:

$$0 < \|b\|_{L^\infty(\Omega)} < \|a\|_{L^\infty(\Omega)}, \text{ and } \tau \|b\|_{L^\infty(\Omega)} < \xi < \tau \left(2\|a\|_{L^\infty(\Omega)} - \|b\|_{L^\infty(\Omega)} \right). \tag{3.4.3}$$

By virtue of the auxiliary variable $u(x, \rho, t) = y_t(x, t - \tau\rho)$, for $(x, \rho, t) \in \Omega \times (0, 1) \times (\infty, 0)$ (see [184]), the system (3.4.1) can be written as follows:

$$\begin{cases} \Phi_t(t) = \mathscr{A}_{a,b}\Phi(t), \\ \Phi(0) = \Phi_0 = (y_0, z_0, g), \end{cases} \tag{3.4.4}$$

where $\mathscr{A}_{a,b}$ is an unbounded linear operator defined by

$$\begin{aligned} &\mathscr{D}(\mathscr{A}_{a,b}) = \Big\{(y, z, u) \in \left(H^{1,0}(\Omega)\right)^2 \times L^2(\Omega; H^1(0, 1)); \\ &\Delta y \in L^2(\Omega); \ \frac{\partial y}{\partial \nu} = 0 \text{ on } \Gamma; \ z = u(\cdot, 0) \text{ on } \Omega\Big\}, \\ &\mathscr{A}_{a,b}(y, z, u) = \left(z, \Delta y - az - bu, -\tau^{-1}u_\rho\right), \quad \forall (y, z, u) \in \mathscr{D}(\mathscr{A}_{a,b}). \end{aligned} \tag{3.4.5}$$

Assume that the condition (3.4.3) holds. Then using the same arguments of the previous sections, one can show that $\mathscr{A}_{a,b}$ generates a C_0-semigroup of contractions $e^{t\mathscr{A}_{a,b}}$ on $\mathscr{X} = \overline{\mathscr{D}(\mathscr{A}_{a,b})}$ (see also [19]).

This section is intended to establish a result that is between the exponential convergence (see Sect. 3.2.6) and logarithmic convergence (see Sect. 3.2.5) for the system (3.4.1) (see also (3.4.4)). This desirable outcome will be established by considering, as in [162], the trapping geometry on the domain Ω of $\mathbb{R}^3$. Also, we shall adopt most of the notations in [162] and consider the same geometric situation as in [162] by letting

$$D(r_1, r_2) = \Big\{(x_1, x_2) \in \mathbb{R}^2; |x_1| < r_1, |x_2| < r_2\Big\},$$

where $r_1, r_2 > 0$. Moreover, we pick three positive constants m_1, m_2, and μ. Then, we choose Ω a connected open set in $\mathbb{R}^3$ whose boundary is $\partial\Omega := \Gamma = \Gamma_1 \cup \Gamma_2 \cup Y$, where $\Gamma_1 = \overline{D(m_1, m_2)} \times \{\mu\}$, with boundary $\partial\Gamma_1$, $\Gamma_2 = \overline{D(m_1, m_2)} \times \{-\mu\}$, with boundary $\partial\Gamma_2$,, and Y is a surface with boundary $\partial Y = \partial\Gamma_1 \cup \partial\Gamma_2$. On the other hand, $\partial\Omega$ is assumed to be either C^2 with $Y \subset \mathbb{R}^2 \setminus D(m_1, m_2) \times \mathbb{R}$ (in particular, $Y \in C^2$) or Ω is convex (in particular, Y is Lipschitz). Next, let Θ be a small neighborhood of Y in $\mathbb{R}^3$ such that $\Theta \cap D(M_1, M_2) \times [-\mu, \mu] = \emptyset$ for some $M_1 \in (0, m_1)$ and $M_2 \in (0, m_2)$. Lastly, we choose $\omega = \Omega \cap \Theta$.

In such a situation of trapped geometry (Ω, ω), the well-known geometric control condition (BLR) [32] is not fulfilled. Despite this geometric constraint, Phung [162] managed to establish the polynomial decay convergence result of (3.4.1) with $b \equiv 0$. More precisely, the following outcome is proved in [162].

Theorem 3.4.1 ([162]) *There exist positive constants C and δ such that for all $t > 0$, we have*

$$\left\| e^{t\mathscr{A}_{a,0}} \right\|_{\mathscr{L}(\mathscr{D}(\mathscr{A}_{a,0}), \mathscr{X}_a)} \leq \frac{C}{t^\delta},$$

where

$$\mathscr{X}_a := H^{1,0}(\Omega) \times L_0^2(\Omega),$$

and

$$\mathscr{A}_{a,0} := \begin{pmatrix} 0 & I \\ \Delta & -a \end{pmatrix} : \mathscr{D}(\mathscr{A}_{a,0}) \subset \mathscr{X}_a \to \mathscr{X}_a$$

$$\mathscr{D}(\mathscr{A}_{a,0}) := \left\{ (y, z) \in H^{1,0}(\Omega) \times H^{1,0}(\Omega);\ \Delta y \in L^2(\Omega);\ \frac{\partial y}{\partial \nu} = 0 \ on\ \Gamma \right\}.$$

3.4.2 Main Result

The main result of this section is:

Theorem 3.4.2 *Suppose that the assumptions (3.4.3) are fulfilled and ω satisfies the trapped geometrical condition. Then, there exists $C > 0$ such that the semigroup generated by the operator $\mathscr{A}_{a,b}$ (see (3.4.5)) satisfies*

$$\left\| e^{t\mathscr{A}_{a,b}} \right\|_{\mathscr{L}(\mathscr{D}(\mathscr{A}_{a,b}),\mathscr{X})} \leq C/t^{\delta/2}, \quad \forall t > 0,$$

where $\delta > 0$ is the same constant given in Theorem 3.4.1.

Before giving the proof of the main result, it is worth mentioning that our result improves Theorem 3.4.1 to the case where a localized internal damping is present. More importantly, the presence of a delay leads to a smaller rate of $t^{\delta/2}$.

Proof Thanks to the resolvent method for polynomial stability of a C_0-semigroup of contractions on a Hilbert space (see Theorem 2.3.10 and [38]), the task ahead consists in establishing the following results:

$$\rho(\mathscr{A}_{a,b}) \supset \{ i\gamma;\ \gamma \in \mathbb{R} \} \equiv i\mathbb{R}, \tag{3.4.6}$$

$$\limsup_{|\gamma|\to+\infty} \| |\gamma|^{-1/\beta} (i\gamma I - \mathscr{A}_{a,b})^{-1} \|_{\mathscr{L}(\mathscr{D}(\mathscr{A}_{a,b}),\mathscr{X})} < \infty. \tag{3.4.7}$$

First, we need to validate (3.4.6) by showing that the equation

$$\mathscr{A}_{a,b} Z = i\gamma Z, \tag{3.4.8}$$

with $Z = (y, z, u) \in \mathscr{D}(\mathscr{A}_{a,b})$ and $\gamma \in \mathbb{R}$, has only the trivial solution. Clearly, (3.4.8) writes

$$z = i\gamma y, \tag{3.4.9}$$

$$\Delta y - az - bu(1) = i\gamma z, \tag{3.4.10}$$

$$u_\rho = -i\tau\gamma u, \tag{3.4.11}$$

$$\frac{\partial y}{\partial \nu} = 0 \text{ on } \Gamma, \quad z = u(\cdot, 0) \text{ on } \Omega. \tag{3.4.12}$$

If $\gamma = 0$, then (3.4.9)–(3.4.12) lead us to claim that the only solution of (3.4.8) is the trivial one. In turn, if $\gamma \neq 0$, then by taking the inner product of (3.4.8) with Z, we get

$$0 = 2\Re e\langle \mathscr{A}_{a,b} Z, Z\rangle_{\mathscr{X}} \leq \left[\|b\|_{L^\infty(\Omega)} - 2\|a\|_{L^\infty(\Omega)} + \xi\tau^{-1}\right] \int_\Omega |z|^2 \, d\sigma$$
$$+ \left[\|b\|_{L^\infty(\Omega)} - \xi\tau^{-1}\right] \int_\Omega |u(x, 1)|^2 \, d\sigma \leq 0. \tag{3.4.13}$$

Whereupon, we have $z = u(\cdot, 0) = 0$ on Ω. Consequently, the only solution of (3.4.8) is the trivial one.

It remains now to show that (3.4.7) is satisfied with $\beta = \delta/2$. If this were not true, Banach-Steinhaus theorem (see Theorem 2.1.7) gives rise to the existence of a sequence of real numbers $\gamma_n \to \infty$ and a sequence of vectors $Z_n = (y_n, z_n, u_n) \in \mathscr{D}(\mathscr{A}_{a,b})$ with $\|Z_n\|_{\mathscr{X}} = 1$ such that

$$\||\gamma_n|^{2/\delta}(i\gamma_n I - \mathscr{A}_{a,b})Z_n\|_{\mathscr{X}} \to 0, \quad \text{as} \quad n \to \infty, \tag{3.4.14}$$

that is,

$$|\gamma_n|^{2/\delta} (i\gamma_n y_n - z_n) \equiv f_n \to 0 \quad \text{in} \quad H^{1,0}(\Omega), \tag{3.4.15}$$

$$|\gamma_n|^{2/\delta} (i\gamma_n z_n - \Delta y_n + a z_n + b u_n(\cdot, 1)) \equiv g_n \to 0 \quad \text{in} \quad L_0^2(\Omega), \tag{3.4.16}$$

$$|\gamma_n|^{2/\delta} \left(i\gamma_n u_n + \frac{(u_n)_\rho}{\tau}\right) \equiv h_n \to 0 \quad \text{in} \quad L^2(\Omega \times (0, 1)). \tag{3.4.17}$$

The immediate target is to derive from (3.4.14) that $\|Z_n\|_{\mathscr{X}} \to 0$, which contradicts $\|Z_n\|_{\mathscr{X}} = 1$, for all $n \in \mathbb{N}$. According to (3.4.13), we have

$$|\gamma_n|^{1/\delta} \sqrt{a}\, z_n \to 0, \quad |\gamma_n|^{1/\delta} \sqrt{b}\, u_n(\cdot, 1) \to 0 \text{ in } L^2(\Omega).$$

By virtue of (3.4.15)–(3.4.16), we deduce that

$$
\begin{aligned}
|\gamma_n|^{1/\delta}(i\gamma_n y_n - z_n) &\equiv \frac{f_n}{|\gamma_n|^{1/\delta}} \to 0 \quad \text{in} \quad H^{1,0}(\Omega),\\
|\gamma_n|^{1/\delta}(i\gamma_n z_n - \Delta y_n + a z_n) &\equiv -\frac{b u_n(\cdot,1)}{|\gamma_n|^{1/\delta}} + \frac{g_n}{|\gamma_n|^{1/\delta}} \to 0 \quad \text{in} \quad L_0^2(\Omega),
\end{aligned}
$$

which together with the fact that $(y_n, z_n) \in \mathscr{D}(\mathscr{A}_{a,0})$ (see Theorem 3.4.1), we conclude that

$$
y_n \to 0 \text{ in } H^{1,0}(\Omega) \text{ and } z_n \to 0 \text{ in } L^2(\Omega).
$$

Furthermore, (3.4.17) gives

$$
u_n(x,\rho) = z_n(x)e^{-i\tau\gamma_n x} + \tau \int_0^\rho e^{-i\tau\gamma_n(\rho-s)} \frac{h_n(s)}{|\gamma_n|^{2/\delta}}\, ds \to 0 \quad \text{in } L^2(\Omega \times (0,1)).
$$

Hence, $\|Z_n\|_{\mathscr{X}} \to 0$, and so we managed to show that the conditions (3.4.6)–(3.4.7) are fulfilled. This achieves the proof of Theorem 3.4.2. ◇ □

Remark 3.4.1

(1) In the case of a square domain, for instance, $\Omega = (0,1)\times(0,1)$ and the function $a(x) = \chi_{(0,\varepsilon)\times(0,1)}$, where $\varepsilon > 0$ is a constant, we have, thanks to the result of [175], that $\delta = 2/3$ (which is the optimal decay rate). In this situation, we obtain the polynomial decay rate for the delayed system (3.4.1) with rate $t^{-1/3}$.

(2) Using Theorem 3.4.2 and arguing as in Sect. 3.2, one can prove that the solutions of (3.4.1) polynomially converge (with a rate $t^{-\delta/2}$) to their equilibrium solution $(\zeta, 0, 0)$, where

$$
\zeta = \frac{1}{\displaystyle\int_\Omega (a+b)\, dx}\left[\int_\Omega \left(z_0 + (a+b)y_0 - b\tau \int_0^1 f(\rho)\,\rho\right) dx\right].
$$

Particularly, the convergence rate is $t^{-1/3}$ in the case discussed in the first part of this remark.

3.4.3 Conclusion

The interesting case of a three-dimensional delayed wave equation with trapped ray is discussed. Although the BLR condition does not hold in this case, it is shown that the solutions polynomially converge to the equilibrium state as long as a classical smallness condition is satisfied (see (3.4.3)). Thereby, it is normal to check whether the convergence could be lost if (3.4.3) does not hold.

3.5 Interior Memory Versus Interior Delay: Exponential Stabilization

This section is concerned with the asymptotic behavior of solutions to a multidimensional wave equation in a domain $\Omega \subseteq \mathbb{R}^n$. Assuming that there is no displacement term in the system and taking into consideration the presence of distributed memory or time delay, we show that the solutions exponentially converge to their stationary state. The proof mainly consists in utilizing the history approach combined with the resolvent method. Other physical systems are also discussed.

3.5.1 *Introduction*

We consider in this section the following wave equation with distributed memory:

$$u_{tt}(x,t) + Au(x,t) + a(x)u_t(x,t) + \int_0^{+\infty} f(s)u_t(x,t-s)\,ds = 0, \qquad (3.5.1)$$

for any $x \in \Omega$ and $t > 0$, as well as the homogeneous Neumann boundary condition on the whole boundary Γ

$$\partial_A u(x,t) := \sum_{i,j=1}^{N} a_{ij}(x)\nu_j\partial_i u(x,t) = 0, \quad \forall x \in \Gamma,\ \forall t > 0, \qquad (3.5.2)$$

and initial conditions

$$\begin{cases} u(x,-t) = u_0(x,t), & \forall x \in \Omega,\ \forall t \in \mathbb{R}_+, \\ u_t(x,-t) = u_1(x,t), & \forall x \in \Omega,\ \forall t \in \mathbb{R}_+, \end{cases} \qquad (3.5.3)$$

where u is the displacement and the differential operator with domain $D(A) = H^2(\Omega)$ is given by

$$A = -\sum_{i,j=1}^{N} \partial_i(a_{ij}\partial_j), \qquad (3.5.4)$$

in which $\partial_k = \frac{\partial}{\partial x_k}$. In turn, $a_{ij} \in C^1(\bar{\Omega})$ and $a \in C(\bar{\Omega})$ such that there exist $\alpha_0,\ a_0 > 0$ satisfying

$$a_{ij}(x) = a_{ji}(x), \quad \sum_{k,l=1}^{N} a_{kl}(x)\varepsilon_k\varepsilon_l \geq \alpha_0 \sum_{k=1}^{N} \varepsilon_k^2, \quad \forall x \in \Omega,\ \forall \varepsilon_k,\ \varepsilon_l \in \mathbb{R},$$

$$\forall i, j = 1, \cdots, N, \qquad (3.5.5)$$

and

$$a(x) \geq a_0, \qquad \forall x \in \Omega. \tag{3.5.6}$$

Also, note that $\nu = (\nu_1, \cdots, \nu_n)$ is the unit normal of Γ pointing toward the exterior of Ω. Moreover, u_0 and u_1 are given initial data that belong to a suitable space. Lastly, $f : \mathbb{R}_+ \to \mathbb{R}_+$ is a given function.

Let $H = L^2(\Omega)$ equipped with its standard norm $\|\cdot\|$ and inner product

$$\langle p, q \rangle = \int_\Omega p(x)q(x)\, dx.$$

Set $V = H^1(\Omega)$ equipped with its classical norm $\|\cdot\|_V$ and inner product

$$\langle p, q \rangle_V = \int_\Omega \left(p(x)q(x) + \sum_{i=1}^{N} \partial_i p(x)\partial_i q(x) \right) dx.$$

The main concern of this section is to compensate the effect of the memory term in (3.5.1) via the action of the damping $a(x)u_t$.

In the present section, we shall continue in the direction of Sect. 3.2 by considering an infinite memory. Unlike the discrete-time-delay case that does not take into account the inherent memory effects, we shall deal with the distributed time delay as an infinite past history of the solution. Indeed, the discrete case will be treated as a special case that corresponds to the Dirac delta distribution kernel (at some time τ).

The rest of this section is organized as follows. In Sect. 3.5.2, the system (3.5.1)–(3.5.3) is shown to be well-posed in the sense of semigroups theory of linear operators (see Sect. 2.3). Section 3.5.3 is devoted to the analysis of the long-time behavior of (3.5.1)–(3.5.3) via LaSalle's principle (see Theorem 2.3.11). The proof of the exponential convergence of solutions to (3.5.1)–(3.5.3) is provided in Sect. 3.5.4, by using the resolvent method. We treat in Sect. 3.5.5 the discrete-time-delay case. The results are illustrated via a set of applications to other physical system in Sect. 3.5.6. Finally, this section ends with a conclusion.

3.5.2 *Well-Posedness of (3.5.1)–(3.5.3)*

In this subsection, we state our hypotheses and then prove the well-posedness of (3.5.1)–(3.5.3).

The following assumptions on a and f are supposed to be fulfilled:

(A1) f is of class $C^1(\mathbb{R}_+, \mathbb{R}_+)$, nonincreasing, and satisfies, for some positive constants α_1 and α_2,

$$-\alpha_1 f(s) \leq f'(s) \leq -\alpha_2 f(s), \quad \forall s \in \mathbb{R}_+. \tag{3.5.7}$$

(A2) a is a positive function that satisfies (3.5.5)–(3.5.6) and also

$$\int_0^{+\infty} f(s)ds := f_0 < \frac{1}{\alpha_1}(a_0\alpha_2 + f(0)). \tag{3.5.8}$$

Remark 3.5.1 The assumptions (**A1**) and (**A2**) are fulfilled by a wide class of functions f. Such a class also contains the ones that converge to zero exponentially such as

$$f(s) = de^{-qs},$$

where $d,\ q > 0$. Clearly, (3.5.7) is fulfilled with $\alpha_1 = \alpha_2 = q$, whereas (3.5.8) is satisfied since $qa_0 > 0$.

In order to express the problem (3.5.1)–(3.5.3) as a linear first-order system, we consider the new state variable η and its initial data η_0 introduced in [79] and given by

$$\begin{cases} \eta(x,t,s) = u(x,t) - u(x,t-s), & \forall x \in \Omega,\ \forall t,\ s \in \mathbb{R}_+, \\ \eta_0(x,s) = \eta(x,0,s) = u_0(x,0) - u_0(x,s), & \forall x \in \Omega,\ \forall s \in \mathbb{R}_+. \end{cases} \tag{3.5.9}$$

It is easy to verify that η satisfies

$$\begin{cases} \eta_t(x,t,s) + \eta_s(x,t,s) = u_t(x,t), & \forall x \in \Omega,\ \forall t,\ s \in \mathbb{R}_+, \\ \eta_s(x,t,s) = u_t(x,t-s), & \forall x \in \Omega,\ \forall t,\ s \in \mathbb{R}_+, \\ \eta(x,t,0) = 0, & \forall x \in \Omega,\ \forall t \in \mathbb{R}_+. \end{cases} \tag{3.5.10}$$

To simplify the notation, the variables $x,\ t$ and s are omitted whenever there is no ambiguity. Let

$$\begin{cases} y = u_t, \\ \Psi = (u, y, \eta), \\ \Psi_0 = (u_0(\cdot,0), u_1(\cdot,0), \eta_0) \end{cases}$$

and

$$\mathscr{H} = V \times H \times L_f,$$

where L_f is the weighted space defined by

$$L_f = \left\{\eta : \mathbb{R}_+ \to H;\quad \int_0^{+\infty} f(s)\|\eta(s)\|^2\, ds < +\infty\right\}$$

and endowed with the inner product

$$\langle \eta, \tilde{\eta} \rangle_{L_f} = \int_0^{+\infty} f(s) \langle \eta(s), \tilde{\eta}(s) \rangle \, ds.$$

Then (3.5.1)–(3.5.3) has an equivalent form

$$\begin{cases} \Psi_t(x,t) = \mathscr{A}\Psi(x,t), & \forall x \in \Omega, \ \forall t > 0, \\ \Psi(x,0) = \Psi_0(x), & \forall x \in \Omega, \end{cases} \tag{3.5.11}$$

where $\mathscr{A}$ is a linear operator given by

$$\mathscr{A}(u, y, \eta) = \left(y, -Au - ay - \int_0^{+\infty} f(s)\eta_s(s) \, ds, -\eta_s + y \right). \tag{3.5.12}$$

The domain $D(\mathscr{A})$ of $\mathscr{A}$ is given by

$$D(\mathscr{A}) = \left\{ \Psi = (u, y, \eta) \in \mathscr{H}, \ Au \in H, \ y \in V, \ \eta_s \in L_f, \ \eta(0) = 0, \right.$$
$$\left. \partial_A u = 0 \text{ on } \Gamma \right\}. \tag{3.5.13}$$

Due to the Neumann boundary condition on the entire boundary Γ and since the displacement term is considered neither in Ω nor on Γ, the classical energy norm $\| \cdot \|_E$ given by

$$\|(u, y, \eta)\|_E^2 = \int_\Omega \sum_{i=1,j}^{N} a_{ij} \partial_i u \partial_j u \, dx + \|y\|^2 + \|\eta\|_{L_f}^2$$

is only a seminorm in our case. In fact, if $\|(u, y, \eta)\|_E^2 = 0$, then we do not necessarily have $u = 0$. Hence, a new inner product on $\mathscr{H}$ must be defined in order to make it a Hilbert space. To achieve this goal, we integrate (3.5.1) on Ω and obtain (formally)

$$\partial_t \left(\int_\Omega \left(u_t + au + \int_0^{+\infty} f(s) u(t-s) \, ds \right) dx \right) = 0.$$

Using the definition of y, η, and f_0, we get

$$\partial_t \left(\int_\Omega \left((f_0 + a)u + y - \int_0^{+\infty} f(s)\eta(s) \, ds \right) dx \right) = 0. \tag{3.5.14}$$

This time-invariant expression will play a crucial role in introducing an appropriate inner product on $\mathscr{H}$. In fact, we pick up an arbitrary positive constant ε_0 and consider on $\mathscr{H}$ the inner product

$$\langle (u, y, \eta), (\tilde{u}, \tilde{y}, \tilde{\eta}) \rangle_{\mathscr{H}} = \int_\Omega \left(\sum_{i,j=1}^{n} a_{ij} \partial_i u \partial_j \tilde{u} \right) dx + \langle y, \tilde{y} \rangle + \alpha_1 \langle \eta, \tilde{\eta} \rangle_{L_f}$$

$$+ \varepsilon_0 \left(\int_\Omega \left((f_0 + a)u + y - \int_0^{+\infty} f(s)\eta(s)\, ds \right) dx \right)$$
$$\times \int_\Omega \left((f_0 + a)\tilde{u} + \tilde{y} - \int_0^{+\infty} f(s)\tilde{\eta}(s)\, ds \right) dx,$$

where α_1 is the positive constant appearing in (3.5.7). To prove that $(\mathscr{H}, \langle \cdot, \cdot \rangle_{\mathscr{H}})$ is a Hilbert space, it is enough to prove that there exists $\varepsilon_0 > 0$ such that the norm $\| \cdot \|_{V \times H \times L_f}$ and the one $\| \cdot \|_{\mathscr{H}}$ generated by $\langle \cdot, \cdot \rangle_{\mathscr{H}}$ are equivalent. This will be done in the following lemma.

Lemma 3.5.1 *There exists $\varepsilon_1 > 0$ such that for any $\varepsilon_0 \in (0, \varepsilon_1)$, and for any $(u, y, \eta) \in \mathscr{H}$, we have*

$$c_1 \|(u, y, \eta)\|_{V \times H \times L_f}^2 \le \|(u, y, \eta)\|_{\mathscr{H}}^2 \le c_2 \|(u, y, \eta)\|_{V \times H \times L_f}^2, \tag{3.5.15}$$

where c_1, c_2 are positive constants.

Proof First, note that there exists $k_1 > 0$ such that

$$\sum_{i,j=1}^{N} a_{ij} \varepsilon_i \varepsilon_j \le k_1 \sum_{i=1}^{N} \varepsilon_i^2, \quad \forall\ (\varepsilon_1, \cdots, \varepsilon_N) \in \mathbb{R}^N. \tag{3.5.16}$$

It is clear that, using Hölder's and Young's inequalities, the direct inequality in (3.5.15) holds, for any $\varepsilon_0 > 0$ and where c_2 depends on ε_0, k_1, a, f, and Ω.

To prove the reverse inequality in (3.5.15), we note that

$$\left(\int_\Omega \left((f_0 + a)u + y - \int_0^{+\infty} f(s)\eta(s)\, ds \right) dx \right)^2 = \left(\int_\Omega (f_0 + a)u\, dx \right)^2 \tag{3.5.17}$$

$$+ \left(\int_\Omega \left(y - \int_0^{+\infty} f(s)\eta(s)\, ds \right) dx \right)^2 + 2 \left(\int_\Omega (f_0 + a)u\, dx \right)$$
$$\times \int_\Omega \left(y - \int_0^{+\infty} f(s)\eta(s)\, ds \right) dx.$$

Using Young's and Hölder's inequalities, we have

$$2\left(\int_\Omega (f_0+a)u\,dx\right)\left(\int_\Omega \left(y-\int_0^{+\infty} f(s)\eta(s)\,ds\right)dx\right)$$
$$\geq -\frac{1}{2}\left(\int_\Omega (f_0+a)u\,dx\right)^2 - 2\left(\int_\Omega \left(y-\int_0^{+\infty} f(s)\eta(s)\,ds\right)dx\right)^2$$

and

$$\left(\int_\Omega \left(y-\int_0^{+\infty} f(s)\eta(s)\,ds\right)dx\right)^2 \leq c\left(\|y\|^2+\|\eta\|^2_{L_f}\right), \quad \forall (y,\eta)\in H\times L_f,$$

for some positive constant c. Furthermore, in light of (3.5.6), and using a classical compactness argument, one can show the following form of the generalized Poincaré's inequality:

There exists a positive constant ϖ_0 depending on Ω, f_0 and a such that

$$\|u\|^2 \leq \varpi_0\left(\|\nabla u\|^2 + \left(\int_\Omega (f_0+a)u\,dx\right)^2\right), \quad \forall u\in V. \tag{3.5.18}$$

Combining the above three inequalities together with (3.5.17), we get

$$\varepsilon_0\left(\int_\Omega \left((f_0+a)u + y - \int_0^{+\infty} f(s)\eta(s)\,ds\right)dx\right)^2$$
$$\geq \frac{\varepsilon_0}{2\varpi_0}\|u\|^2 - \frac{\varepsilon_0}{2}\|\nabla u\|^2 - \varepsilon_0 c\|y\|^2 - \varepsilon_0 c\|\eta\|^2_{L_f}.$$

Therefore, after recalling (3.5.5), we have

$$\|(u,y,\eta)\|^2_{\mathcal{H}} \geq \left(\alpha_0-\frac{\varepsilon_0}{2}\right)\|\nabla u\|^2 + \frac{\varepsilon_0}{2\varpi_0}\|u\|^2 + (1-\varepsilon_0 c)\|y\|^2 + (\alpha_1-\varepsilon_0 c)\|\eta\|^2_{L_f}.$$

Thereafter, choosing ε_0 as follows:

$$0<\varepsilon_0<\varepsilon_1 := \min\left\{2\alpha_0, \frac{1}{c}, \frac{\alpha_1}{c}\right\},$$

we obtain the reverse inequality in (3.5.15) with

$$c_1 = \min\left\{\alpha_0-\frac{\varepsilon_0}{2}, \frac{\varepsilon_0}{2\varpi_0}, 1-\varepsilon_0 c, \alpha_1-\varepsilon_0 c\right\} > 0.$$

□

The well-posedness of our problem (3.5.11) is ensured by the following result:

Theorem 3.5.1 *Assume that* (A1)–(A2) *hold. Then, the operator* $\mathscr{A}$ *is the infinitesimal generator of a linear* C_0*-semigroup* $S(t)$ *of contractions on* $\mathscr{H}$*, and its domain* $D(\mathscr{A})$ *is dense in* $\mathscr{H}$*. Moreover, for any* $\Psi_0 \in \mathscr{H}$*, the system (3.5.11) has a unique weak solution*

$$\Psi \in C(\mathbb{R}_+, \mathscr{H}). \tag{3.5.19}$$

In turn, if $\Psi_0 \in D(\mathscr{A})$*, then the solution of (3.5.11) satisfies (classical solution)*

$$\Psi \in C^1(\mathbb{R}_+, \mathscr{H}) \cap C(\mathbb{R}_+, D(\mathscr{A})). \tag{3.5.20}$$

Proof Let us show that $\mathscr{A}$ is m-dissipative, that is, $\mathscr{A}$ is dissipative and $I - \mathscr{A}$ is onto. Let $\Psi \in D(\mathscr{A})$. Using the definition of Ψ, $\mathscr{A}$, and $\langle \cdot, \cdot \rangle_{\mathscr{H}}$ and integrating by parts, we find

$$\begin{aligned}
\langle \mathscr{A}\Psi, \Psi \rangle_{\mathscr{H}} = &-\|\sqrt{a}y\|^2 + \alpha_1 \int_\Omega y \int_0^{+\infty} f(s)\eta(s)\, ds\, dx \\
&- \int_\Omega y \int_0^{+\infty} f(s)\eta_s(s)\, ds\, dx \\
&- \alpha_1 \int_\Omega \int_0^{+\infty} f(s)\eta(s)\eta_s(s)\, ds\, dx.
\end{aligned}$$

Then, integrating with respect to s the last two integrals and noting that $\eta(0) = 0$, we obtain

$$\begin{aligned}
\langle \mathscr{A}\Psi, \Psi \rangle_{\mathscr{H}} = &-\|\sqrt{a}y\|^2 + \frac{\alpha_1}{2} \int_0^{+\infty} f'(s)\|\eta(s)\|^2\, ds \\
&+ \int_\Omega y \int_0^{+\infty} (f'(s) + \alpha_1 f(s))\eta(s)\, ds\, dx.
\end{aligned}$$

Applying Young's and Hölder's inequalities to the last integral of the above identity, we deduce that, for any $\delta > 0$,

$$\begin{aligned}
&\langle \mathscr{A}\Psi, \Psi \rangle_{\mathscr{H}} \leq \left(\frac{\delta}{2} - a_0\right) \|y\|^2 \\
&+ \frac{1}{2} \int_0^{+\infty} \left(\int_0^{+\infty} \frac{(f'(\tau) + \alpha_1 f(\tau))^2}{\delta f(\tau)}\, d\tau + \alpha_1 \frac{f'(s)}{f(s)} \right) f(s)\|\eta\|^2\, ds,
\end{aligned}$$

where a_0 is given in (3.5.6). Exploiting (3.5.7) and (3.5.8), we have

$$\int_0^{+\infty} \frac{(f'(\tau) + \alpha_1 f(\tau))^2}{\delta f(\tau)}\, d\tau + \alpha_1 \frac{f'(s)}{f(s)}$$
$$= \frac{1}{\delta} \int_0^{+\infty} \left(\frac{(f'(\tau))^2}{f(\tau)} + 2\alpha_1 f'(\tau) + \alpha_1^2 f(\tau) \right) d\tau + \alpha_1 \frac{f'(s)}{f(s)}$$
$$\leq \frac{2}{\delta} \left(\alpha_1^2 f_0 - \alpha_1 f(0) \right) - \alpha_1 \alpha_2.$$

Therefore,

$$\langle \mathscr{A}\Psi, \Psi \rangle_{\mathscr{H}} \leq \left(\frac{\delta}{2} - a_0 \right) \|y\|^2 + \frac{\alpha_1}{2} \left(\frac{2}{\delta} \left(\alpha_1 f_0 - f(0) \right) - \alpha_2 \right) \|\eta\|_{L_f}^2.$$

Thus, choosing

$$\frac{2(\alpha_1 f_0 - f(0))}{\alpha_2} < \delta < 2a_0,$$

which is possible, thanks to (3.5.8), we get

$$\langle \mathscr{A}\Psi, \Psi \rangle_{\mathscr{H}} \leq -d_0 \left(\|y\|^2 + \|\eta\|_{L_f}^2 \right), \tag{3.5.21}$$

where

$$d_0 = \frac{1}{2} \min \left\{ 2a_0 - \delta, \alpha_1 \left(\alpha_2 - \frac{2}{\delta} \left(\alpha_1 f_0 - f(0) \right) \right) \right\} > 0. \tag{3.5.22}$$

Thereby, $\mathscr{A}$ is dissipative.

Now, we show that $I - \mathscr{A}$ is onto $\mathscr{H}$. To this end, let $F = (f_1, f_2, f_3) \in \mathscr{H}$, and we must show the existence of an element $\Psi = (u, y, \eta) \in D(\mathscr{A})$ so that

$$\Psi - \mathscr{A}\Psi = F. \tag{3.5.23}$$

Clearly, the first equation in (3.5.23) is equivalent to

$$y = u - f_1. \tag{3.5.24}$$

In turn, it follows from (3.5.24) that the last equation in (3.5.23) is reduced to

$$\eta_s + \eta = u - f_1 + f_3. \tag{3.5.25}$$

Integrating with respect to s and noting that η should satisfy $\eta(0) = 0$, we get

$$\eta(s) = (1 - e^{-s})(u - f_1) + \int_0^s e^{\tau - s} f_3(\tau)\, d\tau. \tag{3.5.26}$$

Using (3.5.24) and (3.5.26), we find that the second equation in (3.5.23) is reduced to

$$\begin{aligned} &Au + (a + 1 + \tilde{f}_0)u \\ &\quad = f_2 + (a + 1 + \tilde{f}_0) f_1 + \int_0^{+\infty} f(s) \left(\int_0^s e^{\tau - s} f_3(\tau)\, d\tau - f_3(s) \right) ds, \end{aligned} \tag{3.5.27}$$

where

$$\tilde{f}_0 = \int_0^{+\infty} e^{-s} f(s)\, ds.$$

Note that if (3.5.27) admits a solution $u \in V$, then (3.5.24) implies that $y \in V$. Additionally, $\eta_s \in L_f$, whenever $\eta \in L_f$. Henceforth, it suffices to prove that $\eta \in L_f$ in order to get the required regularity on η in $D(\mathscr{A})$. Indeed,

$$s \mapsto (1 - e^{-s})(u - f_1) \in L_f \tag{3.5.28}$$

since $f_1 \in H$ and $u \in V$. On the other hand, using Fubini theorem along with Hölder inequality, we get

$$\begin{aligned} &\int_0^{+\infty} f(s) \left\| \int_0^s e^{\tau - s} f_3(\tau)\, d\tau \right\|^2 ds \\ &\leq \int_0^{+\infty} e^{-2s} f(s) \left(\int_0^s e^{\tau}\, d\tau \right) \int_0^s e^{\tau} \, \| f_3(\tau) \|^2 \, d\tau\, ds \\ &\leq \int_0^{+\infty} e^{-s} (1 - e^{-s})\, f(s) \int_0^s e^{\tau} \, \| f_3(\tau) \|^2 \, d\tau\, ds \\ &\leq \int_0^{+\infty} e^{-s} f(s) \int_0^s e^{\tau} \, \| f_3(\tau) \|^2 \, d\tau\, ds \\ &\leq \int_0^{+\infty} e^{\tau} \, \| f_3(\tau) \|^2 \int_{\tau}^{+\infty} e^{-s} f(s)\, ds\, d\tau \\ &\leq \int_0^{+\infty} e^{\tau} f(\tau) \, \| f_3(\tau) \|^2 \int_{\tau}^{+\infty} e^{-s}\, ds\, d\tau \\ &\leq \| f_3 \|_{L_f}^2 < +\infty, \end{aligned}$$

in which the fact that $f_3 \in L_f$ is invoked. Thereafter,

$$s \mapsto \int_0^s e^{\tau - s} f_3(\tau)\, d\tau \in L_f. \tag{3.5.29}$$

Thus, (3.5.26), (3.5.28), and (3.5.29) imply that $\eta \in L_f$. On the other hand, since $f_2 \in H$, $f_1 \in V$, $f_3 \in L_f$ and thanks to (3.5.29), we have

$$\tilde{f} := f_2 + (a + 1 + \tilde{f}_0) f_1 + \int_0^{+\infty} f(s) \left(\int_0^s e^{\tau - s} f_3(\tau)\, d\tau - f_3(s) \right) ds \in H. \tag{3.5.30}$$

Consequently, it amounts to demonstrating that (3.5.27) admits a solution $u \in V$ that satisfies $Au \in H$ and $\partial_A u = 0$ on Γ. To do so, we consider the variational formulation of (3.5.27)

$$\int_\Omega \left(\sum_{i,j=1}^N a_{ij} \partial_i u \partial_j v + (a + 1 + \tilde{f}_0) u v \right) dx = \int_\Omega \tilde{f} v\, dx, \quad \forall v \in C_c^\infty(\Omega).$$

Then using (3.5.5), (3.5.6), (3.5.30), as well as Lax-Milgram theorem (see Theorem 2.1.6) and classical elliptic regularity arguments, we deduce that (3.5.27) admits a unique solution $u \in V$ satisfying

$$Au \in H \quad \text{and} \quad \partial_A u = 0 \text{ on } \Gamma.$$

This shows that (3.5.23) has a unique solution $\Psi \in D(\mathscr{A})$. By the resolvent identity, we have $\lambda I - \mathscr{A}$ is surjective, for any $\lambda > 0$ (see (2.3.20)). Consequently, Lumer-Phillips theorem (see Theorem 2.3.4) implies that $\mathscr{A}$ is the infinitesimal generator of a linear C_0-semigroup $S(t)$ of contractions on $\mathscr{H}$, and its domain $D(\mathscr{A})$ is dense in $\mathscr{H}$ (see [158] or Sect. 2.3). Lastly, Theorem 3.5.1 follows from semigroups theory (see Sect. 2.3). □

3.5.3 Asymptotic Convergence of Solutions to (3.5.11)

In this subsection, an asymptotic behavior result is established for the unique solution Ψ of (3.5.11) in $\mathscr{H}$. Specifically, we prove that, as time tends to infinity, Ψ converges to a steady state. We have:

Theorem 3.5.2 *Assume that the assumptions* (A1)–(A2) *are fulfilled. Then, for any initial data* $\Psi_0 = (u_0, u_1, \eta_0) \in \mathscr{H}$*, the solution* $\Psi = (u, u_t, \eta)$ *of (3.5.11)* *converges in $\mathscr{H}$ to the equilibrium state $(\mathscr{C}, 0, 0)$ as $t \longrightarrow +\infty$, where*

$$\mathscr{C} = \left(\int_\Omega (f_0 + a)\, dx \right)^{-1} \int_\Omega \left[(f_0 + a)u_0 + u_1 - \int_0^{+\infty} f(s)\eta_0(s)\, ds \right] dx. \tag{3.5.31}$$

Proof By a standard density argument, together with the contraction of the semigroup $S(t)$ generated by $\mathscr{A}$, it suffices to prove Theorem 3.5.2 for smooth initial data

$$\Psi_0 = (u_0, u_1, \eta_0) \in D(\mathscr{A}).$$

Next, let $\Psi = (u, u_t, \eta) = S(\cdot)\Psi_0$ be the solution of (3.5.11). It follows from the dissipativity of $\mathscr{A}$ that the solutions trajectory set $\{\Psi(t)\}_{t \in \mathbb{R}_+}$ is bounded and precompact. Applying LaSalle's principle (see Theorem 2.3.11), we deduce that the ω-limit set $\omega(\Psi_0)$ is non-empty, compact, and invariant under the semigroup $S(t)$. Additionally, we have

$$S(t)\Psi_0 \longrightarrow \omega(\Psi_0) \quad \text{as } t \to +\infty. \tag{3.5.32}$$

It suffices to show that $\omega(\Psi_0)$ is reduced to $(\mathscr{C}, 0, 0)$. To this end, let

$$\tilde{\Psi}_0 = (\tilde{u}_0, \tilde{u}_1, \tilde{\eta}_0) \in \omega(\Psi_0) \subset D(\mathscr{A})$$

and

$$\tilde{\Psi}(t) = (\tilde{u}(t), \tilde{u}_t(t), \tilde{\eta}(t)) = S(t)\tilde{\Psi}_0 \in D(\mathscr{A})$$

the unique classical solution of (3.5.11). Recalling that $t \mapsto \|\tilde{\Psi}(t)\|_{\mathscr{H}}$ is a constant function [114], we have

$$\frac{d}{dt}\left(\frac{1}{2} \|\tilde{\Psi}(t)\|_{\mathscr{H}}^2 \right) = 0.$$

Then, using (3.5.11) and (3.5.21),

$$0 = \left\langle \tilde{\Psi}_t, \tilde{\Psi} \right\rangle_{\mathscr{H}} = \left\langle \mathscr{A}\tilde{\Psi}, \tilde{\Psi} \right\rangle_{\mathscr{H}} \leq -d_0 \left(\|\tilde{u}_t\|^2 + \|\tilde{\eta}\|_{L_f}^2 \right) \leq 0. \tag{3.5.33}$$

This gives $\tilde{u}_t = \tilde{\eta} = 0$, and therefore, $\tilde{u}$ is a solution of the following problem:

$$\begin{cases} A\tilde{u} = 0, & \forall x \in \Omega, \\ \partial_A \tilde{u} = 0, & \forall x \in \Gamma. \end{cases} \tag{3.5.34}$$

It is simple to check that (3.5.34) yields $\tilde{u}$ that is constant. Thus, we have proved that for any $\tilde{\Psi}_0 = (\tilde{u}_0, \tilde{u}_1, \tilde{\eta}_0) \in \omega\left(\tilde{\Psi}_0\right) \subset D(\mathscr{A})$, the solution $\tilde{\Psi}(t) = (\tilde{u}, \tilde{u}_t, \tilde{\eta}) = S(t)\tilde{\Psi}_0 \in D(\mathscr{A})$ satisfies

$$\tilde{\Psi}(t) = (\mathscr{C}, 0, 0), \quad \forall t \in \mathbb{R}_+,$$

where $\mathscr{C}$ is a constant. In particular, $\tilde{\Psi}_0 = (\mathscr{C}, 0, 0)$, and hence, the ω-limit set $\omega(\Psi_0)$ consists of constant elements $(\mathscr{C}, 0, 0)$. Now, let us prove that $\mathscr{C}$ is unique and is given by (3.5.31). To proceed, let $(\mathscr{C}, 0, 0) \in \omega(\Psi_0)$. This implies that there exists $\{t_n\} \to +\infty$, as $n \to +\infty$ such that

$$\Psi(t_n) = (u(t_n), u_t(t_n), \eta(t_n)) = S(t_n)\Psi_0 \longrightarrow (\mathscr{C}, 0, 0) \tag{3.5.35}$$

in the state space $\mathscr{H}$. Furthermore, using (3.5.14), the function

$$t \mapsto \int_\Omega \left((f_0 + a)u + u_t - \int_0^{+\infty} f(s)\eta(s)\,ds \right) dx$$

is a constant function, and hence,

$$\begin{aligned} &\int_\Omega \left((f_0 + a)u + u_t - \int_0^{+\infty} f(s)\eta(s)\,ds \right) dx \\ &\quad = \int_\Omega \left((f_0 + a)u_0 + u_1 - \int_0^{+\infty} f(s)\eta_0(s)\,ds \right) dx. \end{aligned} \tag{3.5.36}$$

Finally, let $t = t_n$ in (3.5.36) with $n \to +\infty$ and use (3.5.35) to obtain

$$\int_\Omega (f_0 + a)\mathscr{C}\,dx = \int_\Omega \left((f_0 + a)u_0 + u_1 - \int_0^{+\infty} f(s)\eta_0(s)\,ds \right) dx,$$

which gives the unique value in (3.5.31). This achieves the proof of Theorem 3.5.2. □

3.5.4 Exponential Convergence of Solutions to (3.5.11)

This subsection is consecrated to obtaining the convergence rate of solutions to our system. Indeed, we will prove that the solutions exponentially converge to their equilibrium state given in Theorem 3.5.2. To do so, we shall use the resolvent method (see Theorem 2.3.8).

First, let us consider the closed subspace $\hat{\mathscr{H}}$ of $\mathscr{H}$ given by

$$\hat{\mathscr{H}} = \left\{ (u, y, \eta) \in \mathscr{H};\ \int_\Omega \left((f_0 + a)u + y - \int_0^{+\infty} f(s)\eta(s)\, ds \right) dx = 0 \right\}$$

and equipped with the inner product

$$\langle (u, y, \eta), (\tilde{u}, \tilde{y}, \tilde{\eta}) \rangle_{\hat{\mathscr{H}}} = \int_\Omega \left(\sum_{i,j=1}^{n} a_{ij} \partial_i u \partial_j \tilde{u} \right) dx + \langle y, \tilde{y} \rangle + \alpha_1 \langle \eta, \tilde{\eta} \rangle_{L_f},$$

where α_1 is the positive constant defined in (3.5.7). Also, $\hat{\mathscr{A}}$ denotes a new operator defined by

$$\begin{aligned} &\hat{\mathscr{A}} : D(\hat{\mathscr{A}}) := D(\mathscr{A}) \cap \hat{\mathscr{H}} \subset \hat{\mathscr{H}} \to \hat{\mathscr{H}}, \\ &\hat{\mathscr{A}}(u, y, \eta) = \mathscr{A}(u, y, \eta),\ \forall (u, y, \eta) \in D(\hat{\mathscr{A}}). \end{aligned} \tag{3.5.37}$$

According to the previous subsections, one can claim the operator $\hat{\mathscr{A}}$ defined by (3.5.37) generates on $\hat{\mathscr{H}}$ a C_0-semigroup of contractions $e^{t\hat{\mathscr{A}}}$. Moreover, $\sigma(\hat{\mathscr{A}})$, the spectrum of $\hat{\mathscr{A}}$, consists of isolated eigenvalues of finite algebraic multiplicity only (see Remark 2.1.2).

Now we are in a position to state and prove the exponential stability of the semigroup $e^{t\hat{\mathscr{A}}}$ on $\hat{\mathscr{H}}$, and hence, the solutions of the system exponentially converge to the steady state.

Theorem 3.5.3 *Suppose that the assumptions* (A1)–(A2) *hold. Then, there exist two positive constants* κ *and* ω *such that we have*

$$\left\| e^{t\hat{\mathscr{A}}} \right\|_{\mathscr{L}(\hat{\mathscr{H}})} \leq \kappa e^{-\omega t}, \quad \forall t > 0. \tag{3.5.38}$$

Proof For the sake of clarity, we will proceed by steps.
Step 1: The immediate objective is to show (2.3.16) is fulfilled by the operator $\hat{\mathscr{A}}$, that is, given a real number γ, the equation

$$\hat{\mathscr{A}}\Psi = i\gamma\Psi, \quad \Psi = (u, y, \eta) \in D(\hat{\mathscr{A}}) \tag{3.5.39}$$

has only a trivial solution. Note that (3.5.39) yields

$$y = i\gamma u, \tag{3.5.40}$$

$$-Au - ay - \int_0^{+\infty} f(s)\eta_s(s)\, ds = i\gamma y, \tag{3.5.41}$$

$$-\eta_s + y = i\gamma\eta, \tag{3.5.42}$$

$$\int_\Omega \left((f_0 + a)u + y - \int_0^{+\infty} f(s)\eta(s)\, ds \right) dx = 0, \tag{3.5.43}$$

$$\partial_A u = 0 \quad \text{on}\, \Gamma, \tag{3.5.44}$$

$$\eta(0) = 0. \tag{3.5.45}$$

Obviously, if $\gamma = 0$, then (3.5.40) implies that $y = 0$ in $L^2(\Omega)$. This, together with (3.5.42) and (3.5.45), leads to $\eta = 0$ in L_f. Going back to (3.5.41) and using (3.5.43)–(3.5.44), we deduce that $u = 0$ in $H^1(\Omega)$. Thereby, (3.5.39) has only the trivial solution.

In the event that $\gamma \neq 0$, taking the inner product of (3.5.39) with Ψ and using (3.5.21), we obtain

$$\begin{aligned} 0 = \Re e \left(i\gamma \|\Psi\|^2_{\hat{\mathscr{H}}} \right) &= \Re e \left(\langle i\gamma\Psi, \Psi\rangle_{\hat{\mathscr{H}}} \right) = \Re e \left(\langle \hat{\mathscr{A}}\Psi, \Psi\rangle_{\hat{\mathscr{H}}} \right) \\ &\leq -d_0 \left(\|y\|^2 + \|\eta\|^2_{L_f} \right), \end{aligned} \tag{3.5.46}$$

where d_0 is a positive constant (see (3.5.22)). Whence $y = 0$ in $L^2(\Omega)$ and $\eta = 0$ in L_f. Finally, invoking (3.5.40)–(3.5.45), we conclude once again that the only solution of (3.5.39) is the trivial one.

Step 2: The objective is to show that the resolvent operator of $\hat{\mathscr{A}}$ satisfies (2.3.17). If this were not true, then, Banach-Steinhaus theorem (see Theorem 2.1.7) leads us to claim that there exist a sequence of real numbers

$$\gamma_n \to +\infty \tag{3.5.47}$$

and a sequence of vectors $\Psi_n = (u_n, y_n, \eta_n) \in D(\hat{\mathscr{A}})$ with

$$\|\Psi_n\|_{\hat{\mathscr{H}}} = 1, \quad \forall n \in \mathbb{N} \tag{3.5.48}$$

such that

$$\left\| (i\gamma_n I - \hat{\mathscr{A}})\Psi_n \right\|_{\hat{\mathscr{H}}} \to 0 \quad \text{as} \quad n \to +\infty, \tag{3.5.49}$$

that is,

$$i\gamma_n u_n - y_n \equiv P_n \to 0 \quad \text{in} \quad H^1(\Omega), \tag{3.5.50}$$

$$i\gamma_n y_n + Au_n + ay_n + \int_0^{+\infty} f(s)(\eta_n)_s(s)\, ds \equiv Q_n \to 0 \quad \text{in} \quad L^2(\Omega), \tag{3.5.51}$$

$$i\gamma_n \eta_n + (\eta_n)_s - y_n \equiv R_n \to 0 \quad \text{in} \quad L_f \tag{3.5.52}$$

and

$$\int_{\Omega}\left((f_0+a)u_n+y_n-\int_0^{+\infty} f(s)\eta_n(s)\,ds\right)dx=0, \tag{3.5.53}$$

$$\partial_A u_n = 0 \ \text{ on } \Gamma, \tag{3.5.54}$$

$$\eta_n(0)=0. \tag{3.5.55}$$

Exploring (3.5.48) and the fact that

$$\begin{aligned}\|\Psi_n\|_{\hat{\mathscr{H}}}\left\|(i\gamma_n I-\hat{\mathscr{A}})\Psi_n\right\|_{\hat{\mathscr{H}}} &\geq \Re e\left(\langle (i\gamma_n I-\hat{\mathscr{A}})\Psi_n,\Psi_n\rangle_{\hat{\mathscr{H}}}\right)\\ &=-\Re e\left(\langle \hat{\mathscr{A}}\Psi_n,\Psi_n\rangle_{\hat{\mathscr{H}}}\right),\end{aligned} \tag{3.5.56}$$

and recalling (3.5.46), as well as, (3.5.49), we reach

$$y_n \to 0 \text{ in } L^2(\Omega) \quad \text{and} \quad \eta_n \to 0 \text{ in } L_f. \tag{3.5.57}$$

Thenceforth,

$$\gamma_n u_n \to 0 \quad \text{and} \quad u_n \to 0 \text{ in } L^2(\Omega), \tag{3.5.58}$$

by means of (3.5.47), (3.5.50), and (3.5.57).

Next, taking the inner product of (3.5.51) with u_n in $L^2(\Omega)$ and applying Green's formula, it follows that

$$\begin{aligned}&\int_{\Omega}(i\gamma_n+a)y_n u_n\,dx+\int_{\Omega}\sum_{i,j=1}^{N} a_{ij}\partial_i u_n\partial_j u_n\,dx\\ &\quad+\int_{\Omega}u_n\int_0^{+\infty} f(s)(\eta_n)_s(s)\,ds\,dx\to 0.\end{aligned}$$

This, together with (3.5.57) and (3.5.58), implies that

$$\int_{\Omega}\sum_{i,j=1}^{N} a_{ij}\partial_i u_n\partial_j u_n\,dx+\int_{\Omega}u_n\int_0^{+\infty} f(s)(\eta_n)_s(s)\,ds\,dx\to 0. \tag{3.5.59}$$

Now, taking the inner product of (3.5.52) with u_n in L_f, we get

$$\begin{aligned}&i\gamma_n\int_{\Omega}u_n\int_0^{+\infty} f_n(s)\eta_n(s)\,ds\,dx+\int_{\Omega}u_n\int_0^{+\infty} f_n(s)(\eta_n)_s(s)\,ds\,dx\\ &\quad-f_0\int_{\Omega}u_n y_n\,dx\to 0.\end{aligned}$$

Thereby, from (3.5.57) and (3.5.58), we obtain

$$\int_{\Omega} u_n \int_0^{+\infty} f_n(s)(\eta_n)_s(s)\,ds\,dx \to 0. \tag{3.5.60}$$

The limits (3.5.59) and (3.5.60) lead to

$$\sum_{i,j=1}^{N} a_{ij}\partial_i u_n \partial_j u_n \to 0 \text{ in } L^2(\Omega).$$

Then (3.5.5) and (3.5.16) yield

$$\nabla u_n \to 0 \text{ in } L^2(\Omega). \tag{3.5.61}$$

Combining (3.5.57), (3.5.58), and (3.5.61), we finally find that $\|\Psi_n\|_{\hat{\mathscr{H}}}$ converges to zero in $\hat{\mathscr{H}}$, which contradicts (3.5.48).

Thereby, the two conditions of Theorem 2.3.8 are verified, and the proof of Theorem 3.5.3 is achieved. □

Remark 3.5.2

(i) Let $\Psi_0 = (u_0, u_1, \eta_0) \in \mathscr{H}$ and $\Psi = (u, u_t, \eta)$ be the solution of (3.5.11). We set

$$\hat{\Psi} = \Psi - (\mathscr{C}, 0, 0) = (u - \mathscr{C}, u_t, \eta) \quad \text{and} \quad \hat{\Psi}_0 = \Psi_0 - (\mathscr{C}, 0, 0)$$
$$= (u_0 - \mathscr{C}, u_1, \eta_0),$$

where $\mathscr{C}$ is defined in (3.5.31). Then $\hat{\Psi}$ is a solution of

$$\begin{cases} \hat{\Psi}_t(x,t) = \mathscr{A}\hat{\Psi}(x,t), & \forall x \in \Omega,\ \forall t > 0, \\ \hat{\Psi}(x,0) = \hat{\Psi}_0(x), & \forall x \in \Omega. \end{cases}$$

On the other hand, we note

$$\int_{\Omega}\left((f_0 + a)(u_0 - \mathscr{C}) + u_1 - \int_0^{+\infty} f(s)\eta_0(s)\,ds\right)dx$$

$$= \int_{\Omega}\left((f_0 + a)u_0 + u_1 - \int_0^{+\infty} f(s)\eta_0(s)\,ds\right)dx - \mathscr{C}\int_0^{+\infty}(f_0 + a)\,dx = 0,$$

and hence, $\hat{\Psi}_0 \in \hat{\mathscr{H}}$. Lastly, Theorem 3.5.3 implies that there exist two positive constants C and ω such that

$$\|\Psi - (\mathscr{C}, 0, 0)\|_{\hat{\mathscr{H}}} \le \kappa e^{-\omega t}, \quad \forall t > 0. \tag{3.5.62}$$

(ii) Note that the inequality condition (3.5.8) has a similar form for ordinary differential equations (ODEs) with distributed delay (see, for instance, [125, equation (1.13), p. 107] for scalar ODEs). Note also that the delayed terms in [125] were treated as perturbations. Moreover, the nondelayed system is assumed asymptotically stable. This result has been extended in [176] to the case of linear systems of ODEs by means of linear matrix inequalities [176]. In turn, unlike this study, the authors in [125] and [176] derived the stability results by using augmented Lyapunov functionals.

3.5.5 *Case of Discrete Time Delay*

In this subsection, we consider the case of discrete time delay

$$u_{tt}(x,t) + Au(x,t) + a(x)u_t(x,t) + b(x)u_t(x,t-\tau) = 0, \quad \forall x \in \Omega,\ \forall t > 0, \tag{3.5.63}$$

with, as in the previous subsections, the homogeneous Neumann boundary condition on the whole boundary Γ

$$\partial_A u(x,t) := \sum_{i,j=1}^{N} a_{ij}(x)\nu_j \partial_i u(x,t) = 0, \quad \forall x \in \Gamma,\ \forall t > 0 \tag{3.5.64}$$

and initial conditions

$$\begin{cases} u(x,0) = u_0(x), & \forall x \in \Omega, \\ u_t(x,0) = u_1(x), & \forall x \in \Omega, \\ u_t(x,t) = u_2(x,t), & \forall x \in \Omega,\ \forall t \in (-\tau, 0), \end{cases} \tag{3.5.65}$$

where A and a are defined in (3.5.4), (3.5.5), and (3.5.6). Moreover, $\tau \in (0, +\infty)$, and (u_0, u_1, u_2) are the given initial data. Finally, b is a real space variable function of $C(\bar{\Omega})$ so that

$$\|b\|_\infty < a_0. \tag{3.5.66}$$

3.5.5.1 Well-Posedness of (3.5.63)–(3.5.65)

Arguing as usual (see [149, 184]), we consider the variable z and its initial data z_0 given by

$$\begin{cases} z(x,t,p) = u_t(x, t-\tau p), & \forall x \in \Omega,\ \forall t \in \mathbb{R}_+,\ \forall p \in (0,1), \\ z_0(x,p) = z(x,0,p) = u_2(x, -\tau p), & \forall x \in \Omega,\ \forall p \in (0,1). \end{cases} \tag{3.5.67}$$

Then z satisfies

$$\begin{cases} \tau z_t(x,t,p) + z_p(x,t,p) = 0, & \forall x \in \Omega,\ \forall t \in \mathbb{R}_+,\ \forall p \in (0,1), \\ z(x,t,0) = u_t(x,t), & \forall x \in \Omega,\ \forall t \in \mathbb{R}_+. \end{cases}$$

Next, let $y = u_t$, $\Phi = (u, y, z)$, and $\Phi_0 = (u_0, u_1, z_0) \in \mathscr{X}$, where

$$\mathscr{X} = V \times H \times L^2((0,1), H)$$

and

$$L^2((0,1), H) = \left\{ w : (0,1) \to H; \quad \int_0^1 \|w(p)\|^2 \, dp < +\infty \right\}$$

endowed with the inner product

$$\langle w_1, w_2 \rangle_{L^2((0,1),H)} = \int_0^1 \langle w_1(p), w_2(p) \rangle \, dp.$$

The system (3.5.63)–(3.5.65) can be written in the abstract form

$$\begin{cases} \Phi_t(x,t) = \mathscr{B}\Phi(x,t), & \forall x \in \Omega,\ \forall t > 0, \\ \Phi(x,0) = \Phi_0(x), & \forall x \in \Omega, \end{cases} \tag{3.5.68}$$

in which the operator $\mathscr{B}$ is given by

$$\mathscr{B}(u, y, z) = \left(y, -Au - ay - bz(1), -\frac{z_p}{\tau} \right), \tag{3.5.69}$$

and the domain

$$D(\mathscr{B}) = \Big\{ (u, y, z) \in \mathscr{X};\ Au \in H,\ y \in V,\ z_p \in L^2((0,1), H), \\ z(0) = y,\ \partial_A u = 0 \text{ on } \Gamma \Big\}.$$

Whereupon, the systems (3.5.63)–(3.5.65) and (3.5.68) are equivalent. As in the case of distributed delay (see, for instance, the previous subsection), one should introduce an inner product on $\mathscr{X}$, which makes $\mathscr{X}$ a Hilbert space. As Neumann boundary condition is considered on the whole boundary Γ and no displacement term is considered neither on Ω nor on Γ, the classical energy norm $\|\cdot\|_E$ given by

$$\|(u, y, z)\|_E^2 = \int_\Omega \sum_{i,j=1}^N a_{ij} \partial_i u \partial_j u \, dx + \|y\|^2 + \|z\|^2_{L^2((0,1),H)}$$

is only a seminorm in our case, since $\|(u, y, z)\|_E^2 = 0$ does not imply that $u = 0$. By integrating (3.5.63) on Ω, we get

$$\partial_t \left(\int_\Omega (u_t + au + bu(x, t - \tau))\, dx \right) = 0.$$

Using the definition of y and z, we get

$$\partial_t \left(\int_\Omega \left((a + b)u + y - \tau b \int_0^1 z(x, p)\, dp \right) dx \right) = 0. \tag{3.5.70}$$

Now, we fix a positive constant ε_2 and consider on $\mathscr{X}$ the inner product

$$\begin{aligned}
\langle (u, y, z), (\tilde{u}, \tilde{y}, \tilde{z}) \rangle_{\mathscr{X}} &= \int_\Omega \left(\sum_{i,j=1}^n a_{ij} \partial_i u \partial_j \tilde{u} \right) dx + \langle y, \tilde{y} \rangle + \xi \, \langle z, \tilde{z} \rangle_{L^2((0,1),H)} \\
&+ \varepsilon_2 \left(\int_\Omega \left((a + b)u + y - \tau b \int_0^1 z(x, p)\, dp \right) dx \right) \\
&\times \int_\Omega \left((a + b)\tilde{u} + \tilde{y} - \tau b \int_0^1 \tilde{z}(x, p)\, dp \right) dx,
\end{aligned}$$

where ξ is a positive constant satisfying

$$\tau \|b\|_\infty < \xi < \tau\, (2a_0 - \|b\|_\infty) \tag{3.5.71}$$

and a_0 is given in (3.5.6). Note that ξ exists in light of (3.5.66). The space $(\mathscr{X}, \langle \cdot, \cdot \rangle_{\mathscr{X}})$ is a Hilbert space if $\| \cdot \|_{V \times H \times L^2((0,1),H)}$ and $\| \cdot \|_{\mathscr{X}}$ are equivalent. Hence, it is sufficient to prove the next lemma.

Lemma 3.5.2 *There exists $\varepsilon_3 > 0$ such that for any $\varepsilon_2 \in (0, \varepsilon_3)$ and for each $(u, y, z) \in \mathscr{X}$, we have*

$$\zeta_1 \|(u, y, z)\|^2_{V \times H \times L^2((0,1),H)} \le \|(u, y, z)\|^2_{\mathscr{X}} \le \zeta_2 \|(u, y, z)\|^2_{V \times H \times L^2((0,1),H)}, \tag{3.5.72}$$

in which ζ_1 and ζ_2 are positive constants.

Proof Invoking Hölder's and Young's inequalities, and using (3.5.16), one can show that the right inequality in (3.5.72) holds for any $\varepsilon_2 > 0$, and for some ζ_2 depending on ε_2, k_1, ξ, Ω, a, and b. Concerning the proof of the left inequality in (3.5.72), one can proceed in a similar way as for (3.5.15). Indeed, we see that

$$\left(\int_\Omega \left((a + b)u + y - \tau b \int_0^1 z(p)\, dp \right) dx \right)^2 = \left(\int_\Omega (a + b)u\, dx \right)^2 \tag{3.5.73}$$

$$+\left(\int_{\Omega}\left(y-\tau b\int_0^1 z(p)\,dp\right)dx\right)^2+2\left(\int_{\Omega}(a+b)u\,dx\right)$$
$$\times\int_{\Omega}\left(y-\tau b\int_0^1 z(p)\,dp\right)dx.$$

Using Young's and Hölder's inequalities, we have

$$2\left(\int_{\Omega}(a+b)u\,dx\right)\left(\int_{\Omega}\left(y-\tau b\int_0^1 z(p)\,dp\right)dx\right)$$
$$\geq -\frac{1}{2}\left(\int_{\Omega}(a+b)u\,dx\right)^2-2\left(\int_{\Omega}\left(y-\tau b\int_0^1 z(p)\,dp\right)dx\right)^2,$$

and there exists a positive constant c such that

$$\left(\int_{\Omega}\left(y-\tau b\int_0^1 z(p)\,dp\right)dx\right)^2$$
$$\leq c\left(\|y\|^2+\|z\|^2_{L^2((0,1),H)}\right),\quad \forall (y,z)\in H\times L^2((0,1),H).$$

Combining the above two inequalities, (3.5.73), together with the generalized Poincaré's inequality (thanks to (3.5.66) and similar to (3.5.18))

$$\|u\|^2\leq \varpi_1\left(\|\nabla u\|^2+\left(\int_{\Omega}(a+b)u\,dx\right)^2\right),\quad \forall u\in V,$$

we reach

$$\varepsilon_2\left(\int_{\Omega}\left((a+b)u+y-\tau b\int_0^1 z(p)\,dp\right)dx\right)^2$$
$$\geq \frac{\varepsilon_2}{2\varpi_1}\|u\|^2-\frac{\varepsilon_2}{2}\|\nabla u\|^2-\varepsilon_2 c\|y\|^2-\varepsilon_2 c\|z\|^2_{L^2_{((0,1),H)}}.$$

Here, ϖ_1 depends on Ω, a and b. Therefore

$$\|(u,y,z)\|^2_{\mathscr{X}}\geq\left(\alpha_0-\frac{\varepsilon_2}{2}\right)\|\nabla u\|^2+\frac{\varepsilon_2}{2\varpi_1}\|u\|^2+(1-\varepsilon_2 c)\|y\|^2$$
$$+(\xi-\varepsilon_2 c)\|z\|^2_{L^2((0,1),H)}.$$

Whereupon, for ε_2, satisfying

$$0 < \varepsilon_2 < \varepsilon_3 := \min\left\{2\alpha_0, \frac{1}{c}, \frac{\xi}{c}\right\},$$

we obtain the reverse inequality in (3.5.72) with

$$\zeta_1 = \min\left\{\alpha_0 - \frac{\varepsilon_2}{2}, \frac{\varepsilon_2}{2\varpi_1}, 1 - \varepsilon_2 c, \xi - \varepsilon_2 c\right\} > 0.$$

□

Our well-posedness result is announced in the following theorem:

Theorem 3.5.4 *Assume that (3.5.66) holds. Then, for any $\Phi_0 \in \mathscr{X}$, the system (3.5.68) has a unique weak solution*

$$\Phi \in C(\mathbb{R}_+, \mathscr{X}). \tag{3.5.74}$$

Moreover, if $\Phi_0 \in D(\mathscr{B})$, then (3.5.68) has a classical solution and satisfies

$$\Phi \in C^1(\mathbb{R}_+, \mathscr{X}) \cap C(\mathbb{R}_+, D(\mathscr{B})). \tag{3.5.75}$$

Proof As in Sect. 3.5.2, we will prove that $\mathscr{B}$ is m-dissipative. Direct computations yield

$$\langle \mathscr{B}\Phi, \Phi\rangle_{\mathscr{X}} = -\|\sqrt{a}y\|^2 - \int_\Omega byz(1)\,dx - \frac{\xi}{\tau}\int_\Omega\int_0^1 zz_p\,dp\,dx. \tag{3.5.76}$$

Integrating with respect to p the last integral of (3.5.76), recalling that $z(0) = y$, and using Young's inequality for the term $-byz(1)$, we obtain

$$\langle \mathscr{B}\Phi, \Phi\rangle_{\mathscr{X}} \leq -\frac{1}{2}\left(2a_0 - \|b\|_\infty - \frac{\xi}{\tau}\right)\|y\|^2 - \frac{1}{2}\left(\frac{\xi}{\tau} - \|b\|_\infty\right)\|z(1)\|^2,$$

and then

$$\langle \mathscr{B}\Phi, \Phi\rangle_{\mathscr{X}} \leq -r_0\left(\|y\|^2 + \|z(1)\|^2\right), \tag{3.5.77}$$

where

$$r_0 = \frac{1}{2}\min\left\{2a_0 - \|b\|_\infty - \frac{\xi}{\tau}, \frac{\xi}{\tau} - \|b\|_\infty\right\} > 0. \tag{3.5.78}$$

Note that $r_0 > 0$ in view of (3.5.71). This proves that $\mathscr{B}$ is dissipative.

In order to prove that $I - \mathscr{B}$ is surjective, let $F = (f_1, f_2, f_3) \in \mathscr{X}$ and we must find $\Phi = (u, y, z) \in D(\mathscr{B})$ satisfying

$$\Phi - \mathscr{B}\Phi = F. \tag{3.5.79}$$

The first equation in (3.5.79) is equivalent to

$$y = u - f_1. \tag{3.5.80}$$

The last equation in (3.5.79) is reduced to

$$z_p + \tau z = \tau f_3. \tag{3.5.81}$$

Integrating with respect to p and noting that z should satisfy $z(0) = y = u - f_1$, we get

$$z(p) = e^{-\tau p}(u - f_1) + \tau \int_0^p e^{\tau(\theta - p)} f_3(\theta)\, d\theta. \tag{3.5.82}$$

Using (3.5.80) and (3.5.82), we find that the second equation in (3.5.79) becomes

$$Au + (a + 1 + be^{-\tau})u = \tilde{f}, \tag{3.5.83}$$

where

$$\tilde{f} = f_2 + (a + 1 + be^{-\tau}) f_1 - \tau e^{-\tau} \int_0^1 e^{\tau\theta} f_3(\theta)\, d\theta \in H.$$

Note that $\tilde{f} \in H$ since $f_1 \in V$, $f_2 \in H$, $f_3 \in L^2((0,1), H)$ and $e^{\tau\theta} \leq e^{\tau}$, for $\theta \in (0,1)$. It is also worth mentioning that, according to (3.5.6) and (3.5.66), we have

$$a + 1 + be^{-\tau} \geq a_0 + 1 - \|b\|_\infty > 1.$$

Then, considering the variational formulation of (3.5.83) and using Lax-Milgram theorem along with classical elliptic regularity arguments, we deduce that (3.5.83) has a unique solution $u \in V$ satisfying

$$Au \in H \quad \text{and} \quad \partial_A u = 0 \text{ on } \Gamma.$$

Therefore, (3.5.80) and (3.5.82) imply that

$$y \in V, \quad z \in L^2((0,1), H) \quad \text{and} \quad z(0) = y.$$

From (3.5.81), we have $z_p \in L^2((0,1), H)$. This proves that (3.5.79) has a unique solution $\Phi \in D(\mathscr{B})$; that is, $I - \mathscr{B}$ is onto $\mathscr{X}$. Consequently, one can conclude that $\mathscr{B}$ is the infinitesimal generator of a linear C_0-semigroup $S(t)$ of contractions on $\mathscr{X}$ and its domain $D(\mathscr{B})$ is dense in $\mathscr{X}$. Hence, Theorem 3.5.4 is shown. □

3.5.5.2 Asymptotic Convergence of Solutions to (3.5.68)

As in Sect. 3.5.3, we can prove the following convergence result:

Theorem 3.5.5 *Suppose that (3.5.66) is satisfied. Then for any initial data* $\Phi_0 = (u_0, u_1, z_0) \in \mathscr{X}$, *the solution* $\Phi = (u, y, z)$ *of (3.5.68) tends in* $\mathscr{X}$ *to* $(\mathscr{K}, 0, 0)$, *as* $t \longrightarrow +\infty$, *where*

$$\mathscr{K} = \left(\int_\Omega (a+b)\,dx\right)^{-1} \int_\Omega \left[(a+b)u_0 + u_1 - \tau b \int_0^1 z_0(p)\,dp\right] dx. \tag{3.5.84}$$

Proof The proof of (3.5.84) is identical to the one of (3.5.31) given in Sect. 3.5.3 (using (3.5.70) and (3.5.77)); we omit the details here. □

3.5.5.3 Exponential Convergence of Solutions to (3.5.68)

Once again, we will use Theorem 2.3.8 to prove that the solutions of (3.5.68) converge to $(\mathscr{K}, 0, 0)$ (see Theorem 3.5.5) with an exponential rate. Therefore, it suffices to proceed as in Sect. 3.5.4. Thus, consider the subspace $\hat{\mathscr{X}}$ of $\mathscr{X}$ defined by

$$\hat{\mathscr{X}} = \left\{(u, y, z) \in \mathscr{X};\ \int_\Omega \left((a+b)u + y - \tau b \int_0^1 z(x,p)\,dp\right) dx = 0\right\} \tag{3.5.85}$$

with inner product

$$\langle (u,y,z), (\tilde{u}, \tilde{y}, \tilde{z})\rangle_{\hat{\mathscr{X}}} = \int_\Omega \left(\sum_{i,j=1}^n a_{ij}\partial_i u \partial_j \tilde{u}\right) dx + \langle y, \tilde{y}\rangle + \xi \langle z, \tilde{z}\rangle_{L^2((0,1),H)}, \tag{3.5.86}$$

where ξ is a positive constant that satisfies (3.5.71). Then, define the operator $\hat{\mathscr{B}}$ by

$$\begin{aligned} &\hat{\mathscr{B}} : D(\hat{\mathscr{B}}) := D(\mathscr{B}) \cap \hat{\mathscr{X}} \subset \hat{\mathscr{X}} \to \hat{\mathscr{X}}, \\ &\hat{\mathscr{B}}(u,y,z) = \mathscr{B}(u,y,z),\ \forall\, (u,y,z) \in D(\hat{\mathscr{B}}), \end{aligned} \tag{3.5.87}$$

which generates a C_0-semigroup of contractions $e^{t\hat{\mathscr{B}}}$ on $\hat{\mathscr{X}}$.

Theorem 3.5.6 *Suppose that the condition (3.5.66) holds. Then, the solutions of (3.5.68) exponentially converge to their equilibrium state $(\mathscr{K}, 0, 0)$ in $\mathscr{X}$.*

Proof Using Remark 3.5.2 with $\mathscr{K}$ instead of $\mathscr{C}$, it is clear that, in order to prove the desired result, it amounts to showing the exponential stability of the semigroup operator $e^{t\hat{\mathscr{B}}}$ on $\hat{\mathscr{X}}$. To do so, let us first check that $\rho(\hat{\mathscr{B}}) \supset i\mathbb{R}$, which means that $\mathscr{B}$ satisfies (2.3.16). In fact, given $\gamma \in \mathbb{R}$, let us seek $\Phi = (u, y, z) \in D(\mathscr{B})$ such that $\mathscr{B}\Phi = i\gamma\Phi$, that is,

$$y = i\gamma u, \tag{3.5.88}$$

$$-Au - ay - bz(1) = i\gamma y, \tag{3.5.89}$$

$$z_p = -i\tau\gamma z, \tag{3.5.90}$$

$$\int_\Omega \left((a+b)u + y - \tau b \int_0^1 z(x,p)\,dp\right) dx = 0, \tag{3.5.91}$$

$$\partial_A u = 0 \ \text{ on } \Gamma, \tag{3.5.92}$$

$$z(0) = y. \tag{3.5.93}$$

If $\gamma = 0$, then $y = 0 = z(0)$ in $L^2(\Omega)$ by (3.5.88) and (3.5.93), which, together with (3.5.90), implies that $z = 0$ in $L^2((0,1), H)$. Using (3.5.89) and (3.5.91)–(3.5.92), we can conclude that $u = 0$ in $H^1(\Omega)$. Hence, (2.3.16) is verified when $\gamma = 0$.

If $\gamma \neq 0$, then taking the inner product of $\mathscr{B}\Phi = i\gamma\Phi$ with Φ and using (3.5.77), we obtain

$$0 = \Re e\left(\langle \hat{\mathscr{B}}\Phi, \Phi\rangle_{\hat{\mathscr{X}}}\right) \leq -r_0\left(\|y\|^2 + \|z(1)\|^2\right), \tag{3.5.94}$$

where r_0 is a positive constant (see (3.5.78)). Thus

$$y = z(1) = 0 \quad \text{in } L^2(\Omega),$$

and therefrom $u = 0$ in $L^2(\Omega)$ thanks to (3.5.88). Next, solving (3.5.90), we get

$$z_n(p) = e^{-i\tau\gamma_n p}\, z_n(0).$$

Because $z(1) = 0$, then $z_n(0) = 0$, and hence, $z_n = 0$ in $L^2((0,1), H)$. Lastly, amalgamating these findings with (3.5.89) and (3.5.91)–(3.5.92), we obtain $u = 0$ in $H^1(\Omega)$. Accordingly, (2.3.16) is also satisfied if $\gamma \neq 0$.

We turn now to the proof of (2.3.17). Suppose that this property is not satisfied by the operator $\hat{\mathscr{B}}$. Thereby, there exist a sequence of real numbers

$$\gamma_n \to +\infty \tag{3.5.95}$$

and a sequence of vectors $\Phi_n = (u_n, y_n, z_n,) \in D(\hat{\mathscr{B}})$ with

$$\|\Phi_n\|_{\hat{\mathscr{X}}} = 1, \quad \forall n \in \mathbb{N}, \tag{3.5.96}$$

such that

$$\left\|(i\gamma_n I - \hat{\mathscr{B}})\Phi_n\right\|_{\hat{\mathscr{X}}} \to 0 \quad \text{as} \quad n \to +\infty; \tag{3.5.97}$$

that is,

$$i\gamma_n u_n - y_n \equiv V_n \to 0 \quad \text{in} \quad H^1(\Omega), \tag{3.5.98}$$

$$i\gamma_n y_n + Au_n + ay_n + bz_n(1) \equiv W_n \to 0 \quad \text{in} \quad L^2(\Omega), \tag{3.5.99}$$

$$i\gamma_n z_n + \frac{1}{\tau}(z_n)_p \equiv X_n \to 0 \quad \text{in} \quad L^2((0,1), H), \tag{3.5.100}$$

together with

$$\int_\Omega \left((a+b)u_n + y - \tau b \int_0^1 z_n(x,p)\,dp\right) dx = 0, \tag{3.5.101}$$

$$\partial_A u_n = 0 \ \text{on}\, \Gamma, \tag{3.5.102}$$

$$z_n(0) = y_n. \tag{3.5.103}$$

Combining

$$\begin{aligned}\|\Phi_n\|_{\hat{\mathscr{X}}} \left\|(i\gamma_n I - \hat{\mathscr{B}})\Phi_n\right\|_{\hat{\mathscr{X}}} &\geq \Re e\left(\langle(i\gamma_n I - \hat{\mathscr{B}})\Phi_n, \Phi_n\rangle_{\hat{\mathscr{X}}}\right) \\ &= -\Re e\left(\langle\hat{\mathscr{B}}\Phi_n, \Phi_n\rangle_{\hat{\mathscr{X}}}\right),\end{aligned} \tag{3.5.104}$$

with (3.5.94), (3.5.96), (3.5.97), and (3.5.103), we obtain

$$y_n = z_n(0) \to 0 \quad \text{and} \quad z_n(1) \to 0 \quad \text{in}\ L^2(\Omega). \tag{3.5.105}$$

This immediately yields after using (3.5.95) and (3.5.98)

$$u_n \to 0 \quad \text{and} \quad \gamma_n u_n \to 0 \quad \text{in}\ L^2(\Omega). \tag{3.5.106}$$

In light of (3.5.100), we have

$$z_n(p) = z_n(0)\, e^{-i\tau\gamma_n p} + \tau \int_0^p e^{-i\tau\gamma_n(p-\theta)}\, X_n(\theta)\, d\theta,$$

and thereby $z_n \to 0$ in $L^2((0,1), H)$. In turn, arguing as before, that is, taking the inner product of (3.5.99) with u_n in $L^2(\Omega)$, we get (3.5.61), and then we conclude that $u_n \to 0$ in $H^1(\Omega)$. Thus $\|\Phi_n\|_{\hat{\mathscr{X}}} \to 0$ in $\hat{\mathscr{X}}$, which contradicts (3.5.96). This completes the proof of Theorem 3.5.6. □

Remark 3.5.3 Theorems 3.5.3 and 3.5.6 ensure the exponential convergence of the solutions to their steady state, but no estimate of the exponential rate ω is provided. In order to obtain such an estimate, one should adopt another method (instead of the resolvent method) such as energy method (or Lyapunov method). Notwithstanding, this method requires an appropriate choice of an energy norm. Let us consider the system (3.5.63)–(3.5.65). The case of system (3.5.1)–(3.5.3) can be treated in a similar way.

For simplicity and without loss of generality, assume that $A = -\Delta$ (that is, $a_{ij} = \delta_{ij}$) and a and b are positive constants satisfying the condition (3.5.66), that is, $b < a$. In this case, (3.5.71) becomes

$$\tau b < \xi < \tau (2a - b), \tag{3.5.107}$$

and the norm induced by the inner product (3.5.86) in $\hat{\mathscr{X}}$ is

$$\|(u, y, z)\|^2_{\hat{\mathscr{X}}} = \|\nabla u\|^2 + \|y\|^2 + \xi \|z\|^2_{L^2((0,1),H)}. \tag{3.5.108}$$

We can also suppose that

$$\int_\Omega u(x,t)\, dx = 0, \tag{3.5.109}$$

which implies that there exists a positive constant χ such that

$$\int_\Omega u^2(x,t)\, dx \le \chi \int_\Omega |\nabla u(x,t)|^2 dx. \tag{3.5.110}$$

Obviously, if (3.5.109) is not satisfied, we can set, as in [[107], Remark 2.1-3]

$$\hat{u}(x,t) = u(x,t) - \frac{1}{|\Omega|} g(t)$$

and

$$(\hat{u}_0(x), \hat{u}_1(x), \hat{u}_2(x,t)) = (u_0(x), u_1(x), u_2(x,t)) - \frac{1}{|\Omega|}(g(0), g'(0), g'(t)),$$

where $g : \mathbb{R}_+ \to \mathbb{R}$ is defined as in [107] (see Remark 2.1-3 with $\psi = u$, $\rho_2 = 1$, $\lambda_2 = a$, $k_1 = 0$, $\mu_2 = b$, $\tau_2 = \tau$, $f_2 = u_2$, $\psi_0 = u_0$, $\psi_1 = u_1$, and Ω instead of $(0, L)$), and we see that $\hat{u}$ is also a solution of (3.5.63)–(3.5.65) (instead of u) for the initial data $(\hat{u}_0, \hat{u}_1, \hat{u}_2)$ (instead of (u_0, u_1, u_2)). Moreover, (3.5.109) is satisfied for $\hat{u}$ (instead of u).

Thereafter, consider the following energy functional:

$$\begin{aligned}\mathscr{E}(t) &= \frac{1}{2}\int_\Omega \left(u_t^2(x,t) + |\nabla u(x,t)|^2\right) dx + \frac{\xi}{2}\int_\Omega\int_0^1 u_t^2(x, t-\tau p)\, dp\, dx \\ &+ K\int_\Omega u(x,t)u_t(x,t)dx, \qquad (3.5.111)\end{aligned}$$

where ξ satisfies (3.5.107), while K is a positive constant to be determined. Then consider the solutions in $\hat{\mathscr{X}}$ (see (3.5.85)) of (3.5.63)–(3.5.65). One can readily check that $\mathscr{E}(t)$ defines a new norm of (u, y, z) as long as K is sufficiently small, where $y(x,t) = u_t(x,t)$ and $z(x,t,p) = u_t(x, t-\tau p)$. Indeed, using Young's inequality and using (3.5.110), we note that there exists a positive constant c_* such that

$$\left|\mathscr{E}(t) - \frac{1}{2}\|(u,y,z)\|^2_{\hat{\mathscr{X}}}\right| \le K\int_\Omega |u||y|dx \le Kc_*\|(u,y,z)\|^2_{\hat{\mathscr{X}}},$$

and hence, for $0 < K < \frac{1}{2c_*}$, we get, for $e_1 = \frac{1}{2} - Kc_*$ and $e_2 = \frac{1}{2} + Kc_*$,

$$e_1\|(u,y,z)\|^2_{\hat{\mathscr{X}}} \le \mathscr{E}(t) \le e_2\|(u,y,z)\|^2_{\hat{\mathscr{X}}}. \qquad (3.5.112)$$

Next, differentiating $\mathscr{E}(t)$ and integrating by parts, we obtain, analogously to (3.5.76), the following result:

$$\begin{aligned}\mathscr{E}_t(t) &= -a\int_\Omega u_t^2 dx - b\int_\Omega u_t u_t(x, t-\tau)\, dx - \frac{\xi}{2\tau}\int_\Omega u_t^2(x, t-\tau)\, dx \\ &+ \frac{\xi}{2\tau}\int_\Omega u_t^2\, dx + K\int_\Omega u_t^2\, dx - K\int_\Omega |\nabla u|^2\, dx \\ &- Ka\int_\Omega u u_t\, dx - Kb\int_\Omega u u_t(x, t-\tau)dx. \qquad (3.5.113)\end{aligned}$$

Applying once again Young's inequality and using (3.5.110), we deduce from (3.5.113) that

$$\begin{aligned}\mathscr{E}_t(t) &\le -\left(a - \frac{b}{2} - \frac{\xi}{2\tau}\right)\int_\Omega u_t^2\, dx + K\left(a\frac{\delta}{2} + 1\right)\int_\Omega u_t^2\, dx \\ &- \frac{1}{2}\left(\frac{\xi}{\tau} - b\right)\int_\Omega u_t^2(x, t-\tau)\, dx + Kb\frac{\delta}{2}\int_\Omega\int_0^1 u_t^2(x, t-\tau)\, dx \\ &- \frac{K}{2}\int_\Omega |\nabla u|^2\, dx - \frac{K}{2}\int_\Omega |\nabla u|^2\, dx + \frac{K(a+b)}{2\delta}\int_\Omega u^2\, dx\end{aligned}$$

$$\le -\left(a - \frac{b}{2} - \frac{\xi}{2\tau}\right)\int_\Omega u_t^2\,dx + K\left(a\frac{\delta}{2}+1\right)\int_\Omega u_t^2\,dx$$
$$-\frac{1}{2}\left(\frac{\xi}{\tau} - b\right)\int_\Omega u_t^2(x,t-\tau)\,dx + Kb\frac{\delta}{2}\int_\Omega\int_0^1 u_t^2(x,t-\tau)\,dx$$
$$-\frac{K}{2}\int_\Omega |\nabla u|^2\,dx + \frac{K}{2}\left(\frac{\chi(a+b)}{\delta} - 1\right)\int_\Omega |\nabla u|^2\,dx, \tag{3.5.114}$$

where δ is a positive constant to be fixed.

Now, choose δ large enough so that

$$\frac{\chi(a+b)}{\delta} - 1 \le 0.$$

Next, bearing in mind (3.5.107), we pick up K small so that

$$0 < K < \frac{1}{2c_*},\quad K\left(a\frac{\delta}{2}+1\right) - \left(a - \frac{b}{2} - \frac{\xi}{2\tau}\right) < 0 \quad \text{and}$$
$$Kb\frac{\delta}{2} - \frac{1}{2}\left(\frac{\xi}{\tau} - b\right) < 0.$$

Amalgamating these choices with (3.5.112), it follows the existence of a positive constant ω such that

$$\mathscr{E}_t(t) \le -2\omega\mathscr{E}(t).$$

By integrating with respect to t, we get $\mathscr{E}(t) \le \mathscr{E}(0)e^{-2\omega t}$, and hence, using again (3.5.112),

$$\|(u, y, z)\|_{\hat{\mathscr{X}}} \le \sqrt{\frac{e_2}{e_1}}\,\|(u_0, u_1, z_0)\|_{\hat{\mathscr{X}}}\, e^{-\omega t}.$$

This gives the desired result.

3.5.6 Extensions

The well-posedness and stability results can be extended to an abstract form. To do so, consider in (3.5.1) and (3.5.63) a self-adjoint linear positive definite operator $A : D(A) \to H$ and a Hilbert space H with dense and compact embedding $D(A) \subset H$. Then, with this abstract form, numerous hyperbolic systems are included. We shall present some of them.

3.5.6.1 Petrovsky Equation

It is possible to consider the case

$$A = \beta \Delta^2, \tag{3.5.115}$$

where $\beta > 0$ such that (3.5.2) is replaced by

$$u_\nu(x,t) = (\Delta u)_\nu(x,t) = 0, \quad \forall x \in \Gamma, \ \forall t > 0, \tag{3.5.116}$$

and Γ is assumed to be of class C^4. Here, $V = H^2(\Omega)$ and the equilibrium points $(\mathscr{C}, 0, 0)$ and $(\mathscr{K}, 0, 0)$, as well as the exponential decay, are the same as in Sects. 3.5.3 and 3.5.5, respectively.

3.5.6.2 Coupled Equations

The following two coupled wave and/or Petrovsky equations can also be considered

$$\begin{cases} u_{tt}(x,t) + Au(x,t) + au_t(x,t) + d\tilde{u}_{tt}(x,t) & \\ +G(x,t) = 0, & \forall\, x \in \Omega, \ \forall\, t > 0, \\ \tilde{u}_{tt}(x,t) + \tilde{A}\tilde{u}(x,t) + \tilde{a}\tilde{u}_t(x,t) & \\ +du_{tt}(x,t) + \tilde{G}(x,t) = 0, & \forall\, x \in \Omega, \ \forall\, t > 0, \end{cases} \tag{3.5.117}$$

and

$$\begin{cases} u_{tt}(x,t) + Au(x,t) + au_t(x,t) + d\tilde{u}_t(x,t) & \\ +G(x,t) = 0, & \forall\, x \in \Omega, \ \forall\, t > 0, \\ \tilde{u}_{tt}(x,t) + \tilde{A}\tilde{u}(x,t) + \tilde{a}\tilde{u}_t(x,t) + du_t(x,t) & \\ +\tilde{G}(x,t) = 0, & \forall\, x \in \Omega, \ \forall\, t > 0, \end{cases} \tag{3.5.118}$$

where A and $\tilde{A}$ are of the form (3.5.4) and/or (3.5.115), $a, \ \tilde{a} \in C(\bar{\Omega})$ such that

$$\inf_\Omega a > 0 \quad \text{and} \quad \inf_\Omega \tilde{a} > 0,$$

$d \in C(\bar{\Omega})$ satisfying $\|d\|_\infty < 1$ in case (3.5.117), and

$$\inf_\Omega \{a - |d|\} > 0 \quad \text{and} \quad \inf_\Omega \{\tilde{a} - |d|\} > 0$$

in case (3.5.118), the boundary condition for u is of the form (3.5.2) if A is of the form (3.5.4), of the form (3.5.116) if A is of the form (3.5.115), and similarly for $\tilde{u}$ depending on the form of $\tilde{A}$, and G and $\tilde{G}$ are distributed and/or discrete time delays; that is,

$$G(x,t) = \int_0^{+\infty} f(s)u_t(x,t-s)ds \quad \text{and} \quad \tilde{G}(x,t) = \int_0^{+\infty} \tilde{f}(s)\tilde{u}_t(x,t-s)ds \tag{3.5.119}$$

or

$$G(x,t) = bu_t(x,t-\tau) \quad \text{and} \quad \tilde{G}(x,t) = \tilde{b}\tilde{u}_t(x,t-\tilde{\tau}) \tag{3.5.120}$$

or

$$G(x,t) = \int_0^{+\infty} f(s)u_t(x,t-s)ds \quad \text{and} \quad \tilde{G}(x,t) = \tilde{b}\tilde{u}_t(x,t-\tilde{\tau}), \tag{3.5.121}$$

where $\tau,\ \tilde{\tau} > 0$ and $b,\ \tilde{b} \in C(\bar{\Omega})$ such that

$$\|b\|_\infty < \inf_\Omega a \quad \text{and} \quad \|\tilde{b}\|_\infty < \inf_\Omega \tilde{a},$$

and $\tilde{f}$ is as f in Sect. 3.5.2. For the convergence, we have $(y,\ \tilde{y},\ \eta, \tilde{\eta})$, $(y,\ \tilde{y},\ z, \tilde{z})$ and $(y,\ \tilde{y},\ \eta, \tilde{z})$ converge exponentially to zero, respectively, in

$$H \times H \times L_f \times L_{\tilde{f}}, \quad H \times H \times L^2((0,1),H) \times L^2((0,1),H) \quad \text{and}$$
$$H \times H \times L_f \times L^2((0,1),H)$$

(corresponding to (3.5.119), (3.5.120), and (3.5.121), respectively), where $y = u_t$, $\tilde{y} = \tilde{u}_t$, $\tilde{\eta}$, and $\tilde{z}$ are defined in terms of $\tilde{u}$ as η and z in Sects. 3.5.2 and 3.5.5, respectively. On the other hand, $(u, \tilde{u})$ converges exponentially to $(\mathscr{C}, \tilde{\mathscr{C}})$ in $H^i(\Omega) \times H^j(\Omega)$, $i = 1$ if A is of the form (3.5.4), $i = 2$ if A is of the form (3.5.115), $j = 1$ if $\tilde{A}$ is of the form (3.5.4), and $j = 2$ if $\tilde{A}$ is of the form (3.5.115), where $\mathscr{C}$ and $\tilde{\mathscr{C}}$ are constants satisfying

$$\begin{aligned}&\mathscr{C} \int_\Omega (a+\lambda)\,dx + \tilde{\mathscr{C}} \int_\Omega (\tilde{a}+\tilde{\lambda})\,dx \\ &\quad = \int_\Omega \Big[(a+\lambda)u_0 + (\tilde{a}+\tilde{\lambda})\tilde{u}_0 + (1+d)(u_1+\tilde{u}_1) - F_0(x)\Big]\,dx\end{aligned}$$

in case (3.5.117), and

$$\begin{aligned}&\mathscr{C} \int_\Omega (a+d+\lambda)\,dx + \tilde{\mathscr{C}} \int_\Omega (\tilde{a}+d+\tilde{\lambda})\,dx \\ &\quad = \int_\Omega \Big[(a+d+\lambda)u_0 + (\tilde{a}+d+\tilde{\lambda})\tilde{u}_0 + u_1 + \tilde{u}_1 - F_0(x)\Big]\,dx\end{aligned}$$

in case (3.5.118), where $\lambda = f_0$ and $\tilde{\lambda} = \tilde{f_0}$ in case (3.5.119), $\lambda = b$ and $\tilde{\lambda} = \tilde{b}$ in case (3.5.120), $\lambda = f_0$ and $\tilde{\lambda} = \tilde{b}$ in case (3.5.121), $\tilde{u}_0$ and $\tilde{u}_1$ are the initial data of $\tilde{u}$, and

$$\tilde{f_0} = \int_0^{+\infty} \tilde{f}(s)\,ds \quad \text{and}$$

$$F_0(x) = \begin{cases} \displaystyle\int_0^{+\infty} (f(s)\eta_0(s) + \tilde{f}(s)\tilde{\eta}_0(s))\,ds & \text{in case (3.5.119)}, \\ \displaystyle\int_0^{1} (\tau b z_0(p) + \tilde{\tau}\tilde{b}\tilde{z}_0(p))\,dp & \text{in case (3.5.120)}, \\ \displaystyle\int_0^{+\infty} f(s)\eta_0(s)\,ds + \tilde{\tau}\tilde{b}\int_0^1 \tilde{z}_0(p)\,dp & \text{in case (3.5.121)}. \end{cases}$$

Moreover, the convergence of $(u, \tilde{u}, y, \tilde{y}, \eta, \tilde{\eta})$, $(u, \tilde{u}, y, \tilde{y}, z, \tilde{z})$, and $(u, \tilde{u}, y, \tilde{y}, \eta, \tilde{z})$ to zero is of exponential type in $\hat{\mathscr{H}}$, where $\hat{\mathscr{H}}$ is the space containing the elements of

$$\mathscr{H} := H^i(\Omega) \times H^j(\Omega) \times H \times H$$
$$\times \begin{cases} L_f \times L_{\tilde{f}} & \text{in case (3.5.119)}, \\ L^2((0,1), H) \times L^2((0,1), H) & \text{in case (3.5.120)}, \\ H \times H \times L_f \times L^2((0,1), H) & \text{in case (3.5.121)}, \end{cases}$$

satisfying

$$\int_\Omega \left[(a+\lambda)u + y + d\tilde{y} - L(x)\right] dx = \int_\Omega \left[(\tilde{a}+\tilde{\lambda})\tilde{u} + \tilde{y} + dy - \tilde{L}(x)\right] dx = 0$$

in case (3.5.117), and

$$\int_\Omega \left[(a+\lambda)u + y + d\tilde{u} - L(x)\right] dx = \int_\Omega \left[(\tilde{a}+\tilde{\lambda})\tilde{u} + \tilde{y} + du - \tilde{L}(x)\right] dx = 0$$

in case (3.5.118), where L and $\tilde{L}$ are defined by

$$\begin{cases} L(x) = \displaystyle\int_0^{+\infty} f(s)\eta(s)\,ds, \quad \tilde{L}(x) = \int_0^{+\infty} \tilde{f}(s)\tilde{\eta}(s)\,ds & \text{in case (3.5.119)}, \\ L(x) = \tau b \displaystyle\int_0^1 z(p)\,dp, \quad \tilde{L}(x) = \tilde{\tau}\tilde{b}\int_0^1 \tilde{z}(p)\,dp & \text{in case (3.5.120)}, \\ L(x) = \displaystyle\int_0^{+\infty} f(s)\eta(s)\,ds, \quad \tilde{L}(x) = \tilde{\tau}\tilde{b}\int_0^1 \tilde{z}(p)\,dp & \text{in case (3.5.121)}. \end{cases}$$

If $A = \tilde{A}, a = \tilde{a}, G = \tilde{G}$ (so (3.5.121) is not concerned) and $(u_0, u_1) = (\tilde{u}_0, \tilde{u}_1)$, then

$$\mathscr{C} = \tilde{\mathscr{C}} = \left(\int_\Omega (a+\lambda)\,dx\right)^{-1} \int_\Omega [(a+\lambda)u_0 + (1+d)u_1 - L_0(x)]\,dx$$

in case (3.5.117), and

$$\mathscr{C} = \tilde{\mathscr{C}} = \left(\int_\Omega (a+d+\lambda)\,dx\right)^{-1} \int_\Omega [(a+d+\lambda)u_0 + u_1 - L_0(x)]\,dx$$

in case (3.5.118), where $\lambda = f_0$ in case (3.5.119), $\lambda = b$ in case (3.5.120), and

$$L_0(x) = \begin{cases} \int_0^{+\infty} f(s)\eta_0(s)\,ds & \text{in case (3.5.119)}, \\ \tau b \int_0^1 z_0(p)\,dp & \text{in case (3.5.120)}. \end{cases}$$

3.5.6.3 Elasticity Systems

We can also consider the following elasticity system:

$$\begin{cases} (u_i)_{tt}(x,t) - \sum_{j=1}^{N} \sigma_{ij,j}(x,t) & \forall x \in \Omega,\ \forall t > 0,\ \forall i = 1, \cdots, N, \\ +a_i(u_i)_t(x,t) + G_i(x,t) = 0, & \\ \sum_{j=1}^{N} \sigma_{ij}(x,t)\nu_j = 0, & \forall x \in \Gamma,\ \forall t > 0,\ \forall i = 1, \cdots, N, \end{cases}$$

where $u = (u_1, \cdots, u_N)$ is the solution with the initial condition

$$u_0 = (u_1^0, \cdots, u_N^0) \quad \text{and} \quad u_1 = (u_1^1, \cdots, u_N^1),$$

$a_i \in C(\bar{\Omega})$ such that $\inf_\Omega a_i > 0$,

$$\sigma_{ij,j} = \frac{\partial \sigma_{ij}}{\partial x_j},\ \sigma_{ij} = \sum_{k,l=1}^{N} a_{ijkl}\varepsilon_{kl},\ \varepsilon_{ij} = \frac{1}{2}(u_{i,j}+u_{j,i}),\ u_{i,j} = \frac{\partial u_i}{\partial x_j},\ u_{j,i} = \frac{\partial u_j}{\partial x_i},$$

where $a_{ijkl} \in C^1(\bar{\Omega})$ satisfying

$$a_{ijkl}(x) = a_{klij}(x) = a_{jikl}(x), \quad \forall x \in \Omega,\ \forall i, j, k, l = 1, \cdots, N,$$

and there exists $\alpha_0 > 0$ such that

$$\sum_{i,j,k,l=1}^{N} a_{ijkl}(x)\varepsilon_{ij}\varepsilon_{kl} \geq \alpha_0 \sum_{i,j=1}^{N} \varepsilon_{ij}\varepsilon_{ij}, \quad \forall x \in \Omega$$

and for all symmetric tensor ε_{ij}, and G_i is a distributed time delay

$$G_i(x,t) = \int_0^{+\infty} f_i(s)(u_i)_t(x, t-s)ds \tag{3.5.122}$$

or discrete-time one

$$G_i(x,t) = b_i(u_i)_t(x, t-\tau_i) \tag{3.5.123}$$

with $\tau_i > 0$, f_i is as f in Sect. 3.5.2, and $b_i \in L^\infty(\Omega)$ such that $\|b_i\|_\infty < \inf_\Omega a_i$. Here $(u_i)_t$ converges exponentially in H to zero and u_i converges exponentially in V to $\mathscr{C}_i$, where $\mathscr{C}_i$ is a constant such that

$$\sum_{i=1}^{N} \mathscr{C}_i \int_\Omega (a_i + \lambda_i)\, dx = \sum_{i=1}^{N} \int_\Omega \left[(a_i + \lambda_i)u_i^0 + u_i^1 - F_i^0(x)\right] dx,$$

where

$$\lambda_i = \int_0^{+\infty} f_i(s)\, ds \quad \text{and} \quad F_i^0(x) = \int_0^{+\infty} f_i(s)\eta_i^0(s)\, ds$$

in case (3.5.122), and

$$\lambda_i = b_i \quad \text{and} \quad F_i^0(x) = \tau_i b_i \int_0^1 z_i^0(p)\, dp$$

in case (3.5.123), η_i^0 and z_i^0 are the initial data of η_i and z_i, respectively, and η_i and z_i are defined in terms of u_i as η and z in Sects. 3.5.2 and 3.5.5, respectively.

In addition, (u, u_t, η) exponentially converges to zero in $\hat{\mathscr{H}}$, where $\hat{\mathscr{H}}$ is the space containing the elements

$$(u_1, \cdots, u_N, y_1, \cdots, y_N, \theta_1, \cdots, \theta_N) \in \mathscr{H} := V^N \times H^N \times L_1 \times \cdots \times L_N$$

satisfying

$$\int_\Omega [(a_i + \lambda_i)u_i + y_i - F_i(x)]\, dx = 0, \quad \forall i = 1, \cdots, N,$$

where $y_i = (u_i)_t$, $\theta_i = \eta_i$ in case (3.5.122), $\theta_i = z_i$ in case (3.5.123), $L_i = L_{f_i}$ in case (3.5.122), $L_i = L^2((0,1), H)$ in case (3.5.123), L_{f_i} is defined as L_f in Sect. 3.5.2, and F_i is defined as F_i^0 with η_i and z_i instead of η_i^0 and z_i^0, respectively.

If $a_i = a_j$, $G_i = G_j$, $u_i^0 = u_j^0$, and $u_i^1 = u_j^1$, for all $i, j = 1, \cdots, N$, then

$$\mathscr{C}_i = \left(\int_\Omega (a_1 + \lambda_1)\, dx \right)^{-1} \int_\Omega \left[(a_1 + \lambda_1) u_1^0 + u_1^1 - F_1^0(x) \right] dx,$$

$$\forall i = 1, \cdots, N.$$

3.5.7 Comments and Discussion

We would like to point out that there are many promising research avenues. For instance:

1. It would be very desirable to extend our results to the case of nonlinear damping and nonlinear time delay, that is, replace $a(x)u_t(x,t)$, $u_t(x, t-s)$, and $b(x)u_t(x, t-\tau)$ by $a(x)h_1(u_t(x,t))$, $h_2(u_t(x, t-s))$, and $b(x)h_3(u_t(x, t-\tau))$, respectively, where $h_i : \mathbb{R} \to \mathbb{R}$, $i = 1, 2, 3$, are nonlinear functions. When no delay is considered, the case of nonlinear damping was studied in [61]. Note that a nonlinear situation has been treated in [177] for semilinear diffusion parabolic PDEs by using Lyapunov-Krasovskii method.
2. It would be interesting to extend our results to a larger class of functions f than the one we considered in this section.
3. It would be of great importance to prove our results in the case where the linear damping $au_t(x,t)$ is replaced by infinite memory of the form

$$\int_0^{+\infty} g(s) B u(x, t-s)\, ds, \tag{3.5.124}$$

   where $g : \mathbb{R}_+ \to \mathbb{R}_+$ is a given function and $B : D(B) \subset H \to H$ is a given operator. When $\mathscr{H}$ is a Hilbert space with respect to the inner product that generates the classical energy norm, the case (3.5.124) was treated in [106] and [108].
4. Some numerical analysis of the theoretical results obtained could be performed. This will illustrate the exponential convergence of solutions by having an idea, in particular, on the decay rate ω (given in (3.5.38)) in terms of f, a, and b, and the constant τ.

3.6 Interior Delay: Boundary Exponential Stabilization

In this section, we consider a boundary stabilization problem for the wave equation with interior delay. We prove an exponential stability result under some Lions Geometric Condition. The proof of the main result is based on an identity with multipliers that allows us to obtain a uniform decay estimate for a suitable Lyapunov functional.

3.6.1 Introduction and Main Result

As in the previous sections (except Section (3.4)), we consider a wave equation in an open bounded domain Ω of $\mathbb{R}^n$, $n \geq 2$, with a smooth boundary $\partial\Omega = \Gamma_0 \cup \Gamma_1$, where $\Gamma_0,\ \Gamma_1$ are closed subsets of $\partial\Omega$ such that $\Gamma_0 \cap \Gamma_1 = \emptyset$. Moreover, we assume $meas\Gamma_0 > 0$. The system is given by

$$u_{tt}(x,t) - \Delta u(x,t) + a u_t(x,t-\tau) = 0, \qquad x \in \Omega,\ t > 0, \tag{3.6.1}$$

$$u(x,t) = 0, \qquad x \in \Gamma_0,\ t > 0 \tag{3.6.2}$$

$$\frac{\partial u}{\partial \nu}(x,t) = -k u_t(x,t), \qquad x \in \Gamma_1,\ t > 0 \tag{3.6.3}$$

$$u(x,0) = u_0(x), \quad u_t(x,0) = u_1(x), \qquad x \in \Omega, \tag{3.6.4}$$

$$u_t(x,t) = g(x,t), \qquad x \in \Omega,\ t \in (-\tau, 0), \tag{3.6.5}$$

where ν stands for the unit normal vector of $\partial\Omega$ pointing toward the exterior of Ω and $\frac{\partial u}{\partial \nu}$ is the normal derivative. Moreover, the constant $\tau > 0$ is the time delay, a and k are two positive numbers, and the initial data are taken in suitable spaces.

Denote by m the standard multiplier, that is, $m(x) = x - x_0$, and assume

$$m(x) \cdot \nu(x) \leq 0, \quad x \in \Gamma_0, \quad \text{and} \quad m(x) \cdot \nu(x) \geq \delta > 0, \quad x \in \Gamma_1 . \tag{3.6.6}$$

In order to highlight the contribution of this section, we first recall some results that we already discussed in the beginning of this chapter. In particular, we know that the exponential stability of (3.6.1)–(3.6.5) with $a = 0$ has been studied in [32], where it has been shown that the system is exponentially stable if Γ_1 satisfies the BLR condition. Moreover, if $\tau = 0$ (no delay occurs), the above problem is exponentially stable for any $a > 0$, even if $k = 0$ (see, e.g., [132, 191]). Notwithstanding, there are instability phenomena in the presence of a delay term. In fact, it is shown in [149] that it is possible to find a sequence $\{\tau_k\}_k$ of delays with $\tau_k \to 0$ for which the solutions u_k of the above problem in case $k = 0$ have an increasing energy.

In order to contrast such destabilizing effect of the time delay, a "good" (nondelayed) damping term is introduced in (3.6.1) (see [149]). In fact, the problem considered in [149] is

$$u_{tt}(x,t) - \Delta u(x,t) + a u_t(x,t-\tau) + a_0 u_t(x,t) = 0, \qquad x \in \Omega,\ t > 0, \tag{3.6.7}$$

$$u(x,t) = 0, \qquad x \in \Gamma_0,\ t > 0 \tag{3.6.8}$$

$$\frac{\partial u}{\partial \nu}(x,t) = 0, \qquad x \in \Gamma_1,\ t > 0, \tag{3.6.9}$$

with $a_0, a > 0$ and with some initial data in suitable spaces. If $a_0 > a$, then the system (3.6.7)–(3.6.9) is uniformly exponentially stable [149] (see also [5, 151, 154] for related results).

The main motivation of this section is to contrast the effect of the time delay but this time by means of the dissipative boundary feedback in (3.6.3). Specifically, we are able to show that if the condition (3.6.6)is satisfied (geometric Lions condition), then for any $k > 0$, system (3.6.1)–(3.6.5) is exponentially stable for a sufficiently small.

In order to state our main result, various notations will be introduced. First, let $A = -\Delta$ be the unbounded operator in $H = L^2(\Omega)$ with domain

$$H_1 = \mathscr{D}(A) = \left\{ u \in H^2(\Omega),\ u_{|\Gamma_0} = 0,\ \frac{\partial u}{\partial \nu}_{|\Gamma_1} = 0 \right\}.$$

Next, we define by

$$B \in \mathscr{L}(L^2(\Omega); L^2(\Omega)),\ Bu = B^* u = \sqrt{a}\, u, \forall\, u \in L^2(\Omega),$$

and

$$C \in \mathscr{L}(L^2(\Gamma_1); H_{-1}),\ Cu = \sqrt{k}\, A_{-1} N u,\ \forall u \in L^2(\Gamma_1),$$

$$C^* w = \sqrt{k} w_{|\Gamma_1},\ \forall\, w \in \mathscr{D}(A^{\frac{1}{2}}) = H_{\frac{1}{2}},$$

where $H_{-1} = (\mathscr{D}(A))'$ (the duality is in the sense of H), A_{-1} is the extension of A to H, namely for all $h \in H$ and $\varphi \in \mathscr{D}(A)$, $A_{-1}h$ is the unique element in H_{-1} such that (see, for instance, [180])

$$\langle A_{-1} h; \varphi \rangle_{H_{-1}-H_1} = \int_\Omega h A \varphi \, dx.$$

Also, N denotes the operator in $\mathscr{L}(L^2(\Gamma_1); L^2(\Omega))$, such that $\forall\, v \in L^2(\Gamma_1)$, Nv is the unique solution (transposition solution) of

$$\Delta w = 0,\ w_{|\Gamma_0} = 0,\ \frac{\partial w}{\partial \nu}_{|\Gamma_1} = v.$$

Setting $z(t,\theta) = \dot{u}(t+\theta)$, $\theta \in (-\tau, 0)$, the evolution equation (3.6.1)–(3.6.5)is equivalent to

$$\ddot{u}(t) + Au(t) + CC^*\dot{u}(t) + BB^*z(t,-\tau) = 0, \quad t \geq 0, \tag{3.6.10}$$

$$\dot{z}(t,\theta) - z_\theta(t,\theta) = 0,\ \theta \in (-\tau, 0), \quad t \geq 0, \tag{3.6.11}$$

$$z(t,0) = \dot{u}(t), \quad t \geq 0, \tag{3.6.12}$$

$$u(0) = u^0,\ \dot{u}(0) = u^1,\ z(0,\theta) = g(\theta),\ \theta \in (-\tau, 0), \tag{3.6.13}$$

where $z_\theta = \partial_\theta z$ and $g \in L^2(-\tau, 0; H_{\frac{1}{2}})$.

The system (3.6.10)–(3.6.13) will be written in an abstract Cauchy problem in a product space so that the semigroup theory is used (see Sect. 2.3). To this end, consider the Hilbert space

$$\mathscr{H} := H_{\frac{1}{2}} \times H \times L^2(-\tau, 0; H_{\frac{1}{2}})$$

and the unbounded linear operator

$$\mathscr{A}_d : \mathscr{D}(\mathscr{A}_d) \subset \mathscr{H} \longrightarrow \mathscr{H}$$

as follows:

$$\mathscr{A}_d\,(u_1, u_2, z) = \left(u_2, -Au_1 - CC^*u_2 - BB^*z(-\tau), \partial_\theta z\right), \\ \text{for all } (u_1, u_2, z) \in \mathscr{D}(\mathscr{A}), \tag{3.6.14}$$

where

$$\mathscr{D}(\mathscr{A}_d) := \Big\{(u_1, u_2, z) \in H_{\frac{1}{2}} \times H_{\frac{1}{2}} \times H^1(-\tau, 0; H_{\frac{1}{2}}),\ Au_1 \\ + CC^*u_2 + BB^*z(-\tau) \in H, z(0) = u_2\Big\}. \tag{3.6.15}$$

We have the following well-posedness result:

Proposition 3.6.1

1. *The operator $(\mathscr{A}_d, \mathscr{D}(\mathscr{A}_d))$ generates a strongly continuous semigroup $(\mathscr{T}(t))_{t\geq 0}$ on $\mathscr{H}$.*

2. *The system* (3.6.1)–(3.6.5)*is well-posed. More precisely, for every* $(u^0, u^1, g) \in \mathscr{H}$, *the function* w *given by the first component of* $\mathscr{T}(t)\left(u^0, u^1, g\right)$ *is the mild solution of* (3.6.1)–(3.6.5). *In particular, for* $(u^0, u^1, g) \in \mathscr{D}(\mathscr{A}_d)$, *the problem* (3.6.10)–(3.6.13)*admits a unique classical solution*

$$(u, v, z) \in C([0, +\infty), \mathscr{D}(\mathscr{A}_d)) \cap C^1([0, +\infty), \mathscr{H}),$$

and thus the problem (3.6.1)–(3.6.5)*admits a unique classical solution*

$$u \in C^1([-\tau, +\infty), H_{\frac{1}{2}}) \cap C^2([0, +\infty), H).$$

The proof of this outcome is conducted in the next subsection.

With regard to the exponential stability result, given any regular solution of (3.6.1)–(3.6.5), we define the energy

$$E(t) := E_S(t) + \frac{\xi}{2}\int_{t-\tau}^{t}\int_{\Omega} u_t^2(x, s)dxds = \frac{1}{2}\int_{\Omega}\{|\nabla u(x, t)|^2 + u_t^2(x, t)\}dx$$
$$+ \frac{\xi}{2}\int_{t-\tau}^{t}\int_{\Omega} u_t^2(x, s)dxds\,, \tag{3.6.16}$$

where ξ is a strictly positive real number.

The main finding of this section is:

Theorem 3.6.1 *For any* $k > 0$, *there exist positive constants* a_0, C_1, C_2 *such that*

$$E(t) \leq C_1 e^{-C_2 t} E(0), \tag{3.6.17}$$

for any regular solution of problem (3.6.1)–(3.6.5) *with* $0 \leq a < a_0$. *The constants* a_0, C_1, C_2 *are independent of the initial data, but they depend on k and the geometry of* Ω.

To the authors' best knowledge, the opposite problem, that is, to contrast the effect of a time delay in the boundary condition with a velocity interior term in the wave equation, is still open as it seems to be much harder to deal with. As discussed in the beginning of the current chapter, there is already a positive answer in [84] for the one-dimensional case (see (3.1.2) for more details).

This section is organized as follows. First, we deal with the well-posedness of the problem. Then, we prove the exponential stability of the delayed system (3.6.1)–(3.6.5)by introducing a suitable Lyapunov functional.

3.6.2 Well-Posedness of the Problem: Proof of Proposition 3.6.1

For the well-posedness of the equivalent problems (3.6.1)–(3.6.5)and (3.6.10)–(3.6.13), we will show that the operator $(\mathscr{A}_d, \mathscr{D}(\mathscr{A}_d))$ defined by (3.6.14)–(3.6.15) generates a contraction semigroup on the Hilbert space $\mathscr{H} := H_{\frac{1}{2}} \times H \times L^2(-\tau, 0; H_{\frac{1}{2}})$.

First, in $\mathscr{H}$, we introduce the new inner product

$$\langle (u_1, u_2, z_1), (v_1, v_2, z_2) \rangle = \langle u_1, v_1 \rangle_{H_{\frac{1}{2}}} + \langle u_2, v_2 \rangle_H + \xi \int_{-\tau}^{0} \left\langle B^* z_1(\theta), B^* z_2(\theta) \right\rangle_{L^2(\Omega)} d\theta,$$

where ξ is a strictly positive constant.

It can be easily verified that $\mathscr{H}$ endowed with this inner product is a Hilbert space, and its associated norm is equivalent to the canonical norm of $\mathscr{H}$.

Now, we are able to prove Proposition 3.6.1:

Proof Let us first show that there exists a positive constant c such that $\mathscr{A}_d - cI$ is dissipative. To this end, let $(u, v, z) \in \mathscr{D}(\mathscr{A}_d)$. Then using Cauchy-Schwarz inequality, we have

$$\begin{aligned}
&\langle \mathscr{A}_d (u, v, z), (u, v, z) \rangle \\
=&\left\langle \left(v, -Au - CC^* v - BB^* z(-\tau), \partial_\theta z\right), (u, v, z) \right\rangle \\
=&- < C^* v, C^* v >_{L^2(\Gamma_1)} - < B^* z(-\tau), B^* v >_{L^2(\Omega)} \\
&+ \xi \int_{-\tau}^{0} \left\langle B^* \partial_\theta z(\theta), B^* z(\theta) \right\rangle_{L^2(\Omega)} d\theta \\
=&- < B^* z(-\tau), B^* v >_{L^2(\Omega)} - \|C^* v\|^2_{L^2(\Gamma_1)} \\
&+ \xi \, \|B^* v\|^2_{L^2(\Omega)} - \xi \, \|B^* z(-\tau)\|^2_{L^2(\Omega)} \\
\leq& -\frac{\xi}{2} \|B^* z(-\tau)\|^2_{L^2(\Omega)} + \left(\frac{1}{2\xi} + \xi\right) \|B^* v\|^2_{L^2(\Omega)}.
\end{aligned}$$

Since B^* is bounded from $L^2(\Omega)$ into itself, we deduce that there exists $c > 0$ such that

$$\langle \mathscr{A}_d (u, v, z), (u, v, z) \rangle \leq c \|v\|^2_{L^2(\Omega)}.$$

This shows that $\mathscr{A}_d - cI$ is dissipative.

Next, we prove that $(\lambda I - \mathscr{A}_d)$ is onto $\mathscr{H}$ for some $\lambda > 0$. To proceed, let $(f, g, h) \in \mathscr{H}$, and the objective is to find $(u, v, z) \in \mathscr{D}(\mathscr{A}_d)$ such that

$$(\lambda I - \mathscr{A}_d)(u, v, z) = (f, g, h).$$

This is equivalent to writing

$$\lambda u - v = f, \tag{3.6.18}$$

$$Au + (\lambda I + CC^*)v + BB^* z(-\tau) = g, \tag{3.6.19}$$

$$\lambda z - \partial_\theta z = h. \tag{3.6.20}$$

Note that the function $z \in H^1(-\tau, 0; H_{\frac{1}{2}})$ given by

$$z(\theta) = e^{\lambda\theta} v + \int_\theta^0 e^{\lambda(\theta-\sigma)} h d\sigma$$

is a solution of (3.6.20) and verifies $z(0) = v$. By writing $v = \lambda u - f$ from (3.6.18)and using the expression of z, it follows from (3.6.19)that we only need to seek a solution $u \in D(A)$ of the following equation:

$$\left(\lambda^2 I + A + \lambda(CC^* + BB^* e^{-\lambda\tau})\right) u = k, \tag{3.6.21}$$

where k is known and defined by

$$k = g + \left(\lambda I + CC^* + e^{-\lambda\tau} BB^*\right) f + BB^* \int_{-\tau}^0 e^{-\lambda(\tau+\sigma)} h d\sigma.$$

Thus, it remains to solve (3.6.21). To do so, we first note that if $u \in D(A)$ exists and is a solution of (3.6.21), then we have

$$\left\langle (\lambda^2 I + A + \lambda(CC^* + BB^* e^{-\lambda\tau}))u, \zeta \right\rangle = \langle k, \zeta \rangle, \forall \zeta \in H_{\frac{1}{2}}$$

or equivalently

$$\Lambda(u, \zeta) = \langle k, \zeta \rangle, \forall \zeta \in H_{\frac{1}{2}}, \tag{3.6.22}$$

where

$$\Lambda(u, \zeta) := \lambda^2 \langle u, \zeta \rangle + \left\langle A^{\frac{1}{2}} u, A^{\frac{1}{2}} \zeta \right\rangle + \lambda(\left\langle C^* u, C^* \zeta \right\rangle + e^{-\lambda\tau} \left\langle B^* u, B^* \zeta \right\rangle).$$

Since Λ is a bilinear coercive form on $H_{\frac{1}{2}}$, Lax-Milgram theorem (see Theorem 2.1.6) ensures the existence and uniqueness of a solution $u \in H_{\frac{1}{2}}$ to (3.6.22). Thereafter, some integrations by parts allow us to show that $u \in D(A)$ is indeed a solution of (3.6.21). Consequently, $\lambda I - \mathscr{A}_d$ is onto $\mathscr{H}$ and so is $(\lambda I - (\mathscr{A}_d - cI))$. The density of $\mathscr{D}(\mathscr{A}_d)$ is clear. Finally, Theorem 2.3.4 leads us to claim that $\mathscr{A}_d - cI$ generates a strongly continuous semigroup of contraction in $\mathscr{H}$, and hence, $\mathscr{A}_d$ generates a strongly continuous semigroup in $\mathscr{H}$. □

It is easy to see that if $u : [-\tau, \infty) \longrightarrow H_{\frac{1}{2}}$ is a classical solution of (3.6.1)–(3.6.5), then $(u, \dot{u}, \dot{u}(t + \cdot))$ is the classical solution of (3.6.10)–(3.6.13). The first assertion of Proposition 3.6.1 provides the opposite result, and then the well-posedness of the evolution equation (3.6.1)–(3.6.5).

The well-posedness part follows from the first assertion of Proposition 3.6.1.

3.6.3 Exponential Stability: Proof of Theorem 3.6.1

The aim of this subsection is to prove the exponential stability result stated in Theorem 3.6.1. To do so, we need some preliminary results. First, we have:

Proposition 3.6.2 *For any solution of* (3.6.1)–(3.6.5), *the following estimate holds:*

$$E'(t) \leq \frac{a+\xi}{2} \int_{\Omega} u_t^2(x,t)dx + \frac{a-\xi}{2} \int_{\Omega} u_t^2(x,t-\tau)dx - k \int_{\Gamma_1} u_t^2(x,t)d\Gamma . \tag{3.6.23}$$

Proof Differentiating (3.6.16), we obtain

$$E'(t) = \int_{\Omega} \{\nabla u(t) \cdot \nabla u_t(t) + u_t(t)u_{tt}(t)\}dx + \frac{\xi}{2} \int_{\Omega} u_t^2(t)dx - \frac{\xi}{2} \int_{\Omega} u_t^2(t-\tau)dx.$$

Subsequently, integrating by parts and using (3.6.1), (3.6.2), (3.6.3), we find

$$\begin{aligned} E'(t) = &-a \int_{\Omega} u_t(t)u_t(t-\tau)dx - k \int_{\Gamma_1} u_t^2(t)d\Gamma + \frac{\xi}{2} \int_{\Omega} u_t^2(t)dx \\ &- \frac{\xi}{2} \int_{\Omega} u_t^2(t-\tau)dx . \end{aligned} \tag{3.6.24}$$

Applying Cauchy-Schwarz's inequality in (3.6.24), we obtain (3.6.23). □

We also have the following second preliminary result:

Proposition 3.6.3 *For any regular solution of problem* (3.6.1)–(3.6.5) *and for every* $\varepsilon > 0$, *we have*

$$\begin{aligned}
&\frac{d}{dt}\left\{\int_\Omega [2m\cdot\nabla u + (n-1)u]u_t dx\right\}\\
&\le -\int_\Omega u_t^2(t)dx - \left(1 - \tfrac{\varepsilon}{2}C(P)\right)\int_\Omega |\nabla u(t)|^2 dx\\
&\qquad -2a\int_\Omega (m\cdot\nabla u(t))u_t(t-\tau)dx - (n-1)a\int_\Omega u(t)u_t(t-\tau)dx\\
&\qquad +\left[\left(\|m\|_\infty\frac{2}{\delta} + \frac{(n-1)^2}{2\varepsilon}\right)k^2 + \|m\|_\infty\right]\int_{\Gamma_1} u_t^2(t)d\Gamma,
\end{aligned} \tag{3.6.25}$$

where $C(P)$ *is a sort of Poincaré constant, which is a positive constant depending on* Ω *and independent of the solution* u.

Proof Simple differentiation and integrations by parts yield

$$\begin{aligned}
&\frac{d}{dt}\left\{\int_\Omega [2m\cdot\nabla u + (n-1)u]u_t dx\right\}\\
&= -\int_\Omega\{u_t^2 + |\nabla u|^2\}dx - 2a\int_\Omega (m\cdot\nabla u)u_t(t-\tau)dx\\
&\qquad -(n-1)a\int_\Omega u(t)u_t(t-\tau)dx + \int_\Gamma (m\cdot\nu)(u_t^2 - |\nabla u|^2)d\Gamma\\
&\qquad +2\int_\Gamma (m\cdot\nabla u)\frac{\partial u}{\partial\nu}d\Gamma + (n-1)\int_\Gamma u\frac{\partial u}{\partial\nu}d\Gamma.
\end{aligned} \tag{3.6.26}$$

Now, since $u = 0$ on Γ_0 and $m\cdot\nu \le 0$ on Γ_0, we deduce from (3.6.26)

$$\begin{aligned}
&\frac{d}{dt}\left\{\int_\Omega [2m\cdot\nabla u + (n-1)u]u_t dx\right\}\\
&\le -\int_\Omega\{u_t^2 + |\nabla u|^2\}dx\\
&-2a\int_\Omega (m\cdot\nabla u)u_t(t-\tau)dx - (n-1)a\int_\Omega u(t)u_t(t-\tau)dx\\
&\qquad +\|m\|_\infty\int_{\Gamma_1} u_t^2 d\Gamma - \delta\int_{\Gamma_1}|\nabla u|^2 d\Gamma + 2\int_{\Gamma_1}(m\cdot\nabla u)\frac{\partial u}{\partial\nu}d\Gamma\\
&+(n-1)\int_{\Gamma_1} u\frac{\partial u}{\partial\nu}d\Gamma,
\end{aligned} \tag{3.6.27}$$

where we have also used $m\cdot\nu \ge \delta$ on Γ_1.

Furthermore, thanks to Young's inequality, we have

$$\begin{aligned}
2\int_{\Gamma_1}(m\cdot\nabla u)\frac{\partial u}{\partial\nu}d\Gamma &\le \frac{\delta}{2}\int_{\Gamma_1}|\nabla u|^2 d\Gamma + 2\frac{\|m\|_\infty}{\delta}\int_{\Gamma_1}\left|\frac{\partial u}{\partial\nu}\right|^2 d\Gamma\\
&= \frac{\delta}{2}\int_{\Gamma_1}|\nabla u|^2 d\Gamma + 2\frac{\|m\|_\infty}{\delta}k^2\int_{\Gamma_1}u_t^2(t)d\Gamma,
\end{aligned} \tag{3.6.28}$$

as well as

$$
\begin{aligned}
(n-1)\int_{\Gamma_1} u\frac{\partial u}{\partial \nu}d\Gamma &\leq \frac{\varepsilon}{2}\int_{\Gamma_1} u^2 d\Gamma + \frac{(n-1)^2}{2\varepsilon}\int_{\Gamma_1}\left(\frac{\partial u}{\partial \nu}\right)^2 d\Gamma \\
&\leq \frac{\varepsilon}{2}C(P)\int_\Omega |\nabla u|^2 dx + \frac{(n-1)^2}{2\varepsilon}k^2\int_{\Gamma_1} u_t^2(t)d\Gamma\,,
\end{aligned}
\tag{3.6.29}
$$

where we have also used trace inequality and Poincaré's inequality, namely, $\int_\Omega y^2(x)\, d\Sigma \leq C(P)\int_\Omega |\nabla y(x)|^2\, dx.$

Substituting (3.6.28) and (3.6.29) in (3.6.27), we obtain the estimate (3.6.25). □

As a consequence of the previous result, we have:

Corollary 3.6.1 *For any regular solution of* (3.6.1)–(3.6.5),

$$
\begin{aligned}
&\frac{d}{dt}\left\{\int_\Omega [2m\cdot\nabla u + (n-1)u]u_t dx\right\} \leq -\int_\Omega u_t^2(t)dx \\
&\quad -\left(1-\frac{\varepsilon}{2}C(P) - a\|m\|_\infty - \frac{a}{2}(n-1)^2 C_0(P)\right)\int_\Omega |\nabla u(t)|^2 dx \\
&\quad +\frac{3}{2}a\int_\Omega u_t^2(t-\tau)dx + \left[\left(\|m\|_\infty\frac{2}{\delta} + \frac{(n-1)^2}{2\varepsilon}\right)k^2 + \|m\|_\infty\right]\int_{\Gamma_1} u_t^2(t)d\Gamma\,,
\end{aligned}
\tag{3.6.30}
$$

where $C_0(P)$ is the so-called Poincaré constant.

Proof We apply Cauchy-Schwarz's inequality to the integral in the second line of (3.6.25). □

Now, let us introduce the functional

$$
\mathscr{S}(t) := \int_\Omega\int_{t-\tau}^t e^{s-t}u_t^2(x,s)dsdx.
$$

A straightforward computation gives

$$
\begin{aligned}
\mathscr{S}'(t) &= \int_\Omega u_t^2(t)dx - \int_\Omega e^{-\tau}u_t^2(t-\tau)dx - \int_\Omega\int_{t-\tau}^t e^{s-t}u_t^2(x,s)dsdx \\
&\leq \int_\Omega u_t^2(t)dx - e^{-\tau}\int_\Omega u_t^2(t-\tau)dx - e^{-\tau}\int_\Omega\int_{t-\tau}^t u_t^2(x,s)dsdx\,.
\end{aligned}
\tag{3.6.31}
$$

Now, let us introduce the Lyapunov functional

$$
\mathscr{E}(t) := E(t) + \gamma_1\int_\Omega [2m\cdot\nabla u + (n-1)u]u_t dx + \gamma_2\mathscr{S}(t),
\tag{3.6.32}
$$

where γ_1, γ_2 are suitable small positive constants that will be provided later on.

Note that $\mathscr{E}(t)$ is equivalent to the energy $E(t)$ as long as γ_1 is small enough. In particular, there exist a positive constant C_1 and suitable positive constants α_1, α_2 such that

$$\alpha_1 E(t) \leq \mathscr{E}(t) \leq \alpha_2 E(t)\,, \forall\, 0 < \gamma_1,\ \gamma_2 \leq C_1\,. \tag{3.6.33}$$

Finally, we can state the last preliminary result:

Proposition 3.6.4 *For every $k > 0$, there exist positive constants a_0, c_1 and c_2 such that for any solution of problem* (3.6.1)–(3.6.5) *with $a < a_0$, we have*

$$\mathscr{E}(t) \leq c_1 e^{-c_2 t}\mathscr{E}(0), \quad t > 0. \tag{3.6.34}$$

Proof Differentiating the Lyapunov functional $\mathscr{E}$ and using the previous propositions, we deduce

$$\begin{aligned}\mathscr{E}'(t) \leq & \left(\frac{a+\xi}{2} - \gamma_1 + \gamma_2\right)\int_\Omega u_t^2(x,t)dx - \gamma_2 e^{-\tau}\int_\Omega\int_{t-\tau}^t u_t^2(x,s)dsdx \\ & + \left(\frac{a-\xi}{2} + \frac{3}{2}a\gamma_1 - \gamma_2 e^{-\tau}\right)\int_\Omega u_t^2(x,t-\tau)dx \\ & -\gamma_1\left(1 - \frac{\varepsilon}{2}C(P) - a\|m\|_\infty - \frac{a}{2}(n-1)^2 C_0(P)\right)\int_\Omega |\nabla u(t)|^2 dx \\ & + \left\{\gamma_1 k^2\left(\|m\|_\infty\frac{2}{\delta} + \frac{(n-1)^2}{2\varepsilon}\right) + \gamma_1\|m\|_\infty - k\right\}\int_{\Gamma_1} u_t^2(x,t)dx\,.\end{aligned} \tag{3.6.35}$$

For a fixed $k > 0$, the objective is to choose $\varepsilon, \gamma_1, \gamma_2 < C_1$ and a sufficiently small to reach

$$\mathscr{E}'(t) \leq -cE(t). \tag{3.6.36}$$

Applying the first inequality of (3.6.33), the desired estimate (3.6.34) follows easily.

Now, in order to show that (3.6.35) implies (3.6.36), we simply need the following requirements:

$$\begin{aligned}\frac{a+\xi}{2} - \gamma_1 + \gamma_2 &< 0, \\ \frac{a-\xi}{2} + \frac{3}{2}a\gamma_1 - \gamma_2 e^{-\tau} &\leq 0, \\ 1 - \frac{\varepsilon}{2}C(P) - a\|m\|_\infty - \frac{a}{2}(n-1)^2 C_0(P) &> 0, \\ \gamma_1 k^2\left(\|m\|_\infty\frac{2}{\delta} + \frac{(n-1)^2}{2\varepsilon}\right) + \gamma_1\|m\|_\infty - k &\leq 0.\end{aligned}$$

These conditions are equivalent to

$$\frac{a+\xi}{2} < \gamma_1 - \gamma_2, \tag{3.6.37}$$

$$a(\frac{1}{2} + \frac{3}{2}\gamma_1) - \frac{\xi}{2} \leq \gamma_2 e^{-\tau}, \tag{3.6.38}$$

$$a(\|m\|_\infty + \frac{1}{2}(n-1)^2 C_0(P)) < 1 - \frac{\varepsilon}{2} C(P), \tag{3.6.39}$$

$$\gamma_1 \left[k^2 \Big(\|m\|_\infty \frac{2}{\delta} + \frac{(n-1)^2}{2\varepsilon} \Big) + \|m\|_\infty \right] - k \leq 0. \tag{3.6.40}$$

For any $k > 0$, this last condition is satisfied for γ_1 sufficiently small (once ε is fixed). Then, one needs to check the conditions (3.6.37) to (3.6.39). For the first one, we assume that $\gamma_1 > \gamma_2$, while for (3.6.39) we need to fix ε small enough such that

$$1 - \frac{\varepsilon}{2} C(P) > 0.$$

Subsequently, we fix γ_1, γ_2 and ε fulfilling the above requirements and look at (3.6.37) to (3.6.39) as conditions on a and ξ. These conditions are clearly linear constraints, and a simple investigation shows that the set of pairs (a, ξ) that fulfills these constraints is not empty (see Fig. 3.6).

According to this figure, it is clear that for a and ξ small enough, (3.6.37) to (3.6.39) are valid. Note further that due to (3.6.40) if k goes to ∞ or to 0, then γ_1 must tend to zero, and therefore γ_2 as well, and the minimal value a_0 of a goes to zero. □

From Proposition 3.6.4 and the energy equivalence (3.6.33), we deduce the estimate (3.6.17), and hence, the proof of Theorem 3.6.1 is achieved.

3.6.4 Conclusion

The main idea of this section was to propose a boundary control of type velocity in order to compensate any eventual negative impact of an interior delay on the stability of the wave equation. Under a so-called Lions' geometric condition on the domain, the exponential stability result is established. Note also that this outcome is true, provided that the feedback gain of the delayed damping gain is small enough (see the condition $a < a_0$ in Theorem 3.6.1).

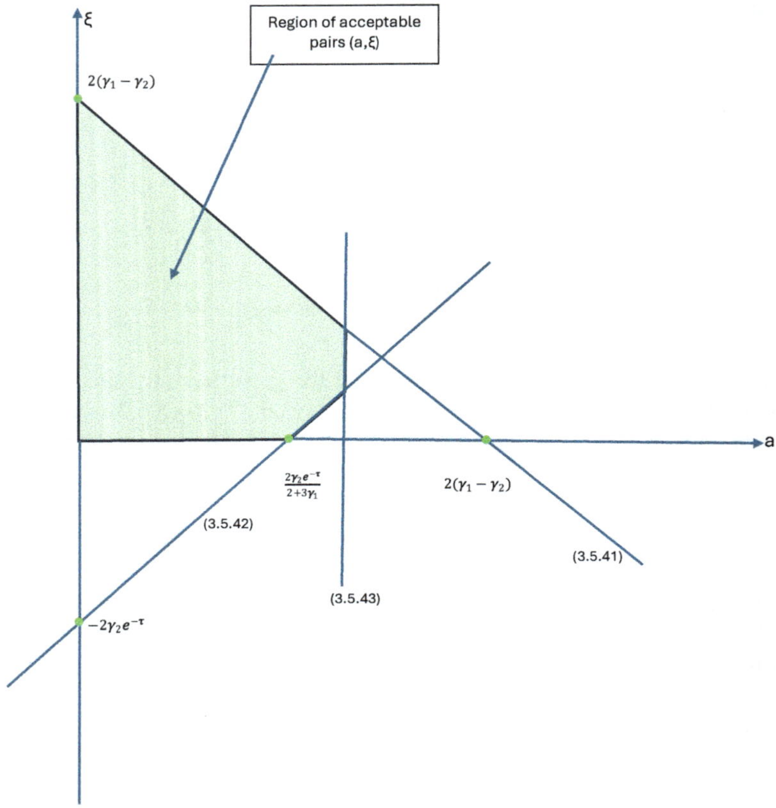

Fig. 3.6 Acceptable solutions (a, ξ)

It would be desirable to consider the same problem when this requirement is not satisfied, that is, a is large. A more challenging problem is to contrast the effect of a time delay in the boundary condition such as

$$\frac{\partial y}{\partial \nu}(x,t) = -ky_t(x,t) + k_0 y_t(x,t-\tau), \quad \text{on } \Gamma_1 \times (0,\infty),$$

with an interior damping (global or local) in the wave equation.

3.7 Abstract-Wave Equation with Delay and Kelvin-Voigt Damping: Exponential Stabilization

In this section, we consider a stabilization problem via a Kelvin-Voigt damping for an abstract-wave equation with delay. We prove an exponential stability result for appropriate damping coefficients. The proof of the main result is based on the resolvent method.

3.7.1 Introduction

Our main goal is to study the internal stabilization of a delayed abstract-wave equation with a Kelvin-Voigt damping. More precisely, given a constant time delay $\tau > 0$, we consider the following system:

$$u''(t) + a\, BB^*u'(t) + BB^*u(t-\tau) = 0, \qquad \text{in} \quad (0, +\infty), \tag{3.7.1}$$

$$u(0) = u_0, \quad u'(0) = u_1, \tag{3.7.2}$$

$$B^*u(t-\tau) = f_0(t-\tau), \qquad \text{in} \quad (0, \tau), \tag{3.7.3}$$

where $a > 0$ is a constant, $B : D(B) \subset H_1 \to H$ is a linear unbounded operator from a Hilbert space H_1 into another Hilbert space H equipped with the respective norms $||\cdot||_{H_1}, ||\cdot||_H$ and inner products $(\cdot, \cdot)_{H_1}, (\cdot, \cdot)_H$, and $B^* : D(B^*) \subset H \to H_1$ is the adjoint of B. The initial datum (u_0, u_1, f_0) belongs to a suitable space to be specified later.

We assume that the operator B^* satisfies the following coercivity assumption: There exists $C > 0$ such that

$$\left\| B^*v \right\|_{H_1} \geq C\, \|v\|_H, \; \forall\, v \in D(B^*). \tag{3.7.4}$$

For shortness, we set $V = D(B^*)$ and then assume that it is closed with the norm $\|v\|_V := \|B^*v\|_{H_1}$ and that it is compactly embedded into H.

Note that the above system is exponentially stable in the absence of time delay, and if $a > 0$. On the other hand, if $a = 0$ and $-BB^*$ corresponds to the Laplace operator with Dirichlet boundary conditions in a bounded domain of $\mathbb{R}^n$, the problem (3.7.1)–(3.7.3) is not well-posed (see [89, 120, 164, 166]). Therefore, in order to restitute the well-posedness character and its stability, we propose, in this section, to add the Kelvin-Voigt damping term $a\, BB^*u'$. Furthermore, the stabilization of (3.7.1)–(3.7.3) is performed using the resolvent method combined with a precise spectral analysis.

This section is organized as follows. The second subsection deals with the well-posedness of our problem, while, in the third subsection, we perform the spectral analysis of the associated operator. In Sect. 3.7.4, we prove the exponential stability of the system (3.7.1)–(3.7.3)if $\tau \leq a$. In the last subsection, we give an example as an application.

3.7.2 Existence Results

In this subsection, we will prove a well-posedness result for the problem (3.7.1)–(3.7.3) via semigroups theory (see Sect. 2.3).

Inspired by [149, 184], we introduce the auxiliary variable

$$z(\rho, t) = B^* u(t - \tau\rho), \qquad \rho \in (0, 1),\ t > 0. \tag{3.7.5}$$

Then, the system (3.7.1)–(3.7.3) has the equivalent form

$$u''(t) + a\, BB^* u'(t) + Bz(1, t) = 0, \qquad \text{in} \quad (0, +\infty), \tag{3.7.6}$$

$$\tau z_t(\rho, t) + z_\rho(\rho, t) = 0 \qquad \text{in} \quad (0, 1) \times (0, +\infty), \tag{3.7.7}$$

$$u(0) = u_0, \quad u'(0) = u_1, \tag{3.7.8}$$

$$z(\rho, 0) = f_0(-\rho\tau), \qquad \text{in} \quad (0, 1), \tag{3.7.9}$$

$$z(0, t) = B^* u(t), \qquad t > 0. \tag{3.7.10}$$

Denoting

$$U := \left(u, u', z\right),$$

we have

$$U' := \left(u', u'', z_t\right) = \left(u', -aBB^*u' - Bz(1, t), -\tau^{-1} z_\rho\right).$$

Now, we define a Hilbert space

$$\mathscr{H} := D(B^*) \times H \times L^2(0, 1; H_1), \tag{3.7.11}$$

equipped with the standard inner product

$$((u, v, z), (u_1, v_1, z_1))_{\mathscr{H}} = (B^* u, B^* u_1)_{H_1} + (v, v_1)_H + \xi \int_0^1 (z, z_1)_{H_1}\, d\rho,$$

where $\xi > 0$ is a parameter that will be fixed later on.

Thereafter, (3.7.6)–(3.7.10) can be rewritten as follows:

$$\begin{cases} U' = \mathscr{A} U, \\ U(0) = (u_0, u_1, f_0(- \cdot \tau)), \end{cases} \tag{3.7.12}$$

where the operator $\mathscr{A}$ is defined by

$$\mathscr{A}\,(u, v, z,) := \Big(v, -aBB^*v - Bz(\cdot, 1), -\tau^{-1}z_\rho\Big),$$

with domain

$$D(\mathscr{A}) := \Big\{ (u, v, z) \in D(B^*) \times D(B^*) \times H^1(0, 1; H_1);\; aB^*v + z(1) \in D(B),\, B^*u = z(0)\Big\}. \tag{3.7.13}$$

We will show that $\mathscr{A}$ generates a C_0-semigroup on $\mathscr{H}$ by proving that $\mathscr{A} - cId$ is maximal dissipative for an appropriate choice of c in terms of ξ, τ and a. Indeed, we prove the following result.

Lemma 3.7.1 *If $\xi > \frac{2\tau}{a}$, then there exists $a^* > 0$ such that $\mathscr{A} - a_*^{-1}Id$ is maximal dissipative in $\mathscr{H}$.*

Proof Take $U = (u, v, z) \in D(\mathscr{A})$. Then we have

$$(\mathscr{A}(u, v, z), (u, v, z))_{\mathscr{H}} = (B^*v, B^*u)_{H_1} - (B(aB^*v + z(1)), v)_H - \xi\tau^{-1}\int_0^1 (z_\rho, z)_{H_1}\, d\rho.$$

Hence, we get

$$(\mathscr{A}(u, v, z), (u, v, z))_{\mathscr{H}} = (B^*v, B^*u)_{H_1} - (aB^*v + z(1), B^*v)_{H_1} - \frac{\xi}{2\tau}\|z(1)\|_{H_1}^2 + \frac{\xi}{2\tau}\|z(0)\|_{H_1}^2.$$

Recalling that $z(0) = B^*u$ and using Young's inequality, we find that

$$\Re e(\mathscr{A}(u, v, z), (u, v, z))_{\mathscr{H}} \le (\varepsilon - a)\|B^*v\|_{H_1}^2 + (\tfrac{1}{2\varepsilon} - \tfrac{\xi}{2\tau})\|z(1)\|_{H_1}^2 + (\tfrac{1}{2\varepsilon} + \tfrac{\xi}{2\tau})\|B^*u\|_{H_1}^2.$$

Choosing $\varepsilon = \frac{a}{2}$, we obtain

$$\Re e(\mathscr{A}(u, v, z), (u, v, z))_{\mathscr{H}} \le -\frac{a}{2}\|B^*v\|_{H_1}^2 + (\frac{1}{a} - \frac{\xi}{2\tau})\|z(1)\|_{H_1}^2 + (\frac{1}{a} + \frac{\xi}{2\tau})\|B^*u\|_{H_1}^2.$$

The choice of ξ is equivalent to $\frac{1}{a} - \frac{\xi}{2\tau} < 0$, and therefore for $a_* = \left(\frac{1}{a} + \frac{\xi}{2\tau}\right)^{-1}$,

$$\Re e(\mathscr{A}(u, v, z), (u, v, z))_{\mathscr{H}} \leq -\frac{a}{2} \|B^* v\|_{H_1}^2 + (\frac{1}{a} - \frac{\xi}{2\tau})\|z(1)\|_{H_1}^2 + a_*^{-1} \|B^* u\|_{H_1}^2. \tag{3.7.14}$$

As $\|B^* u\|_{H_1}^2 \leq \|(u, v, z)\|_{\mathscr{H}}^2$, we get

$$\Re e((\mathscr{A} - a_*^{-1} Id)(u, v, z), (u, v, z))_{\mathscr{H}} \leq -\frac{a}{2} \|B^* v\|_{H_1}^2 + (\frac{1}{a} - \frac{\xi}{2\tau})\|z(1)\|_{H_1}^2 \leq 0,$$

which means that $\mathscr{A} - a_*^{-1} Id$ is dissipative.

Let us turn to the maximality of the operator $\mathscr{A}$, that is, $\lambda I - \mathscr{A}$ is surjective for a fixed $\lambda > 0$. Given $(f, g, h) \in \mathscr{H}$, we seek a solution $U = (u, v, z) \in D(\mathscr{A})$ of the equation

$$(\lambda I - \mathscr{A})(u, v, z) = (f, g, h), \tag{3.7.15}$$

that is,

$$\begin{cases} \lambda u - v = f, \\ \lambda v + B(a B^* v + z(1)) = g, \\ \lambda z + \tau^{-1} z_\rho = h. \end{cases} \tag{3.7.16}$$

Suppose that we have found u with the appropriate regularity. Then

$$v = \lambda u - f, \tag{3.7.17}$$

and we can determine z. Indeed, in view of (3.7.13), we have

$$z(0) = B^* u, \tag{3.7.18}$$

while (3.7.16) produces

$$\lambda z(\rho) + \tau^{-1} z_\rho(\rho) = h(\rho) \quad \text{for } \rho \in (0, 1). \tag{3.7.19}$$

Then, (3.7.18) and (3.7.19) give

$$z(\rho) = B^* u e^{-\lambda\rho\tau} + \tau e^{-\lambda\rho\tau} \int_0^\rho h(\sigma) e^{\lambda\sigma\tau} d\sigma. \tag{3.7.20}$$

In particular, we have

$$z(1) = B^* u e^{-\lambda\tau} + z_0, \tag{3.7.21}$$

with $z_0 \in H_1$ defined by

$$z_0 = \tau e^{-\lambda\tau} \int_0^1 h(\sigma) e^{\lambda\sigma\tau} d\sigma. \tag{3.7.22}$$

The expression in (3.7.16) shows that the function u formally verifies

$$\lambda^2 u + B(aB^*(\lambda u - f) + B^* u e^{-\lambda\tau} + z_0) = g + \lambda f,$$

that is,

$$\lambda^2 u + (\lambda a + e^{-\lambda\tau}) BB^* u = g + \lambda f + B(aB^* f) - Bz_0. \tag{3.7.23}$$

In turn, (3.7.23) can be reformulated as

$$(\lambda^2 u + (\lambda a + e^{-\lambda\tau}) BB^* u, w)_H = (g + \lambda f + B(aB^* f) - Bz_0, w)_H, \quad \forall\, w \in V. \tag{3.7.24}$$

Using the definition of the adjoint of B, the latter writes

$$\begin{aligned} &\lambda^2 (u, w)_H + (\lambda a + e^{-\lambda\tau})(B^* u, B^* w)_{H_1} \\ &\quad = (g + \lambda f, w)_H + (aB^* f - z_0, B^* w)_{H_1}, \ \forall\, w \in V. \end{aligned} \tag{3.7.25}$$

Since the left-hand side of (3.7.25) is coercive on $D(B^*)$, Lax-Milgram theorem (see Theorem 2.1.6) ensures the existence and uniqueness of a solution $u \in V$ of (3.7.25). Once u is obtained, we can define v by (3.7.17), which belongs to V, while z is defined by (3.7.20), which belongs to $H^1(0, 1; H_1)$. Hence, we can set $r = aB^* v + z(1)$. The latter belongs to H_1, but owing to (3.7.25), it fulfills

$$\lambda(v, w)_H + (r, B^* w)_{H_1} = (g, w)_H, \ \forall w \in D(B^*),$$

or equivalently

$$(r, B^* w)_{H_1} = (g - \lambda v, w)_H, \ \forall w \in D(B^*).$$

As $g - \lambda v \in H$, this implies that r belongs to $D(B)$ with

$$Br = g - \lambda v.$$

This shows that the triple $U = (u, v, z)$ belongs to $D(\mathscr{A})$ and satisfies (3.7.15). Thereby, $\lambda I - \mathscr{A}$ is surjective for every $\lambda > 0$. □

We have the following result:

Proposition 3.7.1 *The system* (3.7.1)–(3.7.3)*is well-posed. More precisely, for every* $(u_0, u_1, f_0) \in \mathscr{H}$*, there exists a unique solution* $(u, v, z) \in C(0, +\infty, \mathscr{H})$ *of* (3.7.12)*. Moreover, if* $(u_0, u_1, f_0) \in D(\mathscr{A})$*, then* $(u, v, z) \in C(0, +\infty, D(\mathscr{A})) \cap C^1(0, +\infty, \mathscr{H})$ *with* $v = u'$*, and* u *is indeed a solution of* (3.7.1)–(3.7.3).

3.7.3 The Spectral Analysis

Since $D(B^*)$ is compactly embedded into H, the operator $BB^* : D(BB^*) \subset H \to H$ has a compact resolvent. Hence, let $(\lambda_k)_{k\in\mathbb{N}^*}$ be the set of eigenvalues of BB^* repeated according to their multiplicity (they are positive real numbers and are such that $\lambda_k \to +\infty$ as $k \to +\infty$), and denote by $(\varphi_k)_{k\in\mathbb{N}^*}$ the corresponding eigenvectors that form an orthonormal basis of H (in particular, for all $k \in \mathbb{N}^*$, $BB^*\varphi_k = \lambda_k\varphi_k$).

3.7.3.1 The Discrete Spectrum

We have the following lemma.

Lemma 3.7.2 *If* $\tau \leq a$*, then any eigenvalue* λ *of* $\mathscr{A}$ *satisfies* $\Re e\lambda < 0$.

Proof Let $\lambda \in \mathbb{C}$ and $U = (u, v, z) \in D(\mathscr{A})$ be such that

$$(\lambda I - \mathscr{A})\,(u, v, z) = 0,$$

or equivalently

$$\begin{cases} v = \lambda u, \\ -B(aB^*v + z(\cdot, 1)) = \lambda v, \\ -\tau^{-1} z_\rho = \lambda z. \end{cases} \tag{3.7.26}$$

By (3.7.18), we find that

$$z(\rho) = \lambda^{-1} B^* v e^{-\lambda\rho\tau}. \tag{3.7.27}$$

Using this property in (3.7.26), we find that $u \in D(B^*)$ is solution of

$$\lambda^2 u + (a\lambda + e^{-\lambda\tau}) BB^* u = 0.$$

Hence, a nontrivial solution exists if and only if there exists $k \in \mathbb{N}^*$ such that

$$\frac{\lambda^2}{a\lambda + e^{-\lambda\tau}} = -\lambda_k. \tag{3.7.28}$$

This condition implies that λ does not belong to

$$\Sigma := \{\lambda \in \mathbb{C} : a\lambda + e^{-\lambda\tau} = 0\} \tag{3.7.29}$$

and that

$$e^{-\lambda\tau} + \frac{\lambda^2}{\lambda_k} + a\lambda = 0. \tag{3.7.30}$$

Writing $\lambda = x + iy$, with $x, y \in \mathbb{R}$, we notice that this identity is equivalent to

$$e^{-\tau x}\cos(\tau y) + \frac{x^2 - y^2}{\lambda_k} + ax = 0, \tag{3.7.31}$$

$$-e^{-\tau x}\sin(\tau y) + \frac{2xy}{\lambda_k} + ay = 0. \tag{3.7.32}$$

The second equation (3.7.32) is equivalent to

$$e^{\tau x}\Big(\frac{2x}{\lambda_k} + a\Big)y = \sin(\tau y).$$

Let us consider two cases:
Case 1: $\mathbf{y \neq 0}$. The last equation becomes

$$\frac{e^{\tau x}}{\tau}\Big(\frac{2x}{\lambda_k} + a\Big) = \frac{\sin(\tau y)}{\tau y}.$$

As the modulus of the right-hand side is less than or equal to 1, we obtain

$$\Big|\frac{e^{\tau x}}{\tau}\Big(\frac{2x}{\lambda_k} + a\Big)\Big| \leq 1,$$

or equivalently

$$\Big|\frac{2x}{\lambda_k} + a\Big| \leq \tau e^{-\tau x}.$$

Therefore, if $x \geq 0$, we find that

$$\frac{2x}{\lambda_k} + a \leq \tau e^{-\tau x} \leq \tau,$$

which implies that

$$\frac{2x}{\lambda_k} \leq \tau - a.$$

For $\tau < a$, we reach a contradiction. For $\tau = a$, the sole possibility is $x = 0$, and by (3.7.32), we find that

$$\sin(\tau y) = \tau y,$$

which yields $y = 0$, and once again we obtain a contradiction.
Case 1: $\mathbf{y = 0}$. Clearly, (3.7.32) always holds, and (3.7.31) is equivalent to

$$e^{-\tau x} = -x(\frac{x}{\lambda_k} + a).$$

This equation has no nonnegative solutions x. Indeed, if $x \geq 0$, then the left-hand side is positive, while the right-hand side is nonpositive; hence, again if a solution x exists, it has to be negative.

The proof of the lemma is complete. □

If $a < \tau$, we now show that there exist some pairs of (a, τ) for which the system (3.7.1)–(3.7.3)becomes unstable. Hence, the condition $\tau \leq a$ is optimal for the stability of this system.

Lemma 3.7.3 *There exist pairs of* (a, τ) *such that* $0 < a < \tau$ *and for which the associated operator* $\mathscr{A}$ *has a pure imaginary eigenvalue.*

Proof As we look for a purely imaginary eigenvalue iy of $\mathscr{A}$, the system (3.7.31)–(3.7.32)reduces to

$$\cos(\tau y) = \frac{y^2}{\lambda_k}, \tag{3.7.33}$$

$$\sin(\tau y) = ay. \tag{3.7.34}$$

Such a system has a solution that exists if

$$\frac{y^4}{\lambda_k^2} + a^2 y^2 = 1. \tag{3.7.35}$$

One solution of this equation is

$$y_k = \left(\frac{-a^2\lambda_k^2 + \sqrt{a^4\lambda_k^4 + 4\lambda_k^2}}{2} \right)^{\frac{1}{2}}.$$

We now take any $\tau \in (0, \frac{\pi}{2y_k})$ and $a = \frac{\sin(\tau y_k)}{y_k}$. Then (3.7.34) holds, while (3.7.33) is valid due to (3.7.35)(as $\cos(\tau y_k) > 0$). Finally, $a < \tau$ because

$$\frac{a}{\tau} = \frac{\sin(\tau y_k)}{\tau y_k} < 1.$$

Therefore, with such a choice of a and τ, the operator $\mathscr{A}$ has a purely imaginary eigenvalue equal to $i y_k$. □

3.7.3.2 The Continuous Spectrum

Inspired by Section 3 of [2], we shall use a Fredholm alternative technique to perform the spectral analysis of the operator $\mathscr{A}$.

Recall that an operator T from a Hilbert space X into itself is called singular if there exists a sequence $u_n \in D(T)$ with no convergent subsequence such that $\|u_n\|_X = 1$ and $Tu_n \to 0$ in X, see [183]. According to Theorem 1.14 of [183], T is singular if and only if its kernel is infinite dimensional or its range is not closed. Let Σ be the set defined in (3.7.29). The following results hold:

Theorem 3.7.1

1. *If $\lambda \in \Sigma$, then $\lambda I - \mathscr{A}$ is singular.*
2. *If $\lambda \notin \Sigma$, then $\lambda I - \mathscr{A}$ is a Fredholm operator of index zero.*

Proof For the proof of the first assertion, let us fix $\lambda \in \Sigma$, and for all $k \in \mathbb{N}^*$, set

$$U_k = (u_k, \lambda u_k, B^* u_k e^{-\lambda\tau\cdot}),$$

with $u_k = \frac{1}{\sqrt{\lambda_k}}\varphi_k$. Then U_k belongs to $D(\mathscr{A})$, and easy calculations yield (due to the assumption $\lambda \in \Sigma$)

$$(\lambda I - \mathscr{A})U_k = \lambda^2(0, u_k, 0).$$

Therefore, we deduce that

$$\|(\lambda I - \mathscr{A})U_k\|_{\mathscr{H}} \to 0, \text{ as } k \to \infty.$$

Moreover, due to the property $\|B^* u_k\|_{H_1} = 1$, there exist positive constants c, C, such that

$$c \le \|U_k\|_{\mathscr{H}} \le C, \forall\, k \in \mathbb{N}^*.$$

This shows that $\lambda I - \mathscr{A}$ is singular.

For all $\lambda \in \mathbb{C}$, introduce the (linear and continuous) mapping A_λ from V into its dual by

$$\langle A_\lambda v, w\rangle_{V'-V} = \lambda^2 (v, w)_H + (a\lambda + e^{-\lambda\tau})(B^* v, B^* w)_{H_1}, \ \forall\, v, w \in D(B^*).$$

Then from the proof of Lemma 3.7.1, we know that A_λ is an isomorphism for $\lambda > 0$.

Now for $\lambda \in \mathbb{C} \setminus \Sigma$, we can introduce the operator

$$B_\lambda = (a\lambda + e^{-\lambda\tau})^{-1} A_\lambda.$$

Hence, for $\lambda \in \mathbb{C} \setminus \Sigma$, A_λ is a Fredholm operator of index 0 if and only if B_λ is a Fredholm operator of index 0. Furthermore, for $\lambda, \mu \in \mathbb{C} \setminus \Sigma$ as $B_\lambda - B_\mu$ is a multiple of the identity operator, due to the compact embedding of V into V', and since B_μ is an isomorphism for $\mu > 0$, we finally deduce that A_λ is a Fredholm operator of index 0 for all $\lambda \in \mathbb{C} \setminus \Sigma$.

Subsequently, one can readily check that, for any $\lambda \in \mathbb{C} \setminus \Sigma$, we have the equivalence

$$u \in \ker A_\lambda \iff (u, \lambda u, B^* u e^{-\lambda\tau \cdot}) \in \ker(\lambda I - \mathscr{A}). \tag{3.7.36}$$

This equivalence implies that

$$\dim\, \ker(\lambda I - \mathscr{A}) = \dim\, \ker A_\lambda, \ \forall\, \lambda \in \mathbb{C} \setminus \Sigma. \tag{3.7.37}$$

For the range property for all $\lambda \in \mathbb{C} \setminus \Sigma$, we introduce the inner product

$$(u, z)_{\lambda, V} := \Big((u, \lambda u, B^* u e^{-\lambda\tau \cdot}), (z, \lambda z, B^* z e^{-\lambda\tau \cdot})\Big)_{\mathscr{H}},$$

on V whose associated norm is equivalent to the standard one.

Denote by $\{y^{(i)}\}_{i=1}^N$ an orthonormal basis of $\ker A_\lambda$ for this new inner product (for shortness, the dependence of λ is dropped), i.e.

$$(y^{(i)}, y^{(j)})_{\lambda, V} = \delta_{ij}, \forall\, i, j = 1, \ldots, N.$$

Finally, for all $i = 1, \ldots, N$, we set

$$Z^{(i)} = (y^{(i)}, \lambda y^{(i)}, B^* y^{(i)} e^{-\lambda\tau \cdot}),$$

the element of $\ker(\lambda I - \mathscr{A})$ associated with $y^{(i)}$ that is orthonormal with respect to the inner product of $\mathscr{H}$.

Let us now show that for all $\lambda \in \mathbb{C} \setminus \Sigma$, the range $\Im m(\lambda I - \mathscr{A})$ of $\lambda I - \mathscr{A}$ is closed. Indeed, consider a sequence $U_n = (u_n, v_n, z_n) \in D(\mathscr{A})$ such that

$$(\lambda I - \mathscr{A}) U_n = F_n = (f_n, g_n, h_n) \to F = (f, g, h) \text{ in } \mathscr{H}. \tag{3.7.38}$$

Without loss of generality, we can assume that

$$(U_n, Z^{(i)})_{\mathscr{H}} = -\alpha_{n,i}, \ \forall\, i = 1, \ldots, N, \tag{3.7.39}$$

where

$$\alpha_{n,i} := ((0, f_n, -\tau e^{-\lambda\tau \cdot} \int_0^{\cdot} h_n e^{\lambda\sigma\tau} d\sigma), Z^{(i)})_{\mathscr{H}}.$$

Indeed, if this is not the case, we can consider

$$\tilde{U}_n = U_n - \sum_{i=1}^{N} \beta_i Z^{(i)}$$

that still belongs to $D(\mathscr{A})$ and satisfies

$$(\lambda I - \mathscr{A})\tilde{U}_n = F_n,$$

as well as

$$(\tilde{U}_n, Z^{(i)})_{\mathscr{H}} = -\alpha_{n,i}, \ \forall\, i = 1, \ldots, N,$$

by setting

$$\beta_i = (U_n, Z^{(i)})_{\mathscr{H}} + \alpha_{n,i}, \ \forall\, i = 1, \ldots, N.$$

Note that (3.7.39) is equivalent to

$$(u_n, y^{(i)})_{\lambda,V} = 0, \ \forall\, i = 1, \ldots, N.$$

In other words,

$$u_n \in (\ker A_\lambda)^{\perp_{\lambda,V}}, \tag{3.7.40}$$

where $^{\perp_{\lambda,V}}$ means that the orthogonality is taken with respect to the inner product $(\cdot, \cdot)_{\lambda,V}$.

Returning to (3.7.38), the arguments of the proof of Lemma 3.7.1 imply that

$$A_\lambda u_n = L_{F_n} \text{ in } V',$$

where L_F is defined by

$$L_F(w) := (g, w)_H - \tau e^{-\lambda\tau}(\int_0^1 h(\sigma) e^{\lambda\sigma\tau} d\sigma, B^* w)_{H_1} + (\lambda f + a B^* f, w)_{H_1},$$

when $F = (f, g, h)$. But it is easy to check that

$$L_{F_n} \to L_F \text{ in } V'.$$

Moreover, as $\lambda \in \mathbb{C} \setminus \Sigma$, A_λ is an isomorphism from $(\ker A_\lambda)^{\perp_{\lambda,V}}$ into $\Im m(A_\lambda)$; hence, by (3.7.40), we deduce that there exists a positive constant $C(\lambda)$ such that

$$\|u_n - u_m\|_V \leq C(\lambda)\|L_{F_n} - L_{F_m}\|_{V'}, \forall\, n, m \in \mathbb{N}.$$

Consequently, $(u_n)_n$ is a Cauchy sequence in V, and therefore, there exists $u \in V$ such that

$$u_n \to u \text{ in } V,$$

as well as

$$A_\lambda u = L_F \text{ in } V'.$$

Then defining v by (3.7.17)and z by (3.7.20), we deduce that $U := (u, v, z)$ belongs to $D(\mathscr{A})$ and

$$(\lambda I - \mathscr{A})U = F.$$

In other words, F belongs to $\Im m(\lambda I - \mathscr{A})$. The closedness of $\Im m(\lambda I - \mathscr{A})$ is thus proved.

At this stage, for any $\lambda \in \mathbb{C} \setminus \Sigma$, we show that

$$\operatorname{codim} \Im m(A_\lambda) = \operatorname{codim} \Im m(\lambda I - \mathscr{A}), \tag{3.7.41}$$

where for $W \subset \mathscr{H}$, codim W is the dimension of the orthogonal in $\mathscr{H}$ of W, while for $W' \subset V'$, codim W' is the dimension of the annihilator

$$A := \{v \in V : \langle v, w\rangle_{V-V'} = 0, \ \forall\, w \in W'\},$$

of W' in V.

Indeed, we set $N = \operatorname{codim} \Im m(A_\lambda)$, and then there exist N elements $\varphi_i \in V$, $i = 1, \ldots, N$, such that

$$f \in \Im m(A_\lambda) \iff f \in V' \text{ and } \langle f, \varphi_i\rangle_{V'-V} = 0, \ \forall\, i = 1, \ldots, N.$$

Consequently, for $F \in \mathscr{H}$, if L_F (which belongs to V') satisfies

$$L_F(\varphi_i) = 0, \ \forall\, i = 1, \ldots, N, \tag{3.7.42}$$

then there exists a solution $u \in V$ of

$$A_\lambda u = L_F \text{ in V'},$$

and the arguments of the proof of Lemma 3.7.1 imply that F is in $\Im m(\lambda I - \mathscr{A})$. Hence, the N conditions on $F \in \mathscr{H}$ from (3.7.42) allow to show that it belongs to $\Im m(\lambda I - \mathscr{A})$, and therefore,

$$\text{codim } \Im m(\lambda I - \mathscr{A}) \leq N = \text{ codim } \Im m(A_\lambda). \tag{3.7.43}$$

This shows that $\lambda I - \mathscr{A}$ is a Fredholm operator.

Conversely, we set $M = \text{ codim } \Im m(\lambda I - \mathscr{A})$, and so there exist M elements $\Psi_i = (u_i, v_i, z_i) \in \mathscr{H}$, $i = 1, \ldots, M$, such that

$$F \in \Im m(\lambda I - \mathscr{A}) \iff F \in \mathscr{H} \text{ and } (F, \Psi_i)_{\mathscr{H}} = 0, \forall\, i = 1, \ldots, M.$$

Then, for any $g \in H$, if

$$(g, v_i)_H = ((0, g, 0), \Psi_i)_{\mathscr{H}} = 0, \forall\, i = 1, \ldots, M, \tag{3.7.44}$$

there exists $U = (u, v, z) \in D(\mathscr{A})$ such that

$$(\lambda I - \mathscr{A})U = (0, g, 0),$$

which implies that

$$A_\lambda u = g.$$

This shows that

$$\Im m(A_\lambda) \supset H_0,$$

where $H_0 := \{g \in H \text{ satisfying (3.7.44)}\}$. This inclusion implies that (here $\perp$ means the annihilator of the set in V)

$$\Im m(A_\lambda)^\perp \subset H_0^\perp.$$

Therefore,

$$\begin{aligned}
\Im m(A_\lambda)^\perp &\subset \{v \in V : \langle v, g\rangle_{V-V'} = 0, \forall\, g \in H_0\} \\
&= \{v \in V : (v, g)_H = 0, \forall\, g \in H_0\} \\
&\subset \text{ Span } \{v_i\}_{i=1}^M \cap V.
\end{aligned}$$

Hence,

$$\operatorname{codim} \Im m(A_\lambda) \leq M = \operatorname{codim} \Im m(\lambda I - \mathscr{A}). \tag{3.7.45}$$

The inequalities (3.7.43) and (3.7.45) imply (3.7.41). □

Lemma 3.7.4 *If $\tau \leq a$, then*

$$\Sigma \subset \{\lambda \in \mathbb{C} : \Re e\lambda < 0\}.$$

Proof Let $\lambda = x + iy \in \Sigma$, with $x, y \in \mathbb{R}$, and we deduce that

$$ax + e^{-\tau x}\cos(\tau y) = 0,$$
$$ay - e^{-\tau x}\sin(\tau y) = 0.$$

This corresponds to the system (3.7.31)–(3.7.32) with $k = \infty$, and hence, the arguments as in the proof of Lemma 3.7.2 yield the result. □

Corollary 3.7.1 *It holds*

$$\sigma(\mathscr{A}) = \sigma_p(\mathscr{A}) \cup \Sigma,$$

and therefore, if $\tau \leq a$,

$$\sigma(\mathscr{A}) \subset \{\lambda \in \mathbb{C} : \Re e\lambda < 0\}.$$

Proof By Theorem 3.7.1,

$$\mathbb{C} \setminus \Sigma \subset \sigma_p(\mathscr{A}) \cup \rho(\mathscr{A}).$$

The first assertion is obvious.

The second assertion follows from Lemmas 3.7.2 and 3.7.4. □

3.7.4 Asymptotic Behavior

In this subsection, we show that if $\tau \leq a$ and $\xi > \frac{2\tau}{a}$, the semigroup $e^{t\mathscr{A}}$ decays to the null steady state with an exponential decay rate. To obtain this, our technique is based on a frequency-domain approach and combines a contradiction argument to carry out a special analysis of the resolvent.

Theorem 3.7.2 *If $\xi > \frac{2\tau}{a}$ and $\tau \leq a$, then there exist constants $C, \omega > 0$ such that the semigroup $e^{t\mathscr{A}}$ is exponentially stable, that is,*

$$\left\| e^{t\mathscr{A}} \right\|_{\mathscr{L}(\mathscr{H})} \leq C\, e^{-\omega t}, \ \forall\, t > 0. \tag{3.7.46}$$

Proof of Theorem 3.7.2 We will employ Theorem 2.3.8:

First, according to Corollary 3.7.1, the spectrum of $\mathscr{A}$ is fully included in $\Re e\lambda < 0$, which clearly implies (2.3.16). Then, the proof of Theorem 3.7.2 is based on the following lemma that shows that (2.3.17) holds. □

Lemma 3.7.5 *The resolvent operator of $\mathscr{A}$ satisfies the condition*

$$\sup_{\Re e\lambda \geq 0} \|(\lambda I - \mathscr{A})^{-1}\|_{\mathscr{L}(\mathscr{H})} < \infty. \tag{3.7.47}$$

Proof Suppose that (3.7.47) is false. Thereby, Banach-Steinhaus theorem (see Theorem 2.1.7) implies that there exists a sequence of complex numbers λ_n such that $\Re e\lambda_n \geq 0$, $|\lambda_n| \to +\infty$ and a sequence of vectors $Z_n = (u_n, v_n, z_n) \in \mathscr{D}(\mathscr{A})$ with

$$\|Z_n\|_{\mathscr{H}} = 1 \tag{3.7.48}$$

such that

$$||(\lambda_n I - \mathscr{A})Z_n||_{\mathscr{H}} \to 0 \quad \text{as} \quad n \to \infty. \tag{3.7.49}$$

This is equivalent to writing

$$\lambda_n u_n - v_n \equiv f_n \to 0 \ \text{ in } \ D(B^*), \tag{3.7.50}$$

$$\lambda_n v_n + a\, B(B^* v_n + z_n(1)) \equiv g_n \to 0 \ \text{ in } \ H,$$

$$\lambda_n\, z_n + \tau^{-1}\partial_\rho z_n \equiv h_n \to 0 \ \text{ in } \ L^2((0,1); H_1). \tag{3.7.51}$$

Our goal is to derive from (3.7.49) that $||Z_n||_{\mathscr{H}}$ converges to zero, which furnishes a contradiction.

In view of (3.7.14)and (3.7.50), we have

$$\begin{aligned} ||(\lambda_n I - \mathscr{A})Z_n||_{\mathscr{H}} &\geq |\Re e\, ((\lambda_n I - \mathscr{A})Z_n, Z_n)_{\mathscr{H}}| \\ &\geq \Re e\, \lambda_n - a_*^{-1} \left\| B^* u_n \right\|_{H_1}^2 + \left(\frac{\xi}{2\tau} - \frac{1}{a} \right) \|z_n(1)\|_{H_1}^2 + \frac{a}{2} \left\| B^* v_n \right\|_{H_1}^2 \\ &= \Re e\, \lambda_n - a_*^{-1} \left\| \frac{B^* v_n + B^* f_n}{\lambda_n} \right\|_{H_1}^2 + \left(\frac{\xi}{2\tau} - \frac{1}{a} \right) \|z_n(1)\|_{H_1}^2 + \frac{a}{2} \left\| B^* v_n \right\|_{H_1}^2 . \end{aligned}$$

Hence, using the inequality

$$\left\| B^* v_n + B^* f_n \right\|_{H_1}^2 \leq 2\| B^* v_n \|_{H_1}^2 + 2\| B^* f_n \|_{H_1}^2,$$

we find that

$$||(\lambda_n I - \mathscr{A}) Z_n||_{\mathscr{H}} \geq \Re e \, \lambda_n - 2a_*^{-1} |\lambda_n|^{-2} \left\| B^* f_n \right\|_{H_1}^2 + \left(\frac{\xi}{2\tau} - \frac{1}{a} \right) \|z_n(1)\|_{H_1}^2$$
$$+ (\frac{a}{2} - 2a_*^{-1} |\lambda_n|^{-2}) \left\| B^* v_n \right\|_{H_1}^2 .$$

Therefore, for n large enough, say $n \geq n^*$, we can suppose that

$$\frac{a}{2} - 2a_*^{-1} |\lambda_n|^{-2} \geq \frac{a}{4},$$

and hence for all $n \geq n^*$, we get

$$||(\lambda_n I - \mathscr{A}) Z_n||_{\mathscr{H}} \geq \Re e \, \lambda_n - 2a_*^{-1} |\lambda_n|^{-2} \left\| B^* f_n \right\|_{H_1}^2 + \left(\frac{\xi}{2\tau} - \frac{1}{a} \right) \|z_n(1)\|_{H_1}^2$$
$$+ \frac{a}{4} \left\| B^* v_n \right\|_{H_1}^2 .$$

Thanks to this estimate, together with (3.7.49)and (3.7.50), we deduce that

$$z_n(1) \to 0, \; B^* v_n \to 0, \; \text{ in } H_1, \text{ as } n \to \infty, \tag{3.7.52}$$

and in particular, from the coercivity property (3.7.4), we have

$$v_n \to 0, \text{ in } H, \text{ as } n \to \infty.$$

This implies, according to (3.7.50), that

$$u_n = \frac{1}{\lambda_n} v_n + \frac{1}{\lambda_n} f_n \to 0, \quad \text{in} \quad D(B^*), \text{ as } n \to \infty, \tag{3.7.53}$$

as well as

$$z_n(0) = B^* u_n \to 0, \quad \text{in } H_1, \text{ as } n \to \infty. \tag{3.7.54}$$

Integrating the identity (3.7.51), we get

$$z_n(\rho) = z_n(0) \, e^{-\tau \lambda_n \rho} + \tau \int_0^\rho e^{-\tau \lambda_n (\rho - \gamma)} \, h_n(\gamma) \, d\gamma. \tag{3.7.55}$$

Recalling that $\Re e\lambda_n \geq 0$, we have

$$\int_0^1 \|z_n(\rho)\|_{H_1}^2 \, d\rho \leq 2\|z_n(0)\|_{H_1}^2 + 2\tau^2 \int_0^1 \int_0^\rho \|h_n(\gamma)\|_{H_1}^2 \, d\gamma \rho \, d\rho \to 0, \text{ as } n \to \infty.$$

Whereupon, we have shown that $\|Z_n\|_{\mathscr{H}}$ converges to zero, which clearly contradicts $\|Z_n\|_{\mathscr{H}} = 1$. □

The two hypotheses of Theorem 2.3.8 are proved, and then (3.7.46) holds. The proof of Theorem 3.7.2 is then completed.

3.7.5 *Application to the Stabilization of the Wave Equation with Delay and Kelvin-Voigt Damping*

We study the internal stabilization of a delayed wave equation given by

$$u_{tt}(x,t) - a\,\Delta u_t(x,t) - \Delta u(x,t-\tau) = 0, \quad \text{in} \quad \Omega \times (0,+\infty), \tag{3.7.56}$$

$$u = 0, \quad \text{on} \quad \partial\Omega \times (0,+\infty), \tag{3.7.57}$$

$$u(x,0) = u_0(x), \quad u_t(x,0) = u_1(x), \quad \text{in} \quad \Omega, \tag{3.7.58}$$

$$\nabla u(x,t-\tau) = f_0(t-\tau), \quad \text{in} \quad \Omega \times (0,\tau), \tag{3.7.59}$$

where Ω is a smooth open bounded domain of $\mathbb{R}^n$ and $a, \tau > 0$ are constants.

This problem fits in our abstract framework with $H = L^2(\Omega)$ and

$$B := -div : D(B) = H^1(\Omega)^n \to L^2(\Omega),$$

$$B^* := \nabla : D(B^*) = H_0^1(\Omega) \to H_1 := L^2(\Omega)^n.$$

The assumption (3.7.4)is satisfied due to Poincaré's inequality. Moreover, the operator $\mathscr{A}$ is then given by

$$\mathscr{A}(u,v,z) := \Big(v, a\Delta v + \operatorname{div} z(\cdot,1), -\tau^{-1} z_\rho\Big),$$

with domain

$$D(\mathscr{A}) := \Big\{ (u,v,z) \in H_0^1(\Omega) \times H_0^1(\Omega) \times L^2(\Omega; H^1(0,1));\ a\nabla v + z(\cdot,1) \in H^1(\Omega),\ \nabla u = z(\cdot,0) \text{ in } \Omega \Big\}, \tag{3.7.60}$$

in the Hilbert space

$$\mathscr{H} := H_0^1(\Omega) \times L^2(\Omega) \times L^2(\Omega \times (0,1)). \tag{3.7.61}$$

According to Lemma 3.7.1 and Theorem 3.7.2, we have:

Corollary 3.7.2 *If* $\tau \leq a$, *the system* (3.7.56)–(3.7.59)*is exponentially stable in* $\mathscr{H}$, *namely for* $\xi > \frac{2\tau}{a}$, *the energy*

$$E(t) = \frac{1}{2}\left(\int_\Omega (|\nabla u(x,t)|^2 + |u_t(x,t)|^2)\,dx + \xi \int_\Omega \int_0^1 |\nabla u(x, t-\tau\rho)|^2\,dx d\rho\right),$$

satisfies

$$E(t) \leq M e^{-\omega t} E(0), \ \forall\, t > 0,$$

for some positive constants M *and* ω.

3.7.6 Conclusion

By a careful spectral analysis combined with a frequency-domain approach, we have shown that the system (3.7.1)–(3.7.3)is exponentially stable if $\tau \leq a$ and that this condition is optimal. But from the general form of (3.7.1), we can only consider interior Kelvin-Voigt damping. Hence, an interesting perspective is to consider the wave equation with dynamical Ventcel boundary conditions with a delayed term and a Kelvin-Voigt damping as done for a particular case in Sect. 3.3.

3.8 Conclusion

This chapter was *mainly* concerned with the long-time behavior of the multi-dimensional wave equation under the presence of a boundary or internal delay term, but at the same time without the presence of any displacement term. The problem is shown to have a unique solution in an appropriate functional space. Then, applying LaSalle's invariance principle, the solutions are proved to converge asymptotically to a well-defined equilibrium state. Additionally, combining an interpolation inequality combined with a resolvent method, the convergence rate is shown to be of logarithmic type without any extra geometric assumption in the control zone. In turn, an exponential convergence outcome has been established by means of the BLR condition. Furthermore, an internal delayed wave equation is considered, where the three-dimensional domain possesses trapped ray, and hence, the BLR geometric condition [32, 133] does not hold. Moreover, the internal damping is localized. In such a situation, polynomial convergence results are established. These two findings improve previous results in [162, 175]. We also studied the long-time behavior of the space-varying wave equation with an internal control of either memory-type or delayed-type control. The long-time behavior of numerous physical systems with memory is also briefly discussed. Furthermore, the energy method has been used to show that the effect of an interior delayed control can be nullified by the action of an appropriate boundary feedback. Finally, an abstract-wave equation with delay and a Kelvin-Voigt damping has been considered. Using the resolvent method, an exponential stability result has been shown for appropriate damping coefficients.

Chapter 4
Stabilization of a 2D Overhead Crane

4.1 Background

This chapter is devoted to the stability and stabilization of the so-called 2D overhead crane by taking into account the presence of a time delay or memory. The latter consists of a motorized platform of mass m moving along a horizontal rail. A flexible cable of length ℓ, holding a load mass M, is attached to the platform (see Fig. 4.1).

Overhead cranes are extensively utilized in a wide variety of industrial and construction sites as well as in freight handling. They have different designs and sizes. Some consist of a hook, a hoisting mechanism such as a cable, and a support mechanism such as a girder (trolley) [3]. Very often, the objective of using cranes is to horizontally transport point to point a suspended mass/load. Furthermore, it is well-known that cables possess the inherent flexibility characteristics and can develop only tension [3]. Such natural features inevitably cause deflection in transversal direction of the cable. It is also noted that, on one hand, the suspended load is always subject to swings due to several reasons such as strong winds or waves. Thereby, the behavior of the overhead crane system with flexible cable can generate complex system dynamics (see [3] for more details). On the other hand, the stabilization of a crane in practice is not a simple task and requires the most suitable controls in order to reach the desired result while ensuring safety. It is also important to mention that a wide class of cranes is used depending on the need. This obviously leads to numerous mathematical models and a fortiori to different dynamics.

In this chapter, it is assumed that:

(i) The cable is completely flexible and nonstretching.
(ii) The length of the cable is constant.
(iii) Transversal and angular displacements are small.
(iv) Friction is neglected.
(v) The masses m and M are point masses.
(vi) The angle of the cable with respect to the vertical x-axis is small everywhere.

K. Ammari, B. Chentouf, *Stabilization of Wave-Type Equations with Delay or Memory*, Advances in Mechanics and Mathematics 57,
https://doi.org/10.1007/978-3-032-28496-9_4

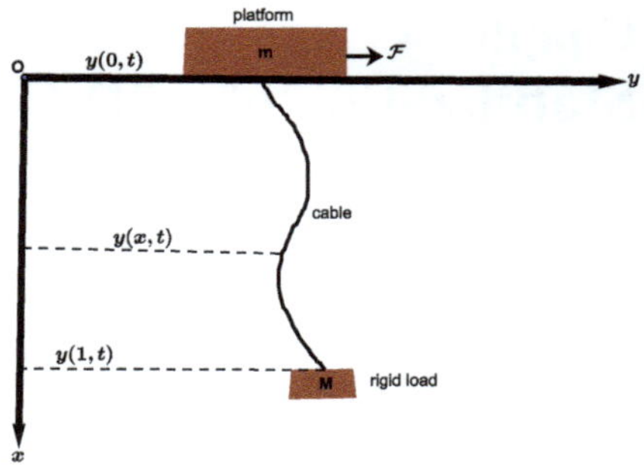

Fig. 4.1 The overhead crane model

Under the above assumptions, the overhead crane is modeled by a hybrid PDE-ODE system (see also for similar models [92, 93] and [172]). For sake of completeness, we shall discuss some details related to the derivation of such a model (the reader is referred to [24] and [167] for further elaborations).

We shall denote the curvilinear coordinate representing the arclength along the cable by x, while $p(x,t)$ is the position of the point at time t and coordinate x. Let $T(x,t)$ and ρ be, respectively, the tension along the cable and the mass density, $\theta(x,t)$ be the angle between T and the x-axis, and consider a portion of the cable of length Δx. Then, second Newton's law leads to the following equation on the y-axis:

$$\rho p_{tt}(x,t)\Delta x = \rho g \Delta x + T(x+\Delta x,t)\theta(x+\Delta x,t) - T(x,t)\theta(x,t).$$

We can write $\theta(x,t) \simeq p_x(x,t)$ as a result of the assumption of smallness of transversal and angular displacements. On the other hand, since the cable tension is essentially due to the action on its lower part, we have

$$y_{tt}(x,t) - (a y_x)_x\,(x,t) = 0,\ 0 < x < \ell,\ t > 0, \tag{4.1.1}$$

in which $y = p - p_{ref}$ (p_{ref} being the equilibrium point) and

$$a(x) = g\left(\frac{M}{\rho} + \ell - x\right).$$

We turn now to the equation of the platform part of the system (see Fig. 4.2). Taking into account the external controlling force $\mathscr{F}(t)$, we have

$$m y_{tt}(0,t) = T(0)\theta(0,t) + \mathscr{F}(t),$$

which can be rewritten as

$$m y_{tt}(0,t) = a(0) y_x(0,t) + \mathscr{F}(t),\ t > 0, \tag{4.1.2}$$

as $T(x) = a(x)$ and $\theta(0,t) \simeq y_x(0,t)$.

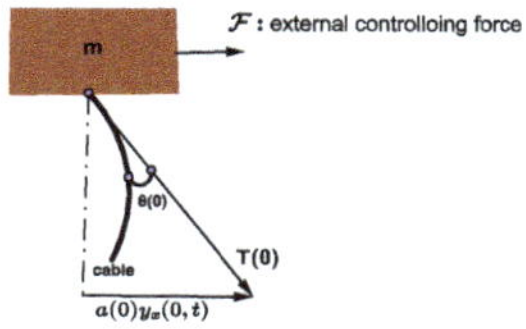

Fig. 4.2 The platform

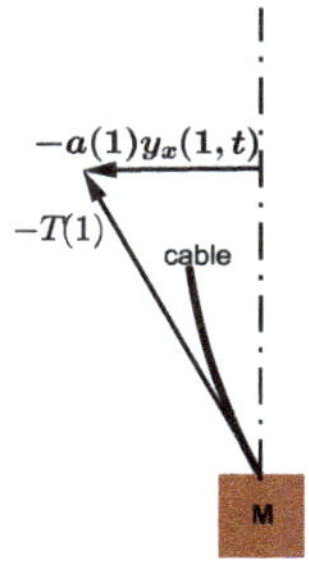

Fig. 4.3 The load mass

Using similar arguments for the load mass (see Fig. 4.3), we obtain

$$M y_{tt}(\ell, t) = -a(\ell) y_x(\ell, t),\ t > 0. \tag{4.1.3}$$

Combining (4.1.1)–(4.1.3), we reach the final form of the system

$$\begin{cases} y_{tt}(x,t) - (a y_x)_x (x,t) = 0, & 0 < x < \ell,\ t > 0, \\ m y_{tt}(0,t) - (a y_x)(0,t) = \mathscr{F}(t), & t > 0, \\ M y_{tt}(\ell,t) + (a y_x)(\ell,t) = 0, & t > 0, \end{cases} \tag{4.1.4}$$

where $a(x)$ is supposed to satisfy the following conditions:

$$\begin{cases} a \in H^1(0,\ell); \\ \text{there exists a positive constant } a_0 \text{ such that } a(x) \geq a_0 > 0 \ \text{ for all } \ x \in [0,\ell]. \end{cases} \tag{4.1.5}$$

It is important to point out that although $a(x)$ is originally equal to $T(x) = g\left(\frac{M}{\rho} + \ell - x\right)$, the latter remains just an approximate expression of $a(x)$. This is because several approximations and assumptions have been made at the beginning of this section. Thereby, it is legitimate to suppose that $a(x)$ obeys the conditions in (4.1.5), which obviously remain fulfilled in the case where $a(x) = g\left(\frac{M}{\rho} + \ell - x\right)$.

As mentioned above, the objective is to seek a delayed control $\mathscr{F}(t)$ depending solely on the velocity so that the solutions of the closed-loop system asymptotically converge to an equilibrium point in a suitable functional space. It is also noteworthy that the open-loop system, i.e., (4.1.4) with $\mathscr{F}(t) = 0$, is unstable. This can be easily verified by taking the energy

$$E(t) = \frac{1}{2}\left\{\int_0^\ell \left(y_t^2(x,t) + a(x) y_x^2(x,t)\right) dx + m y_t^2(0,t) + M y_t^2(\ell,t)\right\}.$$

Then, a formal computation shows that $E_t(t) = 0$, and hence, the energy is conservative. Thereby, it is primordial to design a control $\mathscr{F}(t)$ in order to stabilize the whole structure or at least the cable. For simplicity and without loss of generality, we shall set the length of the cable $\ell = 1$.

The boundary stabilization of the system (4.1.4) has been the object of a considerable mathematical endeavor, and there are basically two categories of research articles: In the first category, at least one of the dynamical terms in the boundary conditions is neglected. In other words, either $my_{tt}(0,t)$ or $My_{tt}(1,t)$ (or even both terms) does not appear in the system. For instance, it has been shown in [167] that the feedback law

$$\mathscr{F}(t) = -cy(0,t) - F(y_t(0,t)),\ c > 0,$$

exponentially stabilizes the system (4.1.4) with $my_{tt}(0,t) = 0$ under appropriate assumptions on the function F. Another stabilization result for the system (4.1.4) with $my_{tt}(0,t) = My_{tt}(1,t) = 0$ has also been established in [77] via the action of the following feedback:

$$\begin{cases} \mathscr{F}(t) = -\alpha y(0,t) - F(y_t(0,t)), \\ \mathscr{U}(t) = -\alpha y(1,t) - F(y_t(1,t)), \quad \alpha > 0, \end{cases}$$

where $\mathscr{U}$ is an additional control that is applied on the load mass. In [24], the asymptotic stabilization has been proved as long as a dynamical control acts on the boundary $y(1,t)$. We also mention that a stabilization result has been obtained in [71] by proposing the feedback law

$$\mathscr{F}(t) = k_p y(0,t) + k_v y_t(0,t) + \int_0^1 G(x)y(x,t)dx + \frac{k_v}{k_p}\int_0^1 G(x)y_t(x,t)dx,$$

with $k_p,\ k_v > 0$ and G is a function in $H^1(0,1)$. Of course, such a result has been established under some conditions on the feedback gains $k_p,\ k_v$ as well as the function G. Similar findings have been proved in [25] for other types of controls containing a displacement term. We conclude this discussion about the first category of articles available in the literature by pointing out (as it has been observed in [70]) that in all references cited above, either the boundary conditions in (4.1.4) or the stabilizing feedback law $\mathscr{F}(t)$ involves the displacement term y. This is mainly due to the fact that most of the authors defined the energy norm of the system by

$$E_0(t) = \frac{1}{2}\int_0^1 \left(y_x^2 + y_t^2\right) dx.$$

This observation has motivated the authors of [70] to add a displacement term in the PDE of (4.1.4) and propose a general class of feedback laws containing only the velocity. In fact, the closed-loop system in [70] has the following form:

$$\begin{cases} y_{tt}(x,t) - (ay_x)_x(x,t) + \alpha y_t(x,t) + \beta y(x,t) = 0, \quad \alpha \geq 0, \ \beta > 0, \ 0 < x < 1, \ t > 0, \\ (ay_x)(0,t) = \epsilon_1 f(y_t(0,t)), \quad t > 0, \\ (ay_x)(1,t) = \epsilon_2 g(y_t(1,t)), \quad t > 0, \end{cases} \tag{4.1.6}$$

in which f and g are two nonlinear functions. The multiplier method has been successfully used in [70] to get the precise decay rate (polynomial or exponential) estimates of the following energy:

$$\hat{E}(t) = \frac{1}{2}\int_0^1 \left(\beta y^2(x,t) + y_t^2(x,t) + a(x)y_x^2(x,t)\right) dx$$

of the system (4.1.6) according to the type of assumptions on the functions f and g. Also, the backstepping approach has been applied to a variant of the system (4.1.4) leading to an exponentially stabilizing boundary feedback controller [25]. In the same spirit, the following feedback law

$$\begin{cases} \mathscr{F}(t) = -\alpha_1 y(0,t) - \beta_1 y_t(0,t), \\ \mathscr{U}(t) = -\alpha_2 y(1,t) - \beta_2 y_t(1,t), \end{cases}$$

has been suggested in [185] in the case where $\alpha_1 + \alpha_2 \neq 0$, $\beta_1 + \beta_2 \neq 0$ and $my_{tt}(0,t) = My_{tt}(1,t) = 0$ and the Riesz basis property has been shown.

The second category of research papers takes into consideration the dynamics of both the load mass and platform mass. Within this context, it has been proved in [74] that the system (4.1.4) can be strongly (but nonexponentially) stabilized by means of the control

$$\mathscr{F}(t) = -\alpha y(0,t) - f(y_t(0,t)), \ \alpha > 0,$$

where f is a suitable function. This means that controls of higher orders are to be designed in order to reach the uniform exponential stability [18, 148] (see also Sect. 3.3). Indeed, the uniform stabilization holds in [148] if

$$\mathscr{F}(t) = -\alpha y(0,t) - \alpha\beta y_t(0,t) + \beta y_{xt}(0,t), \ \alpha > 0, \ \beta > 0.$$

It turned out that the same result can be achieved via the boundary control

$$\mathscr{F}(t) = -\alpha y(0,t) - (\beta + \alpha c)y_t(0,t) + cy_{xt}(0,t),$$

where α, β, and c are positive constants satisfying $\beta c < m$ [148]. Motivated by the work [70], a feedback control depending only on the velocity has been proposed in [76] for the system (4.1.4), and an asymptotic convergence result has been established (see also [1]).

All the papers mentioned above do not take into consideration the presence of a time delay in the system. In turn, it is well-known, as mentioned earlier, that delays are inevitable in practice as they naturally arise in most systems due to numerous reasons such as the time factor needed for the communication among the controllers, the sensors, and the actuators of systems or in some cases due to the dependence of the state variables on past states. Furthermore, recall that it has been noticed that the presence of a delay in a system could be a source of poor performance and instability [81–84] (see also [5, 19, 173]).

The primary concern of this chapter is to investigate the impact of a delay or memory on the stability and stabilization of the crane cable. Different scenarios are considered depending on the type of the control such as boundary, interior, localized, higher order, linear, nonlinear, constant delay, time-dependent delay, etc. Obviously, each scenario leads to a specific type of convergence of solutions to the problem under consideration.

In Sect. 4.2, a delayed boundary control of type velocity is considered. In such an event, we show that the best convergence rate of solutions of the system to the equilibrium state is polynomial. In order to enhance the convergence rate, we add an interior control in Sect. 4.3 and show that the solutions converge exponentially. In turn, the drawback of the additional interior control lies in the fact that it acts on the whole spatial domain. Motivated by practical reasons where the control is often applied on a small portion of the domain, we consider in Sect. 4.4 a localized control damping and show that it is strong enough to reach the desirable exponential stability. In these three sections, the delayed and nondelayed boundary terms are of the same type, namely, of type velocity. This motivated us to treat the case where this situation does not occur. In fact, Sect. 4.5 is dedicated to a study in which the delayed term is still of velocity type, whereas the compensating term is of angular velocity type. To address this scenario, we employ the interior Kelvin-Voigt damping mechanism in addition to another boundary control. Consequently, the feedback law is too costly as it involves several types of controls. In turns out that this stabilizing feedback law may consist of a single boundary control to achieve the exponential stability of the cable. More precisely, in Sect. 4.6, we show that the interior Kelvin-Voigt damping mechanism is not required and neither the higher-order boundary control $(ay_{xt})(1, t)$ nor the velocity type boundary control $y_t(1, t)$. This considerably improves the outcomes of Sect. 4.5. Thereafter, contrary to previous sections, a time-dependent delay is considered in Sect. 4.7. This delay occurs in one of the boundary conditions. We prove that both the cable's dynamics and the coupled dynamics of the cart and suspended load can be exponentially stabilized. Furthermore, an estimate of the decay rate is provided. In Sect. 4.8, we deal with the nonlinear boundary stabilization problem of the crane. Indeed, both delayed control and its compensating one are nonlinear. Due to the mathematical complexity of the problem, we assume that the dynamics of the mass m is neglected. The main result of this section is to provide several decay rates of the solutions depending on the properties of the nonlinearities. Sections 4.9 and 4.10 address the boundary stabilization of the crane system incorporating infinite memory effects. Under suitable assumptions on the memory kernel, we prove uniform exponential stability of the system.

Lastly, note that the long-time behavior of the solutions to the problems treated in this chapter is obtained thanks to the resolvent method, with the exception of Sects. 4.7 and 4.8, where the energy method is utilized. In turn, both methods are invoked in Sect. 4.6.

4.2 Boundary Delay: Polynomial Stabilization

4.2.1 Introduction

This section places primary emphasis on the analysis of the system (4.1.4) under the action of the following input delay:

$$\mathscr{F}(t) = -\beta y_t(0, t) + \alpha y_t(0, t - \tau), \tag{4.2.1}$$

where $\beta > 0$, $\alpha \in \mathbb{R}$, and $\tau > 0$ is the time delay.

It is worth mentioning that the absence of the displacement term in the closed-loop system prevents the utilization of classical Poincaré inequalities. To overcome this difficulty, an appropriate energy norm is suggested as previously done for the multidimensional wave equation.

The main contribution of this section is threefold:

(1) Extend the mathematical outcomes on the overhead crane available in the literature (especially those of [70, 74, 76, 148]), where no delay has been taken into account in the proposed feedback laws.
(2) Demonstrate that, despite the presence of the delay term in the proposed feedback control law, the solutions of the closed-loop system asymptotically converge to a steady state, which depends on the initial conditions.
(3) Provide the rate of convergence of the solutions of the closed-loop system to the equilibrium state, in contrast to the work [76], in which such a result has not been established.

The current section is organized as follows. The next subsection is consecrated to the proof of existence and uniqueness of the solutions to the closed-loop system under a smallness condition on the delay term, which is assumed to be fulfilled throughout the section. Section 4.2.3 deals with the asymptotic behavior of solutions via the use of LaSalle's principle. Section 4.2 is dedicated to the polynomial convergence of solutions. In the event that the smallness condition on the delay term is not satisfied, nonconvergence results are proved in Sect. 4.3.5. This shows that such a smallness condition is optimal. Finally, the section closes with a concluding discussion.

4.2.2 Well-Posedness of the System

With the feedback law in (4.2.1), we consider the closed-loop system

$$
\begin{cases}
y_{tt}(x,t) - (ay_x)_x(x,t) = 0, & 0 < x < 1,\ t > 0,\\
my_{tt}(0,t) - (ay_x)(0,t) = -\beta y_t(0,t) + \alpha y_t(0,t-\tau), & t > 0,\\
My_{tt}(1,t) + (ay_x)(1,t) = 0, & t > 0,\\
y(x,0) = y_0(x),\ \ y_t(x,0) = y_1(x), & x \in (0,1),\\
y_t(0,t-\tau) = f(t-\tau), & t \in (0,\tau),
\end{cases}
\tag{4.2.2}
$$

where a obeys the condition (4.1.5), $\alpha \in \mathbb{R}$, and $\beta > 0$.

Our immediate task is to seek an appropriate energy associated to (4.2.2). To proceed, let

$$
\begin{aligned}
E_0(t) = \frac{1}{2}\Bigg\{ &\int_0^1 \left(y_t^2(x,t) + a(x)y_x^2(x,t)\right) dx + my_t^2(0,t) + My_t^2(1,t)\\
&+K\tau \int_0^1 y_t^2(0,t-x\tau)\, dx \Bigg\},
\end{aligned}
\tag{4.2.3}
$$

where K is a positive constant. Using (4.2.2) and integrating by parts, a formal computation yields

$$
E_0'(t) = -\beta y_t^2(0,t) + \alpha y_t(0,t)y_t(0,t-\tau) - \frac{K}{2}\left(y_t^2(0,t-\tau) - y_t^2(0,t)\right).
\tag{4.2.4}
$$

Applying Young's inequality, the latter becomes

$$
E_0'(t) \le \left(\frac{K}{2} + \frac{|\alpha|}{2c} - \beta\right) y_t^2(0,t) + \frac{1}{2}\left(|\alpha|c - K\right) y_t^2(0,t-\tau),
\tag{4.2.5}
$$

for any positive constant c. Subsequently, we introduce the following additional energy functional (analogously to the multidimensional wave equation in Chap. 3)

$$
E_1(t) = \frac{1}{2}\rho^2(t),
\tag{4.2.6}
$$

where

$$
\begin{aligned}
\rho(t) = &\int_0^1 y_t(x,t)dx + c_1 y_t(0,t) + c_2 y_t(1,t)\\
&+ c_3 \int_0^1 y_t(0,t-x\tau)\, dx + c_4 y(0,t),
\end{aligned}
\tag{4.2.7}
$$

and c_1, c_2, c_3, and c_4 are constants to be determined. Following the same arguments as for $E_0(t)$, we get

$$\begin{aligned} E_1'(t) = \rho(t)\Big((ay_x)(1,t)\left[1-\frac{c_2}{M}\right] + (ay_x)(0,t)\left[\frac{c_1}{m}-1\right] \\ + \beta y_t(0,t)\left[\frac{c_3}{\tau\beta}+\frac{c_4}{\beta}-\frac{c_1}{m}\right] \\ + \alpha y_t(0,t-\tau)\left[\frac{c_1}{m}-\frac{c_3}{\tau\alpha}\right]\Big). \end{aligned} \tag{4.2.8}$$

Thereafter, we define the total energy of the system (4.2.2) as follows:

$$\mathscr{E}(t) = E_0(t) + E_1(t). \tag{4.2.9}$$

This, together with (4.2.5) and (4.2.8), implies that

$$\begin{aligned} \mathscr{E}'(t) \le \left(\frac{K}{2}+\frac{|\alpha|}{2c}-\beta\right) y_t^2(0,t) + \frac{1}{2}\left(|\alpha|c - K\right) y_t^2(0,t-\tau) \\ + \rho(t)\Big\{(ay_x)(1,t)\left[1-\frac{c_2}{M}\right] + (ay_x)(0,t)\left[\frac{c_1}{m}-1\right] \\ + \beta y_t(0,t)\left[\frac{c_3}{\tau\beta}+\frac{c_4}{\beta}-\frac{c_1}{m}\right] \\ + \alpha y_t(0,t-\tau)\left[\frac{c_1}{m}-\frac{c_3}{\tau\alpha}\right]\Big\}. \end{aligned} \tag{4.2.10}$$

In order to make the energy $\mathscr{E}(t)$ decrease (or at least nonincreasing), we shall assume that

$$|\alpha| < \beta \tag{4.2.11}$$

and then choose K such that

$$|\alpha| \le K \le 2\beta - |\alpha|, \tag{4.2.12}$$

whereas the other constants are

$$\begin{cases} c = 1, \\ c_1 = m, \ c_2 = M, \ c_3 = \tau\alpha, \ c_4 = \beta - \alpha. \end{cases} \tag{4.2.13}$$

In light of (4.2.10) and (4.2.11)–(4.2.13), we deduce that

$$\mathscr{E}'(t) \le \frac{1}{2}\left\{(-2\beta + |\alpha| + K)y_t^2(0,t) + (|\alpha| - K)y_t^2(0,t-\tau)\right\} \le 0, \tag{4.2.14}$$

and hence, the energy $\mathscr{E}(t)$ is nonincreasing.

Remark 4.2.1 It is clear from the above choices in (4.2.13) that the additional energy $E_1(t)$ defined by (4.2.6)–(4.2.7) is in fact constant (see (4.2.8)).

Let us now return to our closed-loop system (4.2.2). Using the well-known change of variables [184] (see also [149])

$$u(x,t) = y_t(0, t - x\tau), \quad x \in (0,1), \tag{4.2.15}$$

the system (4.2.2) becomes

$$\begin{cases} y_{tt}(x,t) - (ay_x)_x(x,t) = 0, & (x,t) \in (0,1) \times (0,\infty), \\ \tau u_t(x,t) + u_x(x,t) = 0, & (x,t) \in (0,1) \times (0,\infty), \\ my_{tt}(0,t) - (ay_x)(0,t) = \alpha u(1,t) - \beta u(0,t), & t > 0, \\ My_{tt}(1,t) + (ay_x)(1,t) = 0, & t > 0, \\ y(x,0) = y_0(x), \; y_t(x,0) = y_1(x), & x \in (0,1), \\ u(x,0) = y_t(0, -x\tau) = f(-x\tau), & x \in (0,1). \end{cases} \tag{4.2.16}$$

Let $z(\cdot,t) = y_t(\cdot,t)$, $\xi(t) = y_t(0,t)$, and $\eta(t) = y_t(1,t)$. In the sequel, we shall adopt the standard notation by omitting the time variable t whenever unnecessary. Thereby, consider the state variable $\Phi = (y, z, u, \xi, \eta)$. Then, our state space $\mathscr{X}$ is defined by

$$\mathscr{X} = H^1(0,1) \times L^2(0,1) \times L^2(0,1) \times \mathbb{R}^2,$$

equipped with the following real inner product (the complex case is similar)

$$\begin{array}{l} \langle (y,z,u,\xi,\eta), (\tilde{y},\tilde{z},\tilde{u},\tilde{\xi},\tilde{\eta}) \rangle_{\mathscr{X}} = \displaystyle\int_0^1 (ay_x\tilde{y}_x + z\tilde{z})\,dx \\ +K\tau \int_0^1 u\tilde{u}\,dx + m\xi\tilde{\xi} + M\eta\tilde{\eta} \\ +\varpi \left(\displaystyle\int_0^1 zdx + m\xi + M\eta + \mu y(0) + \tau\alpha \int_0^1 udx \right) \\ \left(\int_0^1 \tilde{z}dx + m\tilde{\xi} + M\tilde{\eta} + \mu\tilde{y}(0) + \tau\alpha \int_0^1 \tilde{u}dx \right) \end{array} \tag{4.2.17}$$

in which $K > 0$ satisfies the condition (4.2.12), while $\mu = \beta - \alpha$ and ϖ is a positive constant to be determined. Note that $\mu = \beta - \alpha$ is positive due to (4.2.11).

The first result, stated below, shows that $\mathscr{X}$ equipped with (4.2.17) is a Hilbert space.

Proposition 4.2.1 *Assume that (4.1.5), (4.2.11), and (4.2.12) hold. Then, the state space $\mathscr{X}$ endowed with the inner product (4.2.17) is a Hilbert space provided that ϖ is small enough.*

Proof It suffices to show the existence of two positive constants A_1 and A_2 such that

$$A_1\|(y, z, u, \xi, \eta)\| \leq \|(y, z, u, \xi, \eta)\|_{\mathscr{X}} \leq A_2\|(y, z, u, \xi, \eta)\|, \tag{4.2.18}$$

where $\|(y, z, u, \xi, \eta)\|$ denotes the usual norm of $H^1(0, 1) \times L^2(0, 1) \times L^2(0, 1) \times \mathbb{R}^2$, that is,

$$\|(y, z, u, \xi, \eta)\|^2 = \int_0^1 \left(y^2 + y_x^2 + z^2 + u^2\right) dx + \xi^2 + \eta^2.$$

The right-hand inequality $\|(y, z, u, \xi, \eta)\|_{\mathscr{X}} \leq A_2\|(y, z, u, \xi, \eta)\|$ is straightforward. Indeed, Young's and Hölder's inequalities yield

$$\|(y, z, u, \xi, \eta)\|_{\mathscr{X}}^2 \leq \int_0^1 \left(ay_x^2 + z^2\right) dx + K\tau \int_0^1 u^2\, dx + m\xi^2 + M\eta^2$$
$$+5\varpi \left(\int_0^1 z^2 dx + m^2\xi^2 + M^2\eta^2 + \mu^2 y^2(0) + \tau^2\alpha^2 \int_0^1 u^2 dx\right).$$

Moreover, by virtue of (4.1.5) and the following trace estimate

$$y^2(0) \leq 2\int_0^1 \left(y^2 + y_x^2\right) dx,$$

the above inequality leads to the desired result with A_2 depending on m, M, $|\alpha|$, τ, β and $||a||_\infty$.

With regard to the other inequality of (4.2.18), we proceed as follows:

$$\begin{aligned}\|(y, z, u, \xi, \eta)\|_{\mathscr{X}}^2 = &\int_0^1 \left(ay_x^2 + z^2\right) dx + K\tau \int_0^1 u^2\, dx + m\xi^2 + M\eta^2\\ &+ \varpi \left(\int_0^1 z\, dx + \tau\alpha \int_0^1 u\, dx + m\xi + M\eta\right)^2\\ &+ \varpi\mu^2 y^2(0) + 2\varpi\mu y(0)\\ &\left[\int_0^1 z\, dx + \tau\alpha \int_0^1 u\, dx + m\xi + M\eta\right].\end{aligned} \tag{4.2.19}$$

It follows from Young's inequality that for any $\kappa > 0$,

$$2y(0)\left[\int_0^1 z\, dx + \tau\alpha \int_0^1 u\, dx + m\xi + M\eta\right] \geq \tag{4.2.20}$$

$$-\frac{4}{\kappa}\left(\left[\int_0^1 z\,dx\right]^2+\tau^2\alpha^2\left[\int_0^1 u\,dx\right]^2+m^2\xi^2+M^2\eta^2\right)-\kappa y^2(0).$$

Combining (4.2.19) and (4.2.20), and choosing $\kappa < \mu = \beta - \alpha$, we obtain

$$\begin{aligned}
&\|(y,z,u,\xi,\eta)\|_{\mathscr{X}}^2 \geq \\
&\int_0^1 ay_x^2dx+\left[1+4\varpi\left(1-\frac{\mu}{\kappa}\right)\right]\int_0^1 z^2dx+\tau\left[K+4\varpi\tau\alpha^2\left(1-\frac{\mu}{\kappa}\right)\right]\int_0^1 u^2\,dx \\
&+m\left[1+4m\varpi\left(1-\frac{\mu}{\kappa}\right)\right]\xi^2+M\left[1+4M\varpi\left(1-\frac{\mu}{\kappa}\right)\right]\eta^2+\varpi\mu(\mu-\kappa)y^2(0).
\end{aligned}\tag{4.2.21}$$

A direct computation gives

$$\begin{aligned}
\int_0^1 y^2dx &= y^2(0)+2\int_0^1\int_0^x yy_s\,ds\,dx \\
&\leq y^2(0)+\varepsilon\int_0^1 y^2dx+\frac{1}{\varepsilon}\int_0^1 y_x^2dx,
\end{aligned}\tag{4.2.22}$$

for any $\varepsilon > 0$. Inserting (4.2.22) into (4.2.21) and using (4.1.5) yield

$$\begin{aligned}
&\|(y,z,u,\xi,\eta)\|_{\mathscr{X}}^2 \geq \left[a_0-\varepsilon^{-1}\varpi\mu(\mu-\kappa)\right]\int_0^1 y_x^2dx \\
&+\varpi\mu(\mu-\kappa)(1-\varepsilon)\int_0^1 y^2dx \\
&+\left[1+4\varpi\left(1-\frac{\mu}{\kappa}\right)\right]\int_0^1 z^2dx+\tau\left[K+4\varpi\tau\alpha^2\left(1-\frac{\mu}{\kappa}\right)\right]\int_0^1 u^2\,dx \\
&+m\left[1+4m\varpi\left(1-\frac{\mu}{\kappa}\right)\right]\xi^2+M\left[1+4M\varpi\left(1-\frac{\mu}{\kappa}\right)\right]\eta^2,
\end{aligned}$$

for any $0 < \kappa < \mu = \beta - \alpha$ and $0 < \varepsilon < 1$. Finally, we choose ϖ so that

$$0 < \varpi < \min\left\{\frac{\varepsilon a_0}{\mu(\mu-\kappa)}, \delta, \frac{\delta}{m}, \frac{\delta}{M}, \frac{K\delta}{\tau\alpha^2}, \right\},$$

where $\delta = \frac{\kappa}{4(\mu-\kappa)} > 0$. Thus, (4.2.18) holds, and the proof of Proposition 4.2.1 is achieved. □

We are now in a position to set our problem in the state space $\mathscr{X}$. Firstly, we define a linear operator $\mathscr{A}$ by

$$\begin{aligned}&\mathscr{D}(\mathscr{A}) = \left\{(y, z, u, \xi, \eta) \in \mathscr{X};\ y \in H^2(0,1),\quad z, u \in H^1(0,1),\right.\\&\quad \left.\xi = u(0) = z(0),\ \eta = z(1)\right\},\\&\mathscr{A}(y, z, u, \xi, \eta) = \left(z, (ay_x)_x, -\frac{u_x}{\tau}, \frac{1}{m}\left[(ay_x)(0) - \beta\xi + \alpha u(1)\right], -\frac{(ay_x)(1)}{M}\right),\\&(y, z, u, \xi, \eta) \in \mathscr{D}(\mathscr{A}).\end{aligned} \tag{4.2.23}$$

Then, the closed-loop system (4.2.2) can now be formulated as an evolution equation in $\mathscr{X}$

$$\begin{cases}\Phi_t(t) = \mathscr{A}\Phi(t),\\ \Phi(0) = \Phi_0,\end{cases} \tag{4.2.24}$$

in which $\Phi = (y, z, u, \xi, \eta)$ and $\Phi_0 = (y_0, y_1, f(-\tau\cdot), \xi_0, \eta_0)$.

The well-posedness result is stated below.

Theorem 4.2.1 *Suppose that (4.1.5), (4.2.11), and (4.2.12) are satisfied. Then, we have:*

(i) The operator $\mathscr{A}$ defined by (4.2.23) is densely defined in $\mathscr{X}$ and generates on $\mathscr{X}$ a C_0-semigroup of contractions $e^{t\mathscr{A}}$. Moreover, $\sigma(\mathscr{A})$, the spectrum of $\mathscr{A}$, consists of isolated eigenvalues of finite algebraic multiplicity only (see Remark 2.1.2).

(ii) For any initial condition $\Phi_0 \in \mathscr{X}$, the system (4.2.24) has a unique mild solution $\Phi \in C([0,\infty); \mathscr{X})$. In turn, if $\Phi_0 \in \mathscr{D}(\mathscr{A})$, then necessarily the solution Φ is strong and belongs to $C([0,\infty); \mathscr{D}(\mathscr{A}) \cap C^1([0,\infty); \mathscr{X})$.

Proof Let $\Phi = (y, z, u, \xi, \eta) \in \mathscr{D}(\mathscr{A})$. Then, in light of (4.2.17) and (4.2.23), a simple integration by parts gives

$$\begin{aligned}\langle \mathscr{A}\Phi, \Phi)\rangle_{\mathscr{X}} &= (ay_x)(1)z(1) - (ay_x)(0)z(0) - \frac{K}{2}(u^2(1) - u^2(0))\\&\quad + \xi(ay_x)(0) - \beta\xi^2 + \alpha\xi u(1)\\&\quad - \eta(ay_x)(1) + \varpi\left(\int_0^1 z\,dx + \tau\alpha\int_0^1 u\,dx + m\xi + M\eta\right)(\alpha u(0)\\&\quad + \beta\xi + (\beta - \alpha)z(0))\end{aligned}$$

$$= \alpha\xi u(1) - \frac{K}{2}u^2(1) + \frac{K}{2}u^2(0) - \beta\xi^2$$

$$\leq \left(-\beta + \frac{K + |\alpha|}{2}\right)\xi^2 + \frac{|\alpha| - K}{2}u^2(1), \tag{4.2.25}$$

and hence, the operator $\mathscr{A}$ is dissipative in light of (4.2.12).

Next, we claim that the operator $\lambda I - \mathscr{A}$ is onto $\mathscr{X}$, for $\lambda > 0$ sufficiently large. To ascertain the correctness of this claim, one has to show that given $(f, g, v, p, q) \in \mathscr{X}$, there exists $(y, z, u, \xi, \eta) \in \mathscr{D}(\mathscr{A})$ for which $(\lambda I - \mathscr{A})(y, z, u, \xi, \eta) = (f, g, v, p, q)$. Although this can be considered as a classical problem, one can easily verify that the latter is equivalent to solving the following system:

$$\begin{cases} \lambda^2 y - (a y_x)_x = \lambda f + g, \\ u_x + \lambda\tau u = \tau v, \\ \lambda(m\lambda + \beta)y(0) - (a y_x)(0) - \alpha u(1) = mp + (m\lambda + \beta)f(0), \\ \lambda^2 M y(1) + (a y_x)(1) = Mq + \lambda M f(1), \\ z = \lambda y - f, \\ \xi = u(0) = z(0) = \lambda y(0) - f(0), \\ \eta = z(1) = \lambda y(1) - f(1). \end{cases} \tag{4.2.26}$$

Solving the equation of u in the above system, one obtains

$$u(x) = e^{-\tau\lambda x}(\lambda y(0) - f(0)) + \tau \int_0^x e^{-\tau\lambda(x-s)} v(s)\, ds, \tag{4.2.27}$$

and hence,

$$u(1) = e^{-\tau\lambda}(\lambda y(0) - f(0)) + \tau \int_0^1 e^{-\tau\lambda(1-s)} v(s)\, ds. \tag{4.2.28}$$

This, together with (4.2.26) and (4.2.27), implies that one has to seek only $y \in H^2(0, 1)$ satisfying

$$\begin{cases} \lambda^2 y - (a y_x)_x = \lambda f + g, \\ \lambda\left[(m\lambda + \beta) - \alpha e^{-\tau\lambda}\right] y(0) - (a y_x)(0) = mp + (m\lambda + \beta - \alpha e^{-\tau\lambda}) f(0) \\ \qquad\qquad + \tau\alpha \displaystyle\int_0^1 e^{-\tau\lambda(1-s)} v(s)\, ds, \\ \lambda^2 M y(1) + (a y_x)(1) = Mq + \lambda M f(1). \end{cases} \tag{4.2.29}$$

Multiplying the first equation in (4.2.29) by $\phi \in H^1(0,1)$, we get the weak formulation

$$\int_0^1 \left(\lambda^2 \phi y + a y_x \phi_x\right) dx + \lambda\left[(m\lambda+\beta) - \alpha e^{-\tau\lambda}\right] y(0)\phi(0) + \lambda^2 M y(1)\phi(1)$$

$$= \int_0^1 (\lambda f + g)\,\phi\, dx + \Big[mp + (m\lambda + \beta - \alpha e^{-\tau\lambda}) f(0)$$

$$+ \tau\alpha \int_0^1 e^{-\tau\lambda(1-s)} v(s)\, ds\Big]\phi(0)$$

$$+ [Mq + \lambda M f(1)]\,\phi(1), \tag{4.2.30}$$

which in turn can be written in the form $\mathcal{L}(y,\phi) = \mathcal{M}(\phi)$, where $\mathcal{L}$ is a bilinear form defined by

$$\mathcal{L} : H^1(0,1) \times H^1(0,1) \longrightarrow \mathbb{R}$$

such that

$$\mathcal{L}(y,\phi) = \int_0^1 \left(\lambda^2 \phi y + a y_x \phi_x\right) dx + \lambda\left[(m\lambda+\beta) - \alpha e^{-\tau\lambda}\right] y(0)\phi(0)$$
$$+ \lambda^2 M y(1)\phi(1),$$

and $\mathcal{M}$ is a linear form given by

$$\mathcal{M} : H^1(0,1) \longrightarrow \mathbb{R}$$

$$\phi \longmapsto \mathcal{M}(\phi) = \int_0^1 (\lambda f + g)\,\phi\, dx +$$

$$\left[mp + (m\lambda + \beta - \alpha e^{-\tau\lambda}) f(0) + \tau\alpha \int_0^1 e^{-\tau\lambda(1-s)} v(s)\, ds\right]\phi(0)$$

$$+ [Mq + \lambda M f(1)]\,\phi(1).$$

Applying Lax-Milgram theorem (see Theorem 2.1.6), one can deduce the existence of a unique solution $y \in H^2(0,1)$ of (4.2.29) as long as $\lambda > 0$. This establishes that the range of $\lambda I - \mathcal{A}$ is $\mathcal{X}$, for $\lambda > 0$ small. Thus, according to semigroup theory [158], the operator $\mathcal{A}$ is densely defined in $\mathcal{X}$ and generates on $\mathcal{X}$ a C_0-semigroup of contractions denoted by $e^{t\mathcal{A}}$. As a direct consequence of the fact that, for $\lambda > 0$, the range of $\lambda I - \mathcal{A}$ is $\mathcal{X}$, and it follows that $(\lambda I - \mathcal{A})^{-1}$ exists and maps $\mathcal{X}$ into $\mathcal{D}(\mathcal{A})$. Finally, using Sobolev embedding [4], one can claim that $(\lambda I - \mathcal{A})^{-1}$

is compact, and hence, the spectrum of $\mathscr{A}$ consists of isolated eigenvalues of finite algebraic multiplicity only (see Remark 2.1.2). This completes the proof of the first assertion (i) in Theorem 4.2.1.

Concerning the proof of the second assertion, it suffices to use (i) and invoke semigroups theory [158]. □

4.2.3 Asymptotic Behavior

It is clear that (4.2.23) implies that $\lambda = 0$ is an eigenvalue of $\mathscr{A}$ whose eigenfunction is $\nu(1, 0, 0, 0, 0)$, where $\nu \in \mathbb{R} \setminus \{0\}$. Thus, the semigroup $e^{t\mathscr{A}}$ generated by $\mathscr{A}$ is not stable by means of Theorem 2.3.12. One can also interpret such a result by noticing that energy $\mathscr{E}(t)$ (see (4.2.9)) of the system (4.2.2) corresponding to $\Phi = \nu(1, 0, 0, 0, 0)$ is constant and therefore will not tend to zero as $t \longrightarrow +\infty$. In turn, we are able to prove the main result of the section that is stated below.

Theorem 4.2.2 *Assume that (4.1.5), (4.2.11) hold and K satisfies $|\alpha| < K < 2\beta - |\alpha|$. Then, for any initial data $\Phi_0 = (y_0, y_1, f, \xi_0, \eta_0) \in \mathscr{X}$, the solution $\Phi(t) = \Big(y, y_t, y_t(0, t - x\tau), y_t(0, t), y_t(1, t)\Big)$ of the system (4.2.2) (or equivalently (4.2.24)) tends in $\mathscr{X}$ to $(\Omega, 0, 0, 0, 0)$ as $t \longrightarrow +\infty$, where*

$$\Omega = \frac{1}{\beta - \alpha}\left[\int_0^1 y_1 dx + \alpha\tau \int_0^1 f(-\tau x)dx + (\beta - \alpha)y_0(0) + m\xi_0 + M\eta_0\right]. \tag{4.2.31}$$

Proof The proof depends on an essential way on the application of LaSalle's invariance principle (see Theorem 2.3.11). Using a standard density argument and the contraction of the semigroup $e^{t\mathscr{A}}$, it suffices to prove Theorem 4.2.2 for smooth initial data $\Phi_0 = (y_0, y_1, f, \xi_0, \eta_0) \in \mathscr{D}(\mathscr{A})$. Let $\Phi(t) = (y(t), y_t(t), u(t), \xi(t), \eta(t)) = e^{t\mathscr{A}}\Phi_0$ be the solution of (4.2.2). It follows from Theorem 4.2.1 that the set of trajectories of solutions $\{\Phi(t)\}_{t \geq 0}$ is bounded for the graph norm and thus precompact by virtue of the compactness of the operator $(\lambda I - \mathscr{A})^{-1}$, for $\lambda > 0$. Invoking LaSalle's principle, we deduce that the ω-limit set

$$\omega(\Phi_0) = \Big\{\Psi \in \mathscr{X};\ \text{there exists a sequence } t_n \to \infty$$
$$\text{as } n \to \infty \text{ such that } \Psi = \lim_{n\to\infty} e^{t_n\mathscr{A}}\Phi_0\Big\}$$

is non-empty, compact, invariant under the semigroup $e^{t\mathscr{A}}$. Additionally, the solution $e^{t\mathscr{A}}\Phi_0 \longrightarrow \omega(\Phi_0)$ as $t \to \infty$ [114]. Clearly, in order to prove the convergence result, it suffices to show that $\omega(\Phi_0)$ reduces to $(\Omega, 0, 0, 0)$. To this end, let $\tilde{\Phi}_0 = \Big(\tilde{y}_0, \tilde{y}_1, \tilde{f}, \tilde{\xi}, \tilde{\eta}\Big) \in \omega(\Phi_0) \subset D(\mathscr{A})$ and consider $\tilde{\Phi}(t) =$

$\left(\tilde{y}(t), \tilde{y}_t(t), \tilde{u}(t), \tilde{\xi}(t), \tilde{\eta}(t)\right) = e^{t\mathscr{A}}\tilde{\Phi}_0 \in D(\mathscr{A})$ as the unique strong solution of (4.2.24). It is well-known that $\|\tilde{\Phi}(t)\|_{\mathscr{X}}$ is constant [114, Theorem 2.1.3 p. 18], and thus $\frac{d}{dt}\left(\|\tilde{\Phi}(t)\|^2_{\mathscr{X}}\right) = 0$. This leads to

$$< \mathscr{A}\tilde{\Phi}, \tilde{\Phi} >_{\mathscr{X}} = 0, \tag{4.2.32}$$

which, together with (4.2.25), implies that $\tilde{\xi} = \tilde{y}_t(0,t) = 0$ and $\tilde{u}(1) = \tilde{y}_t(0, t - \tau) = 0$. Consequently, $\tilde{y}$ is a solution of the system

$$\begin{cases} \tilde{y}_{tt} - (a\tilde{y}_x)_x = 0, & (x,t) \in (0,1) \times (0,\infty), \\ M\tilde{y}_{tt}(1,t) + (a\tilde{y}_x)(1,t) = 0, & t > 0, \\ \tilde{y}_t(0,t) = \tilde{y}_x(0,t) = 0, & t > 0, \\ \tilde{y}(x,0) = \tilde{y}_0(x);\ \tilde{y}_t(x,0) = \tilde{y}_1(x), & x \in (0,1) \\ \tilde{y} \in H^2(0,1). \end{cases} \tag{4.2.33}$$

A straightforward computation shows that $\tilde{z} = \tilde{y}_t$ is a solution of

$$\begin{cases} \tilde{z}_{tt} - (a\tilde{z}_x)_x = 0, & (x,t) \in (0,1) \times (0,\infty), \\ M\tilde{z}_{tt}(1,t) + (a\tilde{z}_x)(1,t) = 0, & t > 0, \\ \tilde{z}(0,t) = \tilde{z}_x(0,t) = 0, & t > 0, \\ \tilde{z}(x,0) = \tilde{y}_1(x);\ \tilde{z}_t(x,0) = (a\tilde{y}_{0x})_x(x), & x \in (0,1). \end{cases} \tag{4.2.34}$$

The problem (4.2.34) admits only the trivial solution $\tilde{z} = 0$. The arguments used to prove this run on much the same lines as in [167] (see also [74]). Consequently, the unique solution $\tilde{y}$ of (4.2.33) is constant. To summarize, we have shown that for any $\tilde{\Phi}_0 = \left(\tilde{y}_0, \tilde{y}_1, \tilde{f}, \tilde{\xi}, \tilde{\eta}\right) \in \omega(\Phi_0) \subset D(\mathscr{A})$, the unique solution $\tilde{\Phi}(t) = \left(\tilde{y}(t), \tilde{y}_t(t), \tilde{u}(t), \tilde{\xi}(t), \tilde{\eta}(t)\right) = e^{t\mathscr{A}}\tilde{\Phi}_0 \in D(\mathscr{A})$ is actually $(\Omega, 0, 0, 0, 0)$, for any $t \geq 0$, where Ω is a constant to be determined. This implies that the initial condition $\tilde{\Phi}_0 = \left(\tilde{y}_0, \tilde{y}_1, \tilde{f}, \tilde{\xi}, \tilde{\eta}\right)$ is also equal to $(\Omega, 0, 0, 0, 0)$. Thereby, the ω-limit set $\omega(\Phi_0)$ consists of only constants $(\Omega, 0, 0, 0, 0)$.

Now, in order to complete the proof, it remains to provide an explicit expression of the constant Ω. To do so, let $(\Omega, 0, 0, 0, 0) \in \omega(\Phi_0)$. This implies that there exists $\{t_n\} \to \infty$, as $n \to \infty$ such that

$$\Phi(t_n) = (y(t_n), y_t(t_n), u(t_n), \xi(t_n), \eta(t_n)) = e^{t_n\mathscr{A}}\Phi_0 \longrightarrow (\Omega, 0, 0, 0, 0) \tag{4.2.35}$$

in the state space $\mathscr{X}$. Furthermore, in view of Remark 4.2.1, any solution of the system (4.2.24) stemmed from $\Phi_0 = (y_0, y_1, f, \xi_0, \eta_0)$ verifies

$$\int_0^1 y_t(x,t)\,dx + my_t(0,t) + My_t(1,t) + \alpha\tau \int_0^1 y_t(0, t - x\tau)\,dx$$

$$+ (\beta - \alpha) y(0, t) = \Upsilon, \ \forall t \geq 0,$$

in which Υ is a constant. Obviously, such a constant can be obtained by taking $t = 0$ in the left-hand side of the last equation. Therefore, we have

$$\begin{aligned}
&\int_0^1 y_t(x, t)\, dx + m y_t(0, t) + M y_t(1, t) \\
&+ \alpha\tau \int_0^1 y_t(0, t - x\tau)\, dx + (\beta - \alpha) y(0, t) \\
&= \int_0^1 y_t(x, 0)\, dx + m y_t(0, 0) + M y_t(1, 0) \\
&+ \alpha\tau \int_0^1 y_t(0, -x\tau)\, dx + (\beta - \alpha) y(0, 0) \\
&= \int_0^1 y_1(x)\, dx + m\xi_0 + M\eta_0 + \alpha\tau \int_0^1 f(-x\tau)\, dx + (\beta - \alpha) y_0(0).
\end{aligned} \tag{4.2.36}$$

Lastly, it suffices to let $t = t_n$ in (4.2.36) with $n \to \infty$ and use (4.2.35) to reach the desired expression of Ω. This achieves the proof of the theorem. □

4.2.4 Polynomial Convergence

The main objective of the current subsection is to show that the convergence result obtained in the previous subsection is in fact polynomial. The proof of such a desired outcome is based on applying Theorem 2.3.10. To proceed, let us first consider the space

$$\dot{\mathscr{X}} = \left\{ (y, z, u, \xi, \eta) \in \mathscr{X};\ \int_0^1 z(x) dx + \alpha\tau \int_0^1 u(x) dx + (\beta - \alpha)\, y(0) \right.$$
$$\left. + m\xi + M\eta = 0 \right\}.$$

Then, a new operator is defined below

$$\dot{\mathscr{A}} : \mathscr{D}(\dot{\mathscr{A}}) := \mathscr{D}(\mathscr{A}) \cap \dot{\mathscr{X}} \subset \dot{\mathscr{X}} \to \dot{\mathscr{X}},$$

$$\dot{\mathscr{A}}(y, z, u, \xi, \eta) = \mathscr{A}(y, z, u, \xi, \eta), \ \forall\, (y, z, u, \xi, \eta) \in \mathscr{D}(\dot{\mathscr{A}}). \tag{4.2.37}$$

Clearly, the operator $\dot{\mathscr{A}}$ defined by (4.2.37) generates on $\dot{\mathscr{X}}$ a C_0-semigroup of contractions $e^{t\dot{\mathscr{A}}}$ provided that the conditions (4.2.11) and (4.2.12) are fulfilled.

Moreover, $\sigma(\dot{\mathscr{A}})$, the spectrum of $\dot{\mathscr{A}}$, consists of isolated eigenvalues of finite algebraic multiplicity only (see Remark 2.1.2).

Now, we are ready to state our result that reflects the polynomial stability in $\dot{\mathscr{X}}$ of the semigroup operator $e^{t\dot{\mathscr{A}}}$.

Theorem 4.2.3 *Assume that (4.1.5) and (4.2.11) hold and K satisfies $|\alpha| < K < 2\beta - |\alpha|$. Then, there exists $C > 0$ such that for all $t > 0$, we have*

$$\left\| e^{t\dot{\mathscr{A}}} \right\|_{\mathscr{L}(\mathscr{D}(\dot{\mathscr{A}}), \dot{\mathscr{X}})} \leq \frac{C}{\sqrt{t}}.$$

Proof of Theorem 4.2.3 The proof is based on the resolvent method and requires two lemmas.

We first look at the point spectrum. □

Lemma 4.2.1 *If γ is a real number, then $i\gamma$ is not an eigenvalue of $\dot{\mathscr{A}}$.*

Proof We will show that the equation

$$\dot{\mathscr{A}} Z = i\gamma Z \tag{4.2.38}$$

with $Z = (y, z, u, \xi, \eta) \in \mathscr{D}(\dot{\mathscr{A}})$ and $\gamma \in \mathbb{R}$ has only the trivial solution. Clearly, the system (4.2.38) writes

$$z = i\gamma y \tag{4.2.39}$$

$$(a\, y_x)_x = i\gamma z \tag{4.2.40}$$

$$-\frac{u_x}{\tau} = i\gamma u \tag{4.2.41}$$

$$\frac{1}{m}\left[(a y_x)(0) - \beta\xi + \alpha u(1)\right] = i\gamma\xi\ . \tag{4.2.42}$$

$$-\frac{(a y_x)(1)}{M} = i\gamma\eta. \tag{4.2.43}$$

Let us first treat the case where $\gamma = 0$. In this case, (4.2.39) gives $z = 0$, and hence, $\xi = \eta = 0$ as $z(0) = \xi$ and $z(1) = \eta$. Using (4.2.41), we deduce that $u = k_1$, where k_1 is a real constant. Since $z = 0$ and $z(0) = u(0)$, we conclude that the constant k_1 is zero, and thus $u = 0$. So far, we have proved that $z = 0$, $u = 0$, and $\xi = \eta = 0$. This, together with (4.2.40) and the fact that $(y, z, u, \xi, \eta) \in \mathscr{D}(\dot{\mathscr{A}})$, implies that $(a\, y_x)_x = 0$ and $y(0) = (a\, y_x)(0) = (a\, y_x)(1) = 0$. This obviously gives $y = 0$, and so (y, z, u, ξ, η) must be the trivial solution of (4.2.38).

Suppose now that $\gamma \neq 0$. Taking the inner product of (4.2.38) with Z, using the inequality (4.2.25), we get

$$0 = \Re e\left(< \dot{\mathscr{A}} Z, Z >_{\dot{\mathscr{X}}}\right)$$

$$\leq \frac{1}{2}\left((-2\beta+|\alpha|+K)|z(0)|^2+(|\alpha|-K)|u(1)|^2\right)(\leq 0). \tag{4.2.44}$$

Thenceforth, we obtain $z(0) = 0$ and $u(1) = 0$, and hence, $\xi = u(0) = 0$. Lastly, we conclude that the only solution of (4.2.38) is the trivial one. □

Lemma 4.2.2 *The resolvent operator of* $\dot{\mathscr{A}}$ *obeys the condition* (2.3.19).

Proof Suppose that condition (2.3.18) is false. By the Banach-Steinhaus theorem (see Theorem 2.1.7), there exist a sequence of real numbers $\gamma_n \to +\infty$ and a sequence of vectors
$Z_n = (y_n, z_n, u_n, \xi_n, \eta_n) \in \mathscr{D}(\dot{\mathscr{A}})$ with $\|Z_n\|_{\dot{\mathscr{X}}} = 1$ such that

$$\|\gamma_n^2 (i\gamma_n I - \dot{\mathscr{A}})Z_n\|_{\dot{\mathscr{X}}} \to 0 \quad \text{as} \quad n \to \infty, \tag{4.2.45}$$

that is, as $n \to \infty$, we have

$$\gamma_n^2 (i\gamma_n y_n - z_n) \equiv \gamma_n^2 f_n \to 0 \quad \text{in} \quad H^1(0,1), \tag{4.2.46}$$

$$\gamma_n^2 (i\gamma_n z_n - (a(y_n)_x)_x) \equiv \gamma_n^2 g_n \to 0 \quad \text{in} \quad L^2(0,1), \tag{4.2.47}$$

$$\gamma_n^2 \left(i\gamma_n u_n + \frac{(u_n)_x}{\tau}\right) \equiv \gamma_n^2 v_n \to 0 \quad \text{in} \quad L^2(0,1), \tag{4.2.48}$$

$$\gamma_n^2 \left(i\gamma_n \xi_n - \frac{1}{m}[(a(y_n)_x)(0) - \beta\xi_n + \alpha u_n(1)]\right) \equiv \gamma_n^2 p_n \to 0 \tag{4.2.49}$$

$$\gamma_n^2 \left(i\gamma_n \eta_n + \frac{(a(y_n)_x)(1)}{M}\right) \equiv \gamma_n^2 q_n \to 0. \tag{4.2.50}$$

Note that (4.2.46)–(4.2.50) also provide the definition of $(f_n, g_n, v_n, p_n, q_n)$.

Our goal is to derive from (3.2.29) that $\|Z_n\|_{\dot{\mathscr{X}}}$ converges to zero; thus there is a contradiction. The proof is divided into three steps.

First Step

We first observe that

$$||\gamma_n^2(i\gamma_n I - \dot{\mathscr{A}})Z_n||_{\dot{\mathscr{X}}} \geq \left|\Re e\left(\langle\gamma_n^2 (i\gamma_n I - \dot{\mathscr{A}})Z_n, Z_n\rangle_{\dot{\mathscr{X}}}\right)\right|. \tag{4.2.51}$$

Amalgamating (4.2.51) with (4.2.44)–(4.2.46), it follows that

$$\gamma_n z_n(0) \to 0,$$

and

$$\gamma_n u_n(1) \to 0.$$

Moreover, since $Z_n \in \mathscr{D}(\dot{\mathscr{A}})$, we deduce that $\xi_n = u_n(0) = z_n(0)$. Thereby

$$\gamma_n \xi_n = \gamma_n u_n(0) \to 0 \text{ and } \xi_n \to 0. \tag{4.2.52}$$

Whereupon, (4.2.49) gives

$$(y_n)_x(0) \to 0. \tag{4.2.53}$$

Solving (4.2.48), we have the following identity

$$u_n(x) = u_n(0)\, e^{-i\tau\gamma_n x} + \tau \int_0^x e^{-i\tau\gamma_n(x-s)} v_n(s)\, ds,$$

which, together with (4.2.52), implies that

$$u_n \to 0 \ \text{ in } \ L^2(0,1). \tag{4.2.54}$$

We have according to (3.2.29)

$$y_n = \frac{1}{i\gamma_n}(z_n + f_n) \to 0 \text{ in } L^2(0,1). \tag{4.2.55}$$

Invoking (4.2.46), (4.2.49), and (4.2.50), we have

$$\begin{aligned}\int_0^1 z_n(x)dx - \frac{1}{i\gamma_n}\int_0^1 g_n(x)dx &= m\frac{p_n}{i\gamma_n} - m\xi_n - M\eta_n + M\frac{q_n}{i\gamma_n} + \circ(1)\\ &= -M\eta_n + \circ(1).\end{aligned}$$

Here and throughout this chapter, $J(n) = \circ(1)$ means that $\lim_{n\to\infty}(J(n)/c) \to 0$ for any positive constant c, and hence, $\lim_{n\to\infty} J(n) = 0$.

Then, since $Z_n \in \dot{\mathscr{X}}$, we have

$$y_n(0) \to 0. \tag{4.2.56}$$

Taking the inner product of (4.2.47) with $a(y_n)_x$ in $L^2(0,1)$, we reach

$$\begin{aligned}&\int_0^1 i\gamma_n\, z_n(x)\, a\,\overline{(y_n)_x}(x)\, dx - \int_0^1 (a(y_n)_x)_x(x) a(x)\overline{(y_n)_x}(x)\, dx\\ &= \int_0^1 g_n(x) a(x)\,\overline{(y_n)_x}(x)\, dx.\end{aligned}$$

Next, we integrate the above identity and then use the following two facts:

(i) $(y_n)_x$ and z_n are bounded in $L^2(0,1)$ (recall that $\|Z_n\|_{\dot{\mathscr{X}}} = 1$).

(ii) $g_n \to 0$ in $L^2(0, 1)$ (see (4.2.47)) and $\gamma_n \to +\infty$.

It follows that

$$\frac{1}{\gamma_n^2}\left(|(a(1)(y_n)_x)(1)|^2 - |(a(0)(y_n)_x)(0)|^2\right) = \circ(1).$$

Therefore, using (4.2.49) and (4.2.50), we have

$$M^2|\eta_n|^2 - m^2|\xi_n|^2 = \circ(1) \Rightarrow \eta_n \to 0. \tag{4.2.57}$$

Since $(f_n, g_n, v_n, p_n, q_n) \in \dot{\mathscr{X}}$, we obtain

$$\int_0^1 g_n(x)\,dx + \alpha\tau \int_0^1 v_n(x)\,dx + (\beta - \alpha)\, f_n(0) + m\,\xi_n + M\,\eta_n = 0, \forall\, n \in \mathbb{N}.$$

Hence,

$$f_n(0) \to 0 \text{ and } f_n(1) = \int_0^1 (f_n)_x(x)\,dx + f_n(0) \to 0. \tag{4.2.58}$$

This also implies, thanks to (4.2.46), that

$$i\gamma_n y_n(1) = z_n(1) + f_n(1) \to 0,\ i\gamma_n y_n(0) = z_n(0) + f_n(0) \to 0.$$

Second Step

Writing z_n in terms of y_n from (4.2.46) and substituting it into (4.2.47), we get

$$-\gamma_n^2 y_n - (a(y_n)_x)_x = i\gamma_n f_n + g_n. \tag{4.2.59}$$

Next, we take the inner product of (4.2.59) with $b(x)(y_n)_x$ in $L^2(0, 1)$, where b is a function in $C^1([0, 1]; \mathbb{R})$ to be specified. We obtain

$$\begin{aligned}
&\int_0^1 \left(-\gamma_n^2 y_n(x) - (a(y_n)_x)_x(x)\right) b(x)\overline{(y_n)_x}(x)\,dx \\
&= \int_0^1 (i\gamma_n f_n(x) + g_n(x))\, b(x)\overline{(y_n)_x}(x)\,dx = \\
&-\int_0^1 i\gamma_n (f_n)_x(x) b(x)\overline{y}_n(x)\,dx + \int_0^1 g_n(x) b(x)(\overline{y_n})_x(x)\,dx \\
&+ i\gamma_n f_n(1) b(1)\overline{y}_n(1) - i\gamma_n f_n(0) b(0)\overline{y}_n(0).
\end{aligned} \tag{4.2.60}$$

It is clear that the right-hand side of (4.2.60) converges to zero since f_n, g_n, $f_n(0)$, $f_n(1)$, $\gamma_n y_n(0)$, and $\gamma_n y_n(1)$ converge to zero in $H^1(0, 1)$, $L^2(0, 1)$ and $\mathbb{C}$, respectively.

On the other hand, a straightforward calculation yields

$$\Re e\left\{\int_0^1 -\gamma_n^2\, y_n b(x)\overline{(y_n)}_x(x)\,dx\right\} = \frac{1}{2}\int_0^1 \gamma_n^2\, b'(x)|y_n(x)|^2\,dx$$
$$-\frac{1}{2}\left(-\gamma_n^2 b(1)|y_n(1)|^2 + \gamma_n^2 b(0)\,|y_n(0)|^2\right)$$
$$= \frac{1}{2}\int_0^1 \gamma_n^2\, b'(x)|y_n(x)|^2\,dx + \circ(1)$$

and

$$\Re e\left\{-\int_0^1 (a(y_n)_x)_x(x)\,b(x)\overline{(y_n)}_x(x)\,dx\right\} = \frac{1}{2}\int_0^1 (ab' - a'b)(x)|(y_n)_x|^2\,dx$$
$$-\frac{1}{2}\left(a(1)b(1)\,|((y_n)_x(1)|^2 - a(0)b(0)\,|(y_n)_x(0)|^2\right)$$
$$= \frac{1}{2}\int_0^1 (ab' - a'b)(x)|(y_n)_x|^2\,dx - \frac{1}{2}a(1)b(1)\,|((y_n)_x(1)|^2 + \circ(1).$$

This, together with (4.2.60), leads to the following

$$\int_0^1 (ab' - a'b)(x)|(y_n)_x|^2\,dx + \int_0^1 \gamma_n^2 b'(x)\,|y_n(x)|^2\,dx - a(1)b(1)|(y_n)_x(1)|^2$$
$$= \circ(1). \tag{4.2.61}$$

Now, let us choose

$$b(x) = \exp\left(\int_x^1 \left|\frac{a'}{a}\right|(s)\,ds\right)\left[-1 - \int_x^1 \exp\left(-\int_s^1 \left|\frac{a'}{a}\right|(r)\,dr\right)\,ds\right],$$
$$\forall x \in [0, 1],$$

and hence,

$$\begin{cases} b(1) = -1, \\ b'(x) \geq 1, \quad \forall x \in [0, 1], \\ (ab' - a'b)(x) \geq a(x), \quad \forall x \in [0, 1], \\ b(x) < 0, \quad \forall x \in [0, 1]. \end{cases} \tag{4.2.62}$$

Combining (4.2.61) and (4.2.62), it follows that

$$\int_0^1 a(x)|(y_n)_x|^2\,dx + \int_0^1 \gamma_n^2\,|y_n(x)|^2\,dx + a(1)|(y_n)_x(1)|^2 = \circ(1),$$

which, together with (4.2.55), yields

$$y_n \to 0 \text{ in } H^1(0,1), \tag{4.2.63}$$

and

$$(y_n)_x(1) = \circ(1).$$

Third Step

Taking the inner product of (4.2.47) with $\frac{z_n}{\gamma_n^2}$ in $L^2(0,1)$, we have

$$\begin{aligned} & i\gamma_n \int_0^1 |z_n(x)|^2\,dx + \int_0^1 (a(y_n))_x(x)\,\overline{i\gamma_n\,(y_n)_x(x) - (f_n)_x(x)}\,dx \\ & - a(1)(y_n)_x(1)\,\overline{i\gamma_n y_n(1) - f_n(1)} + a(0)(y_n)_x(0)\,\overline{i\gamma_n y_n(0) - f_n(0)} \\ & \quad = \int_0^1 g_n(x)\,\overline{i\gamma_n y_n(x) - f_n(x)}\,dx = \circ(1). \end{aligned}$$

Hence,

$$\int_0^1 |z_n(x)|^2\,dx - \int_0^1 a(x)\,|(y_n)_x(x)|^2\,dx = \circ(1),$$

which together with (4.2.63) leads to

$$z_n \to 0 \text{ in } L^2(0,1). \tag{4.2.64}$$

Lastly, the identities (4.2.52), (4.2.54), (4.2.57), (4.2.63), and (4.2.64) clearly contradict the fact that $\|Z_n\|_{\dot{\mathscr{X}}} = 1,\ \forall\, n \in \mathbb{N}$.

Thereby, the two assumptions of Theorem 2.3.9 are shown, and the proof of Theorem 4.2.3 is completed. □

Remark 4.2.2 Combining Theorem 4.2.2 and Theorem 4.2.3, one can claim that the solutions of the system (4.2.2) polynomially tend in $\mathscr{X}$ to $(\Omega, 0, 0, 0, 0)$ as $t \longrightarrow +\infty$, where Ω is given by (4.2.31).

4.2.5 Lack of Exponential Convergence or Lack of Convergence Results

In order to complete our investigation, it remains to see whether the solutions of (4.2.2) can converge exponentially to their equilibrium state (see (4.2.31)). Obviously, this problem is equivalent to checking that the operator $\dot{\mathscr{A}}$ (see (4.2.37) generates an exponentially stable semigroup $e^{t\dot{\mathscr{A}}}$ on $\dot{\mathscr{X}}$. In other words, given an initial condition $\psi_0 \in \dot{\mathscr{X}}$, does the unique solution $e^{t\dot{\mathscr{A}}}\psi_0$ exponentially tend to zero in $\dot{\mathscr{X}}$? To answer such a question, we shall show that the solution $e^{t\dot{\mathscr{A}}}\psi_0$ cannot exponentially decay when $\alpha = 0$ (no delay) and hence use the fact that the presence of a delay in such a system has a destabilizing effect [84]. More precisely, we first consider the system (4.2.2) with $\alpha = 0$

$$\begin{cases} y_{tt}(x,t) - (ay_x)_x(x,t) = 0, & 0 < x < 1,\ t > 0, \\ my_{tt}(0,t) - (ay_x)(0,t) = -\beta y_t(0,t), & t > 0, \\ My_{tt}(1,t) + (ay_x)(1,t) = 0, & t > 0, \\ y(x,0) = y_0(x),\ y_t(x,0) = y_1(x), & x \in (0,1). \end{cases} \tag{4.2.65}$$

The latter can be written as

$$\dot{\mathscr{X}}_0 = \Big\{(y,z,\xi,\eta) \in H^1(0,1) \times L^2(0,1) \times \mathbb{R}^2; \\ \int_0^1 z(x)dx + \beta y(0) + m\xi + M\eta = 0\Big\},$$

equipped with the following real inner product

$$\langle (y,z,\xi,\eta), (\tilde{y},\tilde{z},\tilde{\xi},\tilde{\eta})\rangle_{\dot{\mathscr{X}}_0} = \int_0^1 (ay_x\tilde{y}_x + z\tilde{z})\,dx + m\xi\tilde{\xi} + M\eta\tilde{\eta}, \tag{4.2.66}$$

as follows:

$$\begin{cases} \dot{Q}(t) = (A_0 + P)Q(t), \\ Q(0) = Q_0. \end{cases} \tag{4.2.67}$$

Here, $Q_0 = (y_0, z_0, \xi_0, \eta_0)$, A_0 is the operator defined by

$$\mathscr{D}(A_0) = \{(y,z,\xi,\eta) \in \dot{\mathscr{X}}_0;\ y \in H^2(0,1),\ \ z \in H^1(0,1),\ \ \xi = z(0),\ \eta = z(1)\}, \tag{4.2.68}$$

and for all $(y, z, \xi, \eta) \in \mathscr{D}(A_0)$,

$$A_0(y, z, \xi, \eta) = \left(z, (ay_x)_x, \frac{(ay_x)(0)}{m}, -\frac{(ay_x)(1)}{M} \right). \tag{4.2.69}$$

Additionally, P is the compact operator on $\dot{\mathscr{X}}_0$ given by

$$P(y, z, \xi, \eta) = \left(0, 0, -\frac{\beta}{m}\xi, 0 \right). \tag{4.2.70}$$

Arguing as for $\mathscr{A}$, one can verify that the operator $A_0 + P$ generates on $\dot{\mathscr{X}}_0$ a C_0-semigroup of contractions $S_{A_0+P}(t)$. On the other hand, the operator A_0 defined by (4.2.68)–(4.2.69) is maximal and satisfies $< A_0 Q, Q >= 0$, for all $Q \in \mathscr{D}(A_0)$. Thereby, $< A_0 Q, V >= - < Q, A_0 V >$, for all $Q, V \in \mathscr{D}(A_0)$. Thus, the operator A_0 is antisymmetric. In turn, one can show that $\mathscr{D}(A_0) = \mathscr{D}(A_0^*)$ and $A_0^* = -A_0$, and hence, A_0 is skew adjoint. Finally, it follows that A_0 generates a unitary group $S_1(t)$ of isometries [158].

Now, we are ready to state a result that highlights the optimality of the polynomial convergence:

Theorem 4.2.4 *The solutions of (4.2.65) cannot converge to zero exponentially.*

Proof The proof is based on the result of compact perturbations [170]. Let $S_{A_0+P}(t)$ and $S_{A_0-P}(t)$ denote the groups generated by $A_0 + P$ and $A_0 - P$, respectively. Let $Q(t) = (y, z, \xi, \eta)(t)$ be the solution of (4.2.67) stemmed from $Q_0 \in \mathscr{D}(A_0)$ and define $\tilde{Q}(t) = (\tilde{y}(t), \tilde{z}(t), \tilde{\xi}(t), \tilde{\eta}(t))$, where

$$\tilde{y}(x, t) = y(x, -t), \; \tilde{z}(x, t) = -z(x, -t), \; \tilde{\xi}(t) = -\xi(t), \; \tilde{\eta}(t) = -\eta(-t).$$

It can be verified that $\tilde{Q}(t) = S_{A_0-P}(t)\tilde{Q}_0$. Since P is compact, Russell's result [170] implies that there is no positive constants $0 < \vartheta < 1$ and $T > 0$ such that we have simultaneously

$$||S_{A_0+P}(t)|| \leq \vartheta \quad \text{and} \quad ||S_{A_0-P}(-t)|| \leq \vartheta, \; \forall t \geq T. \tag{4.2.71}$$

Since $\|\tilde{Q}(-t)\| = \|\tilde{Q}(t)\|$, for all $t \in \mathbb{R}$, we have $\|S_{A_0+P}(t)\| = \|S_{A_0-P}(-t)\|$. This, together with (4.2.71), completes the proof of Proposition 4.2.4. □

We turn now to the second issue of this subsection: What is the behavior of the solutions to the system (4.2.2) if $|\alpha| \geq \beta$ (the condition (4.2.11) used to get the convergence results of (4.2.2) is violated)?. The immediate task is to show that in such an event, the solutions of (4.2.2) do not converge to their equilibrium state, which is equivalent to claiming that the semigroup $e^{t\mathscr{A}}$ becomes unstable in $\dot{\mathscr{X}}$, for some delays τ. Therefore, condition (4.2.11) is optimal.

For sake of simplicity, one may assume that $\alpha > 0$ and $a = 1$. Then, under the assumption $\alpha \geq \beta$, we seek a nonzero real number γ and a nonzero function $f \in H^2(0,1)$ so that $y(x,t) = e^{i\gamma t} f(x)$ solves the system associated to the operator $\dot{\mathscr{A}}$ (see (4.2.37) on $\dot{\mathscr{X}}$, that is,

$$\begin{cases} y_{tt}(x,t) - y_{xx}(x,t) = 0, & 0 < x < 1,\ t > 0, \\ my_{tt}(0,t) - y_x(0,t) = \alpha y_t(0,t-\tau) - \beta y_t(0,t), & t > 0, \\ My_{tt}(1,t) + y_x(1,t) = 0, & t > 0, \end{cases} \tag{4.2.72}$$

with $\int_0^1 y_t(x,t)\,dx + \alpha\tau\int_0^1 y_t(0,t-x\tau)\,dx + (\beta-\alpha)\,y(0) + my_t(0,t) + My_t(1,t) = 0$. Once this goal is achieved, we conclude that $\|y\|_{L^2(0,1)} = \|f\|_{L^2(0,1)} \neq 0$, and thereby the solution of (4.2.72) is unstable.

A straightforward computation permits to claim that we just need to solve the following problem:

$$\begin{cases} f_{xx} + \gamma^2 f = 0, \\ f_x(0) + \gamma\left(m\gamma - i\beta + i\alpha e^{-i\gamma\tau}\right) f(0) = 0, \\ f_x(1) - M\gamma^2 f(1) = 0. \end{cases} \tag{4.2.73}$$

As $\alpha \geq \beta$, let us take $\cos(\gamma\tau) = \beta/\alpha \leq 1$, and hence, $\alpha\sin(\gamma\tau) = \sqrt{\alpha^2 - \beta^2}$. This, together with (4.2.73), gives

$$\begin{cases} f_{xx} + \gamma^2 f = 0, \\ f_x(0) + \gamma\left(m\gamma + \sqrt{\alpha^2-\beta^2}\right) f(0) = 0, \\ f_x(1) - M\gamma^2 f(1) = 0. \end{cases} \tag{4.2.74}$$

One can readily verify that (4.2.74) has a nontrivial solution if and only if γ is the solution of the characteristic equation

$$\begin{aligned} \Delta(\gamma) := {} & mM\gamma\sin\gamma + (M\sqrt{\alpha^2-\beta^2} - 1)\sin\gamma - (m+M)\gamma\cos\gamma \\ & - \sqrt{\alpha^2-\beta^2}\cos\gamma = 0. \end{aligned} \tag{4.2.75}$$

We shall discuss two cases:

Case 1: $\alpha = \beta$. Therefore, we have $\gamma\tau = 2n\pi$, with $n \in \mathbb{N}$ as $\cos(\gamma\tau) = \beta/\alpha = 1$. Amalgamating the latter with (4.2.75), we end up with the following equation:

$$\hat{\Delta}(\gamma) := mM\gamma\sin\gamma - \sin\gamma - (m+M)\gamma\cos\gamma = 0, \tag{4.2.76}$$

where $\gamma\tau = 2n\pi$, $n \in \mathbb{N}$. It is a simple task to check that (4.2.74) has a solution $\gamma^* \in (\pi, 2\pi)$. Indeed, note that $\hat{\Delta}(\pi) = (m+M)\pi$, whereas

$\hat{\Delta}(2\pi) = -2(m + M)\pi$. Whereupon, there exists a delay $\tau \in (n, 2n)$, $n \in \mathbb{N}$, for which f satisfies (4.2.74).

Case 2: $\alpha > \beta$. Recalling (4.2.75), we have $\Delta(\pi) = \sqrt{\alpha^2 - \beta^2} + (m+M)\pi > 0$, and $\Delta(2\pi) = -\sqrt{\alpha^2 - \beta^2} - 2(m + M)\pi < 0$. Thereby, (4.2.75) has a solution $\gamma^* \in (\pi, 2\pi)$ and, hence, a positive delay τ such that (4.2.74) has a nontrivial solution.

To sum up, we have the following result:

Theorem 4.2.5 *If the assumption (4.2.11) does not hold, that is, $\alpha \geq \beta$, then the solutions of (4.2.72) are unstable in $\dot{\mathscr{X}}$. In other words, the solutions of the system (4.2.2) do not converge in $\mathscr{X}$ to their equilibrium state* $(\Omega, 0, 0, 0, 0)$ *as* $t \longrightarrow +\infty$, *where Ω is given by (4.2.31).*

4.2.6 Concluding Remarks

To recapitulate, this section dealt with the analysis of the overhead system in the presence of a constant time delay in the boundary velocity control. Assuming that the feedback gain of the delayed term is small, it has been shown that the system is well-posed, whose proof is based on the introduction of a suitable energy norm. Additionally, it has been proved that the solutions of the system asymptotically converge to an equilibrium state, which is explicitly given and depends on the initial conditions. The proof of this result utilized the well-known LaSalle principle. More importantly, the polynomial convergence of solutions has been obtained by means of the resolvent method, and it is proved that this convergence cannot be exponential. Finally, nonconvergence results are put forward in the case where the smallness condition of the delay term is not fulfilled.

We point out that there are many problems, in connection with this section, that deserve to be treated. For instance, it is quite natural to wonder whether the polynomial decay $t^{-1/2}$ obtained is optimal. Moreover, is it possible to extend the results of this section to the case of a nonlinear control? Only a partial answer will be given in Sect. 4.8. It would also be desirable to see if the convergence result still holds when the delay occurring in the boundary control is time dependent. This will be treated in Sect. 4.7 with a special case of boundary conditions. Finally, the modulus of tension of the cable, namely, $a(x)$, is supposed to be a positive function (see (4.1.5)). However, the degenerate case, where $a(x)$ vanishes on a subset of the spatial domain [0, 1], is worth studying as in [11, 109].

4.3 Boundary Delay: Exponential Stabilization via a Global Interior Control

In this section, we go one step further and improve the convergence result obtained in Sect. 4.2 through the action of an *additional interior* (distributed) control

$$\mathcal{G}(x,t) = \sigma w_t(x,t), \tag{4.3.1}$$

where $\sigma > 0$. To be more precise, we show that the solutions of (4.1.4), (4.2.1), and (4.3.1) *exponentially* converge to a steady state, which depends on the initial conditions. Another interesting finding of the current section is to prove that the presence of the boundary velocity (in addition of the interior control) is crucial for the convergence of solutions. Indeed, a nonconvergence result is established when the boundary velocity is omitted ($\beta = 0$ in (4.1.4)) despite the action of the interior control (4.2.1). This shows that the "destabilizing" effect of the delay term in (4.1.4) cannot be compensated by the action of *only* the interior control.

Now, let us briefly provide an overview of this section. In Sect. 4.3.1, we present a comprehensive study of the system (4.1.4), (4.2.1), and (4.3.1) without boundary velocity ($\beta = 0$). The study includes, inter alia, well-posedness and more importantly a nonconvergence result for the system with $\beta = 0$. In turn, an exponential convergence outcome is proved for a shifted system associated to (4.1.4), (4.2.1), and (4.3.1) with $\beta = 0$. In Sect. 4.2.2, we go back to our initial problem (4.1.4), (4.2.1), and (4.3.1) and prove that it is indeed well-posed by means of semigroups theory of linear operators. LaSalle's principle is then used in Sect. 4.3.3 to show that the solutions of the system (4.1.4), (4.2.1), and (4.3.1) converge to an equilibrium state whose expression is explicitly determined. In Sect. 4.3.4, the convergence of solutions to the stationary position is shown to be exponential. The proof is based on the utilization of the frequency-domain method. In Sect. 4.3.5, numerical examples of the spectrum of the closed-loop system (4.1.4), (4.2.1) and (4.3.1) are provided to support our analysis and highlight the relevance of the theoretical outcomes. Finally, this section concludes with some remarks.

4.3.1 *The Closed-Loop System (4.1.4), (4.2.1), and (4.3.1) Without Boundary Velocity Control*

This subsection is concerned with a complete study of the system (4.1.4), (4.2.1), and (4.3.1) with $\beta = 0$, namely,

$$\begin{cases} y_{tt}(x,t) - (ay_x)_x(x,t) + \sigma y_t(x,t) = 0, & 0 < x < 1,\ t > 0, \\ my_{tt}(0,t) - (ay_x)(0,t) = \alpha y_t(0,t-\tau), & t > 0, \\ My_{tt}(1,t) + (ay_x)(1,t) = 0, & t > 0, \\ y(x,0) = y_0(x),\ y_t(x,0) = y_1(x), & x \in (0,1), \\ y_t(0,t-\tau) = \chi(t-\tau), & t \in (0,\tau). \end{cases} \tag{4.3.2}$$

Primary emphasis is placed on examining the long-time behavior of solutions of (4.3.2). In fact, we reveal that the solutions of (4.3.2) will not converge to their equilibrium state. This outcome justifies why the boundary velocity $y_t(0, t)$ has to be present in (4.3.2), if one would desire a convergence result.

4.3.1.1 Well-Posedness of the System (4.3.2)

Thanks to the usual change of variables [149, 184]

$$v(x, t) = y_t(0, t - x\tau),$$

the closed-loop system (4.3.2) takes the form

$$\begin{cases} y_{tt}(x,t) - (ay_x)_x(x,t) + \sigma y_t(x,t) = 0, & (x,t) \in (0,1) \times (0,\infty), \\ \tau v_t(x,t) + v_x(x,t) = 0, & (x,t) \in (0,1) \times (0,\infty), \\ my_{tt}(0,t) - (ay_x)(0,t) = \alpha v(1,t), & t > 0, \\ My_{tt}(1,t) + (ay_x)(1,t) = 0, & t > 0, \\ y(x,0) = y_0(x), \ \ y_t(x,0) = y_1(x), & x \in (0,1), \\ v(x,0) = y_t(0,-x\tau) = \chi(-x\tau), & x \in (0,1). \end{cases} \tag{4.3.3}$$

Then, consider the state variable $\Psi = (y, z, v, \xi, \eta)$, where $z(\cdot, t) = y_t(\cdot, t)$, $\xi(t) = y_t(0, t)$ and $\eta(t) = y_t(1, t)$. Next, the state space of our system is

$$\mathscr{H}_0 = H^1(0,1) \times L^2(0,1) \times L^2(0,1) \times \mathbb{R}^2,$$

equipped with the following real inner product:

$$\begin{aligned} &\langle (y, z, v, \xi, \eta), (\tilde{y}, \tilde{z}, \tilde{v}, \tilde{\xi}, \tilde{\eta}) \rangle_{\mathscr{H}_0} \\ &\quad = \int_0^1 (ay_x\tilde{y}_x + z\tilde{z})\, dx + K\tau \int_0^1 v\tilde{v}\, dx + m\xi\tilde{\xi} + M\eta\tilde{\eta} \\ &\quad + \varpi \left[\int_0^1 (\sigma y + z) dx + m\xi + M\eta - \alpha y(0) + \tau\alpha \int_0^1 v\, dx \right] \\ &\quad \times \left[\int_0^1 (\sigma\tilde{y} + \tilde{z})\, dx + m\tilde{\xi} + M\tilde{\eta} - \alpha\tilde{y}(0) + \tau\alpha \int_0^1 \tilde{v}\, dx \right], \end{aligned} \tag{4.3.4}$$

where $K > 0$ satisfies the condition

$$\alpha \leq K, \tag{4.3.5}$$

and ϖ is a positive constant sufficiently small so that the norm induced by (4.3.4) must be equivalent to the usual norm of $\mathscr{H}_0$. The proof of this claim runs on much the same lines as that of Proposition 4.2.1 (see also Sect. 3.2), and hence, the details are omitted.

Now, one can write the system (4.3.3) as follows:

$$\begin{cases} \Psi_t(t) = \mathscr{A}_0 \Psi(t), \\ \Psi(0) = \Psi_0, \end{cases} \tag{4.3.6}$$

in which $\Psi = (y, z, v, \xi, \eta)$, $\Psi_0 = (y_0, y_1, \chi(-\tau\cdot), \xi_0, \eta_0)$, and $\mathscr{A}_0$ is a linear operator defined by

$$\begin{aligned} &\mathscr{D}(\mathscr{A}_0) = \left\{(y, z, v, \xi, \eta) \in \mathscr{H}_0;\ y \in H^2(0,1),\quad z,\ v \in H^1(0,1),\right. \\ &\quad \left.\xi = v(0) = z(0),\ \eta = z(1)\right\}, \\ &\mathscr{A}_0(y, z, v, \xi, \eta) = \left(z,\ (ay_x)_x - \sigma z,\ -\frac{v_x}{\tau},\ \frac{1}{m}\left[(ay_x)(0) + \alpha v(1)\right],\ -\frac{(ay_x)(1)}{M} \right), \end{aligned} \tag{4.3.7}$$

for each $(y, z, v, \xi, \eta) \in \mathscr{D}(\mathscr{A}_0)$.

The first result is stated below:

Theorem 4.3.1 *If (4.1.5) and (4.3.5) hold, then the operator $\mathscr{A}_0$ defined by (4.3.7) is densely defined in $\mathscr{H}_0$ and generates a C_0-semigroup $e^{t\mathscr{A}_0}$ on $\mathscr{H}_0$. Consequently, if $\Psi_0 \in \mathscr{D}(\mathscr{A}_0)$, then the solution Ψ of (4.3.6) exists in a strong sense and belongs to the space $C([0,\infty); \mathscr{D}(\mathscr{A}_0) \cap C^1([0,\infty); \mathscr{H}_0)$. However, in the case where the initial condition $\Psi_0 \in \mathscr{H}_0$, then the corresponding unique mild solution to (4.3.6) is in $C([0,\infty); \mathscr{H}_0)$.*

Proof Let $\Psi = (y, z, v, \xi, \eta) \in \mathscr{D}(\mathscr{A}_0)$. Using the inner product (4.3.4) as well as (4.3.7), we obtain after simple integrations by parts

$$\begin{aligned} \langle \mathscr{A}_0 \Psi, \Psi) \rangle_{\mathscr{H}_0} = &-\sigma \int_0^1 y^2\,dx + (ay_x)(1)z(1) - (ay_x)(0)z(0) \\ &- \frac{K}{2}[v^2(1) - v^2(0)] + \xi(ay_x)(0) + \alpha\xi v(1) \end{aligned}$$

$$\begin{aligned} &-\eta(ay_x)(1) + \varpi\left(\int_0^1 z\,dx + \tau\alpha \int_0^1 v\,dx + m\xi + M\eta\right)(\alpha v(0) - \alpha z(0)) \\ &= -\sigma \int_0^1 y^2\,dx + \alpha\xi v(1) - \frac{K}{2}v^2(1) + \frac{K}{2}v^2(0). \end{aligned}$$

Applying Young's inequality, we have

$$\langle \mathscr{A}_0 \Psi, \Psi) \rangle_{\mathscr{H}_0} \leq -\sigma \int_0^1 y^2\,dx + \frac{K+\alpha}{2}\xi^2 + \frac{\alpha - K}{2}v^2(1). \tag{4.3.8}$$

Thenceforth, the operator

$$\mathscr{P} := \mathscr{A}_0 - \frac{K+\alpha}{2} I \tag{4.3.9}$$

is dissipative by virtue of the assumption (4.3.5).

Thereafter, let us show that $\lambda I - \mathscr{P}$ is onto $\mathscr{H}_0$, for $\lambda > 0$. To do so, pick any $(f_1, f_2, f_3, f_4, f_5) \in \mathscr{H}_0$, and let us seek $(z, z, u, \xi, \eta) \in \mathscr{D}(\mathscr{A}_0)$ such that $(\lambda I - \mathscr{A}_0)(z, z, u, \xi, \eta) = (f_1, f_2, f_3, f_4, f_5)$. This problem reduces to

$$\begin{cases} \lambda(\lambda+\sigma)z - (ay_x)_x = (\lambda+\sigma)f_1 + f_2, \\ \lambda\left[m\lambda - \alpha e^{-\tau\lambda}\right] y(0) - (ay_x)(0) = mp + (m\lambda - \alpha e^{-\tau\lambda}) f_1(0) \\ \quad +\tau\alpha \displaystyle\int_0^1 e^{-\tau\lambda(1-s)} f_3(s)\,ds, \\ \lambda^2 M y(1) + (ay_x)(1) = Mq + \lambda M f_1(1), \end{cases} \tag{4.3.10}$$

whose weak formulation is

$$\begin{aligned} &\int_0^1 [(\lambda+\sigma)\phi z + a y_x \phi_x]\,dx + \lambda\left[m\lambda - \alpha e^{-\tau\lambda}\right] y(0)\phi(0) + \lambda^2 M y(1)\phi(1) \\ &\quad = \int_0^1 [(\lambda+\sigma) f_1 + f_2]\,\phi\,dx \\ &\quad + \left[mp + (m\lambda - \alpha e^{-\tau\lambda}) f_1(0) + \tau\alpha \int_0^1 e^{-\tau\lambda(1-s)} f_3(s)\,ds \right] \phi(0) \\ &\quad + [Mq + \lambda M f_1(1)]\,\phi(1). \end{aligned}$$

In view of Lax-Milgram theorem (see Theorem 2.1.6), it is easy to check that there exists a unique solution $z \in H^2(0,1)$ of (4.3.10), for $\lambda > 0$, and hence, $\lambda I - \mathscr{A}_0$ is onto $\mathscr{H}_0$. Now, since $\mathscr{P} := \mathscr{A}_0 - \frac{K+\alpha}{2} I$, one can deduce that $\lambda I - \mathscr{P}$ is also onto $\mathscr{H}_0$. Combining the latter with (4.3.8) and evoking semigroups theory [158], the operator $\mathscr{P}$ generates on $\mathscr{H}_0$ a C_0-semigroup of contraction $e^{t\mathscr{P}}$. Lastly, utilizing classical results of bounded perturbations [158], we conclude that $\mathscr{A}_0$ also generates a C_0-semigroup on $\mathscr{H}_0$. The other assertions of the theorem directly follow from semigroups theory of linear operators [158]. □

Remark 4.3.1 Obviously, $\lambda = 0$ is an eigenvalue of $\mathscr{A}_0$, and $(C, 0, 0, 0, 0)$ is the corresponding eigenfunction, where C is a nonzero real number. Nevertheless, one may expect a convergence result of the solutions of (4.3.6) to some equilibrium state $(C, 0, 0, 0, 0)$ as in Sect. 4.2 (see also Sect. 3.2). It turned out that this is not the case. The proof of this claim will be given in the next subsection.

4.3.1.2 Lack of Convergence

As stated before, we shall establish a nonconvergence result of the solutions of (4.3.2) to their steady state. First, consider the following subspace:

$$\tilde{\mathscr{H}_0} = \left\{ (y, z, v, \xi, \eta) \in \mathscr{H}_0; \int_0^1 (\sigma y(x) + z(x) + \alpha \tau v(x))\, dx \right.$$
$$\left. -\alpha\, y(0) + m\xi + M\eta = 0 \right\}, \tag{4.3.11}$$

and the operator

$$\tilde{\mathscr{A}_0} : \mathscr{D}(\tilde{\mathscr{A}_0}) := \mathscr{D}(\mathscr{A}_0) \cap \tilde{\mathscr{H}_0} \subset \tilde{\mathscr{H}_0} \to \tilde{\mathscr{H}_0},$$

$$\tilde{\mathscr{A}_0}(y, z, v, \xi, \eta) = \mathscr{A}_0(y, z, v, \xi, \eta), \text{ for any } (y, z, v, \xi, \eta) \in \mathscr{D}(\tilde{\mathscr{A}_0}). \tag{4.3.12}$$

According to the outcome of Sect. 4.3.1.1, the operator $\tilde{\mathscr{A}_0}$ generates a C_0-semigroup $e^{t\tilde{\mathscr{A}_0}}$. Notwithstanding, 0 is not an eigenvalue of $\tilde{\mathscr{A}_0}$. Now, in order to reach our target (a nonconvergence of solutions of (4.3.2) to their equilibrium state), it amounts to checking that the semigroup $e^{t\tilde{\mathscr{A}_0}}$ generated by the operator $\tilde{\mathscr{A}_0}$ is unstable on $\tilde{\mathscr{H}_0}$. To this end, assume that $a(x) = 1$, for all $x \in [0, 1]$. Then, the immediate task is to seek a solution $y(x, t) = e^{\lambda t} q(x)$ of the system associated to the operator $\tilde{\mathscr{A}_0}$ on $\tilde{\mathscr{H}_0}$, in which λ is a nonzero complex number whose real part is nonnegative and q is a nonzero function $H^2(0, 1)$. This leads to solve the problem

$$\begin{cases} y_{tt}(x,t) - y_{xx}(x,t) + \sigma y_t(x,t) = 0, & 0 < x < 1,\ t \in (0, \infty), \\ m y_{tt}(0,t) - y_x(0,t) = \alpha y_t(0, t-\tau) - \beta y_t(0,t), & t \in (0, \infty), \\ M y_{tt}(1,t) + y_x(1,t) = 0, & t \in (0, \infty), \end{cases} \tag{4.3.13}$$

where $y(x, t) = e^{\lambda t} q(x)$. A straightforward computation gives the following system:

$$\begin{cases} q_{xx} - \left(\lambda^2 + \sigma\lambda\right) q = 0, \\ q_x(0) + \lambda \left(\alpha e^{-\lambda\tau} - m\lambda\right) q(0) = 0, \\ q_x(1) + M\lambda^2 q(1) = 0. \end{cases} \tag{4.3.14}$$

Solving the differential equation of (4.3.14) yields $q(x) = k_1 e^{-\sqrt{\lambda^2+\sigma\lambda}x} + k_2 e^{\sqrt{\lambda^2+\sigma\lambda}}$, where k_1 and k_2 are constants that can be determined via the boundary conditions in (4.3.14). The latter leads us to claim that q is a nontrivial solution of (4.3.14) if and only if λ is a nonzero solution of the following equation:

$$\left[\lambda \left(\alpha e^{-\lambda\tau} - \lambda m\right) - \sqrt{\lambda^2 + \sigma\lambda}\right]\left[M\lambda^2 + \sqrt{\lambda^2 + \sigma\lambda}\right] e^{\sqrt{\lambda^2+\sigma\lambda}}$$

$$- \left[\lambda \left(\alpha e^{-\lambda\tau} - \lambda m\right) + \sqrt{\lambda^2 + \sigma\lambda}\right] \left[M\lambda^2 - \sqrt{\lambda^2 + \sigma\lambda}\right] e^{-\sqrt{\lambda^2+\sigma\lambda}} = 0. \quad (4.3.15)$$

For sake of simplicity, we shall pick up $\lambda = \sigma > 0$. Next, we choose $\tau = \sqrt{2}$ and $M = \sqrt{2}/\sigma$. Thereby, (4.3.15) yields

$$\alpha = \sqrt{2}\left(1 + \frac{m}{M}\right) e^{\frac{2}{M}}.$$

Whereupon, (4.3.15) holds for $\lambda = \sigma > 0$ and the above values of the time delay τ as well as the physical parameters M and α. Thus, $y(x,t) = e^{\sigma t} q(x)$ is a solution of (4.3.13), which clearly does not tend to zero. In summary, we have proved the following result:

Theorem 4.3.2 *Let $\tau = \sqrt{2}$ and $M = \sqrt{2}/\sigma$ and $\alpha = \sqrt{2}\left(1 + \frac{m}{M}\right) e^{\frac{2}{M}}$. Although the assumption (4.3.5) holds, the solutions of (4.3.13) are unstable in $\tilde{\mathcal{H}}_0$. In other words, the solutions to the system (4.3.2) do not tend in $\mathcal{H}_0$ to the steady state, as $t \longrightarrow +\infty$.*

4.3.1.3 Asymptotic Behavior Result of $e^{t\mathscr{P}}$

Now, we are going to study the asymptotic behavior of the semigroup $e^{t\mathscr{P}}$ generated by the operator $\mathscr{P}$ (see (4.3.9)) that has been evoked in the proof of Theorem 4.3.1. First, going back to the proof of Theorem 4.3.1, one can claim that the operator $(\lambda I - \mathscr{P})^{-1}$ exists and maps $\mathcal{H}_0$ into $\mathscr{D}(\mathscr{P})$. Next, Sobolev embedding [4] leads to the assertion that $(\lambda I - \mathscr{P})^{-1}$ is compact. Then, the spectrum of $\mathscr{P}$ only consists of isolated eigenvalues of finite algebraic multiplicity (see Remark 2.1.2). Moreover, the set of trajectories of solutions $\{\Psi(t)\}_{t\geq 0}$ of the shifted system

$$\begin{cases} \Psi_t(t) = \mathscr{P}\Psi(t), \\ \Psi(0) = \Psi_0, \end{cases} \quad (4.3.16)$$

is bounded for the graph norm. Consequently, $\{\Psi(t)\}_{t\geq 0}$ is precompact by means of the compactness of $(\lambda I - \mathscr{P})^{-1}$, for $\lambda > 0$. Now, let $\Psi(t) = (y(t), y_t(t), v(t), \xi(t), \eta(t)) = e^{t\mathscr{P}}\Psi_0$ be the solution of (4.3.16) stemmed from $\Psi_0 = (y_0, y_1, \chi, \xi_0, \eta_0) \in \mathscr{D}(\mathscr{P})$. By virtue of LaSalle's principle (see Theorem 2.3.11), it follows that the ω-limit set

$$\omega(\Psi_0) = \left\{\Psi \in \mathcal{H}_0;\ \Psi = \lim_{n\to\infty} e^{t_n\mathscr{P}}\Psi_0, \text{ for some sequence } t_n \to \infty, \text{ as } n \to \infty\right\}$$

is not empty. Additionally, $\omega(\Psi_0)$ is compact, invariant under $e^{t\mathscr{P}}$, and $e^{t\mathscr{P}}\Psi_0 \longrightarrow \omega(\Psi_0)$ as $t \to \infty$ [114]. Next, let $\tilde{\Psi}_0 = \left(\tilde{y}_0, \tilde{y}_1, \tilde{\chi}, \tilde{\xi}, \tilde{\eta}\right) \in \omega(\Psi_0) \subset D(\mathscr{P})$ and consider

$$\tilde{\Psi}(t) = \left(\tilde{y}(t), \tilde{y}_t(t), \tilde{v}(t), \tilde{\xi}(t), \tilde{\eta}(t)\right) = e^{t\mathscr{P}}\tilde{\Psi}_0 \in D(\mathscr{P})$$

as the strong solution of (4.3.6). Since $\|\tilde{\Psi}(t)\|_{\mathscr{H}_0}$ is constant [114, Theorem 2.1.3 p. 18], we have

$$< \mathscr{P}\tilde{\Psi}, \tilde{\Psi} >_{\mathscr{H}_0} = 0.$$

This, together with (4.3.8), gives $\tilde{z} = \tilde{y}_t = 0$, and $\tilde{v}(1) = \tilde{y}_t(0, t-\tau) = 0$ as long as $K > \alpha$. Consequently, $\tilde{y} = \zeta \in \mathbb{R}$ is a constant function, and hence, $\omega(\Psi_0) = \{(\zeta, 0, 0, 0, 0)\}$. Now, it remains to find ζ. To do so, let $(\zeta, 0, 0, 0, 0) \in \omega(\Psi_0)$, which yields

$$\begin{aligned} \Psi(t_n) = (y(t_n), y_t(t_n), v(t_n), \xi(t_n), \eta(t_n)) = e^{t_n\mathscr{P}}\Psi_0 \longrightarrow (\zeta, 0, 0, 0, 0),\\ \text{for some } t_n \to \infty, \text{as } n \to \infty. \end{aligned} \tag{4.3.17}$$

In turn, any solution of (4.3.16) stemmed from $\Psi_0 = (y_0, y_1, \chi, \xi_0, \eta_0)$ verifies

$$\begin{aligned} \frac{\mathrm{d}}{\mathrm{dt}}\Bigg[\int_0^1 (\sigma y(x,t) + y_t(x,t) + \alpha\tau y_t(0, t - x\tau))\, dx + m y_t(0,t)\\ + M y_t(1,t) - \alpha y(0,t)\Bigg] = 0, \end{aligned}$$

for each $t \geq 0$, and thus

$$\begin{aligned} &\int_0^1 (\sigma y(x,t) + y_t(x,t) + \alpha\tau y_t(0, t - x\tau))\, dx + m y_t(0,t)\\ &+ M y_t(1,t) - \alpha y(0,t)\\ &= \int_0^1 (\sigma y(x,0) + y_t(x,0) + \alpha\tau y_t(0, -x\tau))\, dx + m y_t(0,0)\\ &+ M y_t(1,0) - \alpha y(0,0)\\ &= \int_0^1 (\sigma y_0(x) + y_1(x) + \alpha\tau\chi(-x\tau))\, dx + m\xi_0 + M\eta_0 - \alpha y_0(0). \end{aligned} \tag{4.3.18}$$

Finally, letting $t = t_n$ in (4.3.18) with $n \to \infty$, and then using (4.3.17), one can get the desired value of

$$\zeta = \frac{1}{\sigma - \alpha}\left(\int_0^1 (\sigma y_0(x) + y_1(x) + \alpha\tau\chi(-\tau x))\, dx - \alpha y_0(0) + m\xi_0 + M\eta_0\right),$$

for $\alpha \neq \sigma$.

To sum up, we have shown the following result:

Theorem 4.3.3 *Under the assumptions (4.1.5) and $K > \alpha$, where $\alpha \neq \sigma$, we have: for any initial data $\Psi_0 = (y_0, y_1, \chi, \xi_0, \eta_0) \in \mathscr{H}_0$, the solution $\Psi(t) = \Big(z, y_t, y_t(0, t-x\tau), y_t(0,t), y_t(1,t)\Big)$ of (4.3.16) converges in $\mathscr{H}_0$ to $(\zeta, 0, 0, 0, 0)$, as $t \longrightarrow +\infty$, where*

$$\zeta = \frac{1}{\sigma - \alpha}\left[\int_0^1 (\sigma y_0(x) + y_1(x) + \alpha\tau\chi(-\tau x))\, dx - \alpha y_0(0) + m\xi_0 + M\eta_0\right]. \tag{4.3.19}$$

4.3.1.4 Exponential Convergence Result for $e^{t\mathscr{P}}$

This subsection addresses the problem of determining the convergence rate of the solutions of the shifted system (4.3.16). To proceed, we shall use Theorem 2.3.8.

Let us first define another operator on the subspace $\tilde{\mathscr{H}_0}$ (see (4.3.11)) as follows:

$$\tilde{\mathscr{P}} : \mathscr{D}(\tilde{\mathscr{P}}) := \mathscr{D}(\mathscr{P}) \cap \tilde{\mathscr{H}_0} \subset \tilde{\mathscr{H}_0} \to \tilde{\mathscr{H}_0},$$

$$\tilde{\mathscr{P}}(y, z, v, \xi, \eta) = \mathscr{P}(y, z, v, \xi, \eta),\ \forall\, (y, z, v, \xi, \eta) \in \mathscr{D}(\tilde{\mathscr{P}}). \tag{4.3.20}$$

Recalling the results we previously obtained, one can claim that $\tilde{\mathscr{P}}$ defined by (4.3.20) generates a C_0-semigroup of contractions $e^{t\tilde{\mathscr{P}}}$ in $\tilde{\mathscr{H}_0}$. In turn, the spectrum $\sigma(\tilde{\mathscr{P}})$ only consists of isolated eigenvalues of finite algebraic multiplicity (see Remark 2.1.2).

Next, we have the exponential convergence result:

Theorem 4.3.4 *Assume that (4.1.5) holds and K satisfies $K > \alpha$ with $\alpha \neq \sigma$. Then, there exist $C > 0$ and $\omega > 0$ such that for all $t > 0$, we have*

$$\left\| e^{t\tilde{\mathscr{P}}} \right\|_{\mathscr{L}(\tilde{\mathscr{H}_0})} \leq C e^{-\omega t}.$$

Thereby, the solutions of (4.3.16) exponentially go in $\mathscr{H}_0$ to $(\zeta, 0, 0, 0, 0)$ as $t \longrightarrow +\infty$, where ζ is defined in (4.3.19).

Proof In the first lieu, we are going to check that (2.3.16) holds. Consider the eigenvalue problem

$$\tilde{\mathscr{P}} F = i\gamma F, \tag{4.3.21}$$

where $\gamma \in \mathbb{R}$ and $F = (y, z, v, \xi, \eta) = 0_{\tilde{\mathscr{H}}_0}$. The latter writes

$$z - \frac{K+\alpha}{2} y = i\gamma y \tag{4.3.22}$$

$$(a\, y_x)_x - \left(\sigma + \frac{K+\alpha}{2}\right) z = i\gamma z \tag{4.3.23}$$

$$-\frac{v_x}{\tau} - \frac{K+\alpha}{2} v = i\gamma v \tag{4.3.24}$$

$$\frac{1}{m}\left[(a y_x)(0) + \alpha v(0)\right] - \frac{K+\alpha}{2}\xi = i\gamma\xi\ . \tag{4.3.25}$$

$$-\frac{(a y_x)(1)}{M} - \frac{K+\alpha}{2}\eta = i\gamma\eta. \tag{4.3.26}$$

Next, let us take the inner product of (4.3.21) with F, and recalling (4.3.8), we have

$$0 = \Re e\left(< \tilde{\mathscr{P}} F, F >_{\tilde{\mathscr{H}}_0}\right) \le -\sigma \int_0^1 y^2\, dx - \frac{K-\alpha}{2} v^2(1) (\le 0). \tag{4.3.27}$$

Thereby, $z = 0$ in $L^2(0, 1)$, and $v(1) = 0$. Amalgamating $z = 0$ and (4.3.22) gives $z = 0$, and hence, $z(1) = \eta$ by (4.3.26). Exploiting (4.3.24) together with $v(1) = 0$, we get $u = 0$. Finally, $\xi = z(0) = 0$ by (4.3.25). Whereupon, (2.3.16) is fulfilled by the operator $\tilde{\mathscr{P}}$.

We turn now to the proof of the fact that the resolvent operator of $\tilde{\mathscr{P}}$ satisfies (2.3.17). If this claim were not true, then thanks to Banach-Steinhaus theorem (see Theorem 2.1.7), one can find a sequence of real numbers $\gamma_n \to +\infty$ and a sequence of elements $F^n = (y^n, z^n, v^n, \xi^n, \eta^n) \in \mathscr{D}(\tilde{\mathscr{P}})$ such that

$$\|F^n\|_{\tilde{\mathscr{H}}_0} = \|y^n\|_{H^1(0,1)} + \|z^n\|_{L^2(0,1)} + \|v^n\|_{L^2(0,1)} + |\xi^n|_{\mathbb{C}} + |\eta^n|_{\mathbb{C}} = 1 \tag{4.3.28}$$

and

$$\|(i\gamma_n I - \tilde{\mathscr{P}})F^n\|_{\tilde{\mathscr{H}}_0} \to 0 \quad \text{as} \quad n \to \infty, \tag{4.3.29}$$

that is, as $n \to \infty$, we have

$$\left(i\gamma_n + \frac{K+\alpha}{2}\right) y^n - z^n \equiv I^n \to 0 \ \text{ in } \ H^1(0, 1), \tag{4.3.30}$$

$$\left(\sigma + i\gamma_n + \frac{K+\alpha}{2}\right) z^n - (ay_x^n)_x \equiv J^n \to 0 \quad \text{in } L^2(0,1), \tag{4.3.31}$$

$$\left(i\gamma_n + \frac{K+\alpha}{2}\right) v^n + \frac{v_x^n}{\tau} \equiv K^n \to 0 \quad \text{in } L^2(0,1), \tag{4.3.32}$$

$$\left(i\gamma_n + \frac{K+\alpha}{2}\right) \xi^n - \frac{1}{m}\left[(ay_x^n)(0) + \alpha v^n(1)\right] \equiv Q^n \to 0 \quad \text{in } \mathbb{C}, \tag{4.3.33}$$

$$\left(i\gamma_n + \frac{K+\alpha}{2}\right) \eta^n + \frac{(ay_x^n)(1)}{M} \equiv R^n \to 0 \quad \text{in } \mathbb{C}. \tag{4.3.34}$$

Exploring the fact that

$$\left\|(i\gamma_n I - \tilde{\mathscr{P}})F^n\right\|_{\tilde{\mathscr{H}_0}} \geq \left|\Re e\langle(i\gamma_n I - \tilde{\mathscr{P}})F^n, F^n\rangle_{\tilde{\mathscr{H}_0}}\right|,$$

together with (4.3.27)–(4.3.29), we have

$$z^n \to 0, \quad \gamma_n y^n \to 0, \quad \text{and } y^n \to 0, \quad \text{in } L^2(0,1) \tag{4.3.35}$$

as well as

$$v^n(1) \to 0 \quad \text{in } \mathbb{C}. \tag{4.3.36}$$

Combining (4.3.32) and (4.3.36) yields

$$v^n \longrightarrow 0 \quad \text{in } L^2(0,1). \tag{4.3.37}$$

It follows from (4.3.31) that

$$\frac{(ay_x^n)_x}{\gamma_n} = \left(\frac{\sigma + \dfrac{K+\alpha}{2}}{\gamma_n} + i\right) z^n - \frac{J^n}{\gamma_n}.$$

This, together with (4.3.35), gives

$$\frac{(ay_x^n)_x}{\gamma_n} \longrightarrow 0 \quad \text{in } L^2(0,1). \tag{4.3.38}$$

Applying Gagliardo-Nirenberg interpolation inequality (see Theorem 2.1.4), we have

$$\|ay_x^n\|_2^2 \leq c\,\frac{\|(ay_x^n)_x\|_2}{|\gamma_n|}\,\|\gamma_n y^n\|_2,$$

where $\|\cdot\|_2$ is the usual norm in $L^2(0,1)$, and c is a positive constant independent of n. Thereby,

$$\left\|ay_x^n\right\|_2 \longrightarrow 0 \quad \text{in} \quad L^2(0,1), \tag{4.3.39}$$

due to (4.3.38) and (4.3.35).

Furthermore, we have

$$\frac{(ay_x^n)(1)}{\gamma_n} \longrightarrow 0 \quad \text{in} \quad \mathbb{C}, \tag{4.3.40}$$

by means of (4.3.28) and (4.3.38). Amalgamating (4.3.34) and (4.3.40), we have

$$\eta^n = z^n(1) \to 0, \quad \text{in} \;\; \mathbb{C}. \tag{4.3.41}$$

Similarly, it follows from (4.3.33) that

$$\xi^n = z^n(0) \to 0, \quad \text{in} \;\; \mathbb{C}. \tag{4.3.42}$$

To recapitulate, it follows from (4.3.35)–(4.3.37), (4.3.41), and (4.3.42) that $\|F^n\|_{\tilde{\mathscr{H}}_0} \longrightarrow 0$, which contradicts the fact that $\|F^n\|_{\tilde{\mathscr{H}}_0} = 1, \;\; \forall\, n \in \mathbb{N}$. Thus, conditions (2.3.16) and (2.3.17) are fulfilled. This completes the proof of Theorem 4.3.4. □

Now, we go back to our original system (4.1.4) and (4.2.1) together with (4.3.1) and investigate in the next subsections the existence and uniqueness of its solutions as well as their long-time behavior.

4.3.2 Well-Posedness of the System (4.1.4), (4.2.1), and (4.3.1)

Let $v(x,t) = y_t(0, t - x\tau)$, where $x \in (0,1)$. Then, the closed-loop system (4.1.4), (4.2.1), and (4.3.1) becomes

$$\begin{cases} y_{tt}(x,t) - (ay_x)_x(x,t) + \sigma y_t(x,t) = 0, & x \in (0,1),\ t > 0, \\ \tau v_t(x,t) + v_x(x,t) = 0, & x \in (0,1),\ t > 0, \\ my_{tt}(0,t) - (ay_x)(0,t) = \alpha v(1,t) - \beta v(0,t), & t > 0, \\ My_{tt}(1,t) + (ay_x)(1,t) = 0, & t > 0, \\ y(x,0) = y_0(x),\ \ y_t(x,0) = y_1(x), & x \in (0,1), \\ v(x,0) = y_t(0, -x\tau) = \chi(-x\tau), & x \in (0,1). \end{cases} \tag{4.3.43}$$

Next, consider the state variable $\Phi = (y, z, v, \xi, \eta)$, where $z(\cdot, t) = y_t(\cdot, t)$, $\xi(t) = y_t(0, t)$ and $\eta(t) = y_t(1, t)$. Subsequently, the state space of our system will be

$$\mathscr{H} = H^1(0, 1) \times L^2(0, 1) \times L^2(0, 1) \times \mathbb{R}^2.$$

The immediate task is to equip $\mathscr{H}$ with an appropriate inner product that induces a norm equivalent to the usual one. To do so, let K be a positive constant to be obtained and define the energy associated to (4.3.43) as follows:

$$\mathscr{E}(t) = E_0(t) + E_1(t), \tag{4.3.44}$$

where

$$E_0(t) = \frac{1}{2}\left[\int_0^1 \left[y_t^2(x,t) + a(x)y_x^2(x,t) + K\tau y_t^2(0, t - x\tau)\right] dx \right.$$
$$\left. + m y_t^2(0,t) + M y_t^2(1,t)\right], \tag{4.3.45}$$

and

$$E_1(t) = \frac{1}{2}\left[\int_0^1 [\sigma y(x,t) + y_t(x,t) + \alpha\tau y_t(0, t - x\tau)]\, dx \right.$$
$$\left. + m y_t(0,t) + M y_t(1,t) + (\beta - \alpha) y(0,t)\right]^2. \tag{4.3.46}$$

Differentiating E_1 in a formal way, using (4.3.43), and integrating by parts, we obtain after a straightforward computation

$$\mathscr{E}'(t) \leq -\sigma \int_0^1 y_t^2(x,t)\, dx + \left(\frac{K}{2} + \frac{\alpha}{2} - \beta\right) y_t^2(0,t) + \frac{1}{2}(\alpha - K)\, y_t^2(0, t - \tau). \tag{4.3.47}$$

Clearly, the energy $\mathscr{E}(t)$ is decreasing if we assume that

$$\alpha < \beta \tag{4.3.48}$$

and then choose K such that

$$\alpha \leq K \leq 2\beta - \alpha. \tag{4.3.49}$$

Whereupon, the space $\mathscr{H}$ shall be equipped with the following real inner product:

$$
\begin{aligned}
&\langle (y, z, v, \xi, \eta), (\tilde{y}, \tilde{z}, \tilde{v}, \tilde{\xi}, \tilde{\eta}) \rangle_{\mathscr{H}} = \int_0^1 (a y_x \tilde{y}_x + z\tilde{z})\, dx \\
&+ K\tau \textstyle\int_0^1 v\tilde{v}\, dx + m\xi\tilde{\xi} + M\eta\tilde{\eta} \\
&+ \varpi \left[\int_0^1 (\sigma y + z) dx + m\xi + M\eta + (\beta - \alpha) y(0) + \tau\alpha \int_0^1 v\, dx \right] \\
&\times \left[\int_0^1 (\sigma \tilde{y} + \tilde{z})\, dx + m\tilde{\xi} + M\tilde{\eta} + (\beta - \alpha)\tilde{y}(0) + \tau\alpha \int_0^1 \tilde{v}\, dx \right],
\end{aligned}
\tag{4.3.50}
$$

where $K > 0$ satisfies (4.3.49) and ϖ is a positive constant sufficiently small so that the norm induced by (4.3.50) is equivalent to the usual norm of $\mathscr{H}$. In fact, arguing as in Proposition 4.2.1, one can establish the following result:

Lemma 4.3.1 *The state space $\mathscr{H}$ equipped with the inner product (4.3.50) becomes a Hilbert space as long as ϖ is sufficiently small and also the hypotheses (4.1.5), (4.3.48), and (4.3.49) are fulfilled.*

Now, one can write the closed-loop system (4.3.43) as follows:

$$
\begin{cases}
\Phi_t(t) = \mathscr{A}\Phi(t), \\
\Phi(0) = \Phi_0,
\end{cases}
\tag{4.3.51}
$$

where the linear operator $\mathscr{A}$ is given by

$$
\begin{aligned}
&\mathscr{D}(\mathscr{A}) = \left\{ (y, z, v, \xi, \eta) \in \mathscr{H};\, z \in H^2(0,1), \quad z,\, v \in H^1(0,1), \right. \\
&\quad \left. \xi = v(0) = z(0), \;\; \eta = z(1) \right\}, \\
&\mathscr{A}(y, z, v, \xi, \eta) = \left(z,\, (a y_x)_x - \sigma z,\, -\frac{v_x}{\tau},\, \frac{1}{m} \left[(a y_x)(0) \right.\right. \\
&- \left.\left. \beta\xi + \alpha v(1) \right],\, -\frac{(a y_x)(1)}{M} \right),
\end{aligned}
\tag{4.3.52}
$$

whereas $\Phi = (y, z, v, \xi, \eta)$ and $\Phi_0 = (y_0, y_1, \chi(-\tau\cdot), \xi_0, \eta_0)$.

The well-posedness result is:

Theorem 4.3.5 *Assume that (4.1.5), (4.3.48), and (4.3.49) hold. Then, the operator $\mathscr{A}$ defined by (4.3.52) generates a C_0-semigroup of contractions $e^{t\mathscr{A}}$ on $\mathscr{H}$. Consequently, if $\Phi_0 \in \mathscr{D}(\mathscr{A})$, then the solution Φ is strong and belongs to $C([0,\infty); \mathscr{D}(\mathscr{A}) \cap C^1([0,\infty); \mathscr{H})$. However, for any initial condition $\Phi_0 \in \mathscr{H}$, the system (4.3.51) has a unique mild solution $\Phi \in C([0,\infty); \mathscr{H})$. Finally, the spectrum of $\mathscr{A}$ only consists of isolated eigenvalues of finite algebraic multiplicity.*

Proof Given $\Phi = (y, z, v, \xi, \eta) \in \mathscr{D}(\mathscr{A})$, and recalling (4.3.50) as well as (4.3.52), we obtain after several integrations by parts

$$
\begin{aligned}
\langle \mathscr{A}\Phi, \Phi)\rangle_{\mathscr{H}} &= -\sigma \int_0^1 z^2\,dx + (ay_x)(1)z(1) - (ay_x)(0)z(0) \\
&\quad - \frac{K}{2}(v^2(1) - v^2(0)) + \xi(ay_x)(0) - \beta\xi^2 \\
&\quad + \varpi \left(\int_0^1 z\,dx + \tau\alpha \int_0^1 v\,dx + m\xi + M\eta \right) (\alpha v(0) \\
&\quad + \beta\xi + (\beta - \alpha)z(0)) \\
&\quad + \alpha\xi v(1) - \eta(ay_x)(1) \\
&= -\sigma \int_0^1 z^2\,dx + \alpha\xi v(1) - \frac{K}{2}v^2(1) + \frac{K}{2}v^2(0) - \beta\xi^2.
\end{aligned}
$$

Invoking Young's inequality, we deduce that

$$
\langle \mathscr{A}\Phi, \Phi)\rangle_{\mathscr{H}} \le -\sigma \int_0^1 z^2\,dx + \left[-\beta + \frac{K+\alpha}{2} \right] \xi^2 + \frac{\alpha - K}{2} v^2(1). \tag{4.3.53}
$$

Thenceforth, the operator $\mathscr{A}$ is dissipative due to the assumptions (4.3.48)–(4.3.49).

On the other hand, one can show that the operator $\lambda I - \mathscr{A}$ is onto $\mathscr{H}$, for $\lambda > 0$. Indeed, taking $(g_1, g_2, g_3, g_4, g_5) \in \mathscr{H}$, we seek $(y, z, v, \xi, \eta) \in \mathscr{D}(\mathscr{A})$ such that $(\lambda I - \mathscr{A})(y, z, v, \xi, \eta) = (g_1, g_2, g_3, g_4, g_5)$. A classical argument leads us to deal with the elliptic problem

$$
\begin{cases}
\lambda(\lambda + \sigma)y - (ay_x)_x = (\lambda + \sigma)g_1 + g_2, \\
\lambda\left[(m\lambda + \beta) - \alpha e^{-\tau\lambda}\right] y(0) - (ay_x)(0) = mp + (m\lambda + \beta - \alpha e^{-\tau\lambda})g_1(0) \\
\qquad\qquad\qquad\qquad + \tau\alpha \displaystyle\int_0^1 e^{-\tau\lambda(1-s)} g_3(s)\,ds, \\
\lambda^2 M y(1) + (ay_x)(1) = Mg_5 + \lambda M g_1(1),
\end{cases} \tag{4.3.54}
$$

whose weak formulation is

$$
\begin{aligned}
&\int_0^1 [(\lambda + \sigma)\phi y + ay_x\phi_x]\,dx + \lambda\left[(m\lambda + \beta) - \alpha e^{-\tau\lambda}\right] y(0)\phi(0) + \lambda^2 M y(1)\phi(1) \\
&\quad = \int_0^1 [(\lambda + \sigma)f + g]\,\phi\,dx + \Big[mp + (m\lambda + \beta - \alpha e^{-\tau\lambda})g_1(0) \\
&\qquad + \tau\alpha \int_0^1 e^{-\tau\lambda(1-s)} v(s)\,ds\Big]\phi(0)
\end{aligned}
$$

$$+ [Mg_5 + \lambda M g_1(1)]\,\phi(1). \tag{4.3.55}$$

Applying Lax-Milgram theorem (see Theorem 2.1.6), there exists a unique solution $z \in H^2(0,1)$ of (4.3.54) for $\lambda > 0$, and hence, $\lambda I - \mathscr{A}$ is onto $\mathscr{H}$. Therefore, $(\lambda I - \mathscr{A})^{-1}$ exists and maps $\mathscr{H}$ into $\mathscr{D}(\mathscr{A})$. Thereafter, Sobolev embedding [4] implies that $(\lambda I - \mathscr{A})^{-1}$ is compact and $\sigma(\mathscr{A})$ only has isolated eigenvalues of finite algebraic multiplicity (see Remark 2.1.2). In turn, semigroups theory [158] permits to vindicate the other assertions of our theorem. □

4.3.3 Asymptotic Behavior of the System (4.1.4), (4.2.1), and (4.3.1)

In light of (4.3.52), we have $0 \in \sigma(\mathscr{A})$, and the corresponding eigenfunction is $\Upsilon(1,0,0,0,0)$, where Υ is a nonzero real number. Hence, the corresponding energy $\mathscr{E}(t)$, given by (4.3.44), remains constant and so will never decay to zero as t is large. In turn, the following convergence result will be proved.

Theorem 4.3.6 *Assume that (4.1.5), (4.3.48) hold, and the constant K verifies the inequalities $\alpha < K < 2\beta - \alpha$. Then, for any initial condition $\Phi_0 = (y_0, y_1, \chi, \xi_0, \eta_0) \in \mathscr{H}$, the solution $\Phi(t) = \Big(z, y_t, y_t(0, t - x\tau), y_t(0,t), y_t(1,t)\Big)$ of the system (4.3.51) goes in $\mathscr{H}$ to the state $(\Xi, 0, 0, 0, 0)$ as t tends to $+\infty$, where*

$$\Xi = \frac{1}{\sigma + \beta - \alpha}\left[\int_0^1 (\sigma y_0 x) + y_1 x) + \alpha\tau\chi(-\tau x))\,dx + (\beta - \alpha) y_0(0) + m\xi_0 + M\eta_0\right]. \tag{4.3.56}$$

Proof Let $\Phi(t) = (y(t), y_t(t), v(t), \xi(t), \eta(t)) = e^{t\mathscr{A}}\Phi_0$ be the solution of (4.3.43) whose initial data is $\Phi_0 \in \mathscr{D}(\mathscr{A})$. In light of Theorem 4.3.5, the set of trajectories $\{\Phi(t)\}_{t\geq 0}$ is bounded for the graph norm. Since the operator $(\lambda I - \mathscr{A})^{-1}$, for $\lambda > 0$is compact, we deduce that $\{\Phi(t)\}_{t\geq 0}$ is indeed precompact. Thanks to LaSalle's principle (see Theorem 2.3.11), it follows that the ω-limit set

$$\omega(\Phi_0) = \Big\{\Phi \in \mathscr{H};\quad \Phi = \lim_{n\to\infty} e^{t_n\mathscr{A}}\Phi_0, \text{ for some sequence } t_n \to \infty, \text{ as } n \to \infty\Big\}$$

is non-empty, compact, invariant under the semigroup $e^{t\mathscr{A}}$. Moreover, the solution $e^{t\mathscr{A}}\Phi_0 \longrightarrow \omega(\Phi_0)$ as $t \to \infty$ [114]. Next, let $\tilde{\Phi}_0 = \Big(\tilde{z}_0, \tilde{z}_1, \tilde{f}, \tilde{\xi}, \tilde{\eta}\Big) \in \omega(\Phi_0) \subset D(\mathscr{A})$ and consider the unique strong solution

$$\tilde{\Phi}(t) = \Big(\tilde{z}(t), \tilde{z}_t(t), \tilde{u}(t), \tilde{\xi}(t), \tilde{\eta}(t)\Big) = e^{t\mathscr{A}}\tilde{\Phi}_0 \in D(\mathscr{A})$$

of (4.3.51). Exploiting the fact that $\|\tilde{\Phi}(t)\|_{\mathscr{H}}$ is constant [114, Theorem 2.1.3 p. 18], we have

$$< \mathscr{A}\tilde{\Phi}, \tilde{\Phi} >_{\mathscr{H}} = 0. \tag{4.3.57}$$

Combining (4.3.53) with (4.3.57), we obtain $\tilde{z} = \tilde{y}_t$, $\tilde{\xi} = \tilde{y}_t(0,t) = 0$, and $\tilde{v}(1) = \tilde{y}_t(0, t-\tau) = 0$. Consequently, $\tilde{y}$ is constant, and $\omega(\Phi_0)$ reduces to $(\Xi, 0, 0, 0, 0)$.

Now, the proof will be completed once we find Ξ. This can be done by arguing as in the proof of Theorem 4.3.3. Indeed, let $(\Xi, 0, 0, 0, 0) \in \omega(\Phi_0)$, which yields

$$\Phi(t_n) = (y(t_n), y_t(t_n), v(t_n), \xi(t_n), \eta(t_n)) = e^{t_n \mathscr{A}}\Phi_0 \longrightarrow (\Xi, 0, 0, 0, 0),$$
$$\text{for some } t_n \to \infty, \text{ as } n \to \infty. \tag{4.3.58}$$

In turn, each solution of (4.3.51) stemmed from $\Phi_0 = (y_0, y_1, f, \xi_0, \eta_0)$ verifies

$$\frac{\mathrm{d}}{\mathrm{d}t}\left[\int_0^1 (\sigma y(x,t) + y_t(x,t) + \alpha\tau y_t(0, t - x\tau))\, dx + m y_t(0,t) + M y_t(1,t)\right.$$
$$\left. + (\beta - \alpha) y(0,t)\right] = 0, \tag{4.3.59}$$

for each $t \geq 0$, and thus,

$$\int_0^1 (\sigma y(x,t) + y_t(x,t) + \alpha\tau y_t(0, t - x\tau))\, dx + m y_t(0,t)$$
$$+ M y_t(1,t) + (\beta - \alpha) y(0,t)$$
$$= \int_0^1 (\sigma y_0(x) + y_1(x) + \alpha\tau\chi(-x\tau))\, dx + m\xi_0 + M\eta_0 + (\beta - \alpha) y_0(0). \tag{4.3.60}$$

Lastly, letting $t = t_n$ in (4.3.60) with $n \to \infty$ and then exploring (4.3.58), one can get the expression of Ξ. □

4.3.4 Exponential Convergence Rate for (4.1.4), (4.2.1), and (4.3.1)

The main result of this subsection is to show that the convergence rate of the solutions of the system (4.3.43) is exponential. The proof is essentially based on the resolvent method (see Theorem 2.3.8).

Firstly, consider the space

$$\hat{\mathscr{H}} = \left\{ (y, z, v, \xi, \eta) \in \mathscr{H};\ \int_0^1 (\sigma y(x) + z(x) + \alpha \tau v(x))\, dx + (\beta - \alpha)\, y(0) \right.$$

$$\left. + m\xi + M\eta = 0 \right\}.$$

Clearly, $\hat{\mathscr{H}}$ is a closed subspace of $\mathscr{H}$ and has codimension 1. Subsequently, let us consider a new operator associated to the operator $\mathscr{A}$ (see (4.3.52))

$$\hat{\mathscr{A}} : \mathscr{D}(\hat{\mathscr{A}}) := \mathscr{D}(\mathscr{A}) \cap \hat{\mathscr{H}} \subset \hat{\mathscr{H}} \to \hat{\mathscr{H}},$$

$$\hat{\mathscr{A}}(y, z, v, \xi, \eta) = \mathscr{A}(y, z, v, \xi, \eta),\ \forall\, (y, z, v, \xi, \eta) \in \mathscr{D}(\hat{\mathscr{A}}). \tag{4.3.61}$$

By virtue of Theorem 4.3.5, $\hat{\mathscr{A}}$ generates a C_0-semigroup of contractions $e^{t\hat{\mathscr{A}}}$ on $\hat{\mathscr{H}}$ under the conditions (4.3.48) and (4.3.49). Also, the spectrum $\sigma(\hat{\mathscr{A}})$ of $\hat{\mathscr{A}}$ only possesses isolated eigenvalues of finite algebraic multiplicity (see Remark 2.1.2).

Our main result is:

Theorem 4.3.7 *Assume that the requirements (4.1.5), (4.3.48), and $\alpha < K < 2\beta - \alpha$ are satisfied. Then, there exist two positive constants $C0$ and ω so that for all $t > 0$, we have*

$$\left\| e^{t\hat{\mathscr{A}}} \right\|_{\mathscr{L}(\hat{\mathscr{H}})} \leq C e^{-\omega t}.$$

Proof For sake of clarity, our desired outcome will be proved by proceeding by steps.

Step 1 The task ahead is to check that $\hat{\mathscr{A}}$ satisfies (2.3.16), that is, the equation

$$\hat{\mathscr{A}} \Phi = i\gamma \Phi \tag{4.3.62}$$

with $\Phi = (y, z, v, \xi, \eta) \in \mathscr{D}(\hat{\mathscr{A}})$ and $\gamma \in \mathbb{R}$ cannot have more than one trivial solution. In other words, one should verify from (4.3.62) that the system

$$z = i\gamma y \tag{4.3.63}$$

$$(a\, y_x)_x - \sigma z = i\gamma z \tag{4.3.64}$$

$$-\frac{v_x}{\tau} = i\gamma v \tag{4.3.65}$$

$$\frac{1}{m}\left[(a y_x)(0) - \beta\xi + \alpha v(0)\right] = i\gamma\xi\ . \tag{4.3.66}$$

$$-\frac{(a y_x)(1)}{M} = i\gamma\eta, \tag{4.3.67}$$

has only a trivial solution. To this end, suppose that $\gamma = 0$. Then (4.3.63) gives $z = 0$, and hence, $\xi = \eta = 0$ as $y(0) = \xi$ and $y(1) = \eta$. In turn, if $\gamma \neq 0$, then exploiting (4.3.62) and (4.3.53), we obtain

$$0 = \Re e\left(< \hat{\mathscr{A}}\Phi, \Phi >_{\hat{\mathscr{H}}}\right) \leq -\sigma \int_0^1 y^2\,dx + \left[-\beta + \frac{K+\alpha}{2}\right]\xi^2 + \frac{\alpha - K}{2} v^2(1) (\leq 0). \tag{4.3.68}$$

Hence, $z = 0$ in $L^2(0, 1)$, $y(0) = \xi = 0$, and $v(1) = 0$. Going back to (4.3.62), we deduce our desired result.

Step 2 It amounts to proving that (2.3.17) is satisfied. Otherwise, we deduce from Banach-Steinhaus theorem (see Theorem 2.1.7) the existence of a sequence of real numbers $\gamma_n \to +\infty$ and a sequence of vectors $\Phi^n = (y^n, z^n, v^n, \xi^n, \eta^n) \in \mathscr{D}(\hat{\mathscr{A}})$ with

$$\|\Phi^n\|_{\hat{\mathscr{H}}} = \|y^n\|_{H^1(0,1)} + \|z^n\|_{L^2(0,1)} + \|v^n\|_{L^2(0,1)} + |\xi^n|_{\mathbb{C}} + |\eta^n|_{\mathbb{C}} = 1 \tag{4.3.69}$$

such that

$$\|(i\gamma_n I - \hat{\mathscr{A}})\Phi^n\|_{\hat{\mathscr{H}}} \to 0 \quad \text{as} \quad n \to \infty, \tag{4.3.70}$$

that is, as $n \to \infty$, we have

$$i\gamma_n y^n - z^n \equiv I^n \to 0 \;\; \text{in} \;\; H^1(0, 1), \tag{4.3.71}$$

$$(i\gamma_n + \sigma)z^n - (ay_x^n)_x \equiv J^n \to 0 \;\; \text{in} \;\; L^2(0, 1), \tag{4.3.72}$$

$$i\gamma_n v^n + \frac{v_x^n}{\tau} \equiv K^n \to 0 \;\; \text{in} \;\; L^2(0, 1), \tag{4.3.73}$$

$$i\gamma_n \xi^n - \frac{1}{m}\left[(ay_x^n)(0) - \beta\xi^n + \alpha v^n(1)\right] \equiv Q^n \to 0 \;\; \text{in} \;\; \mathbb{C}, \tag{4.3.74}$$

$$i\gamma_n \eta^n + \frac{(ay_x^n)(1)}{M} \equiv R^n \to 0 \;\; \text{in} \;\; \mathbb{C}. \tag{4.3.75}$$

Exploring the fact that

$$\left\|(i\gamma_n I - \hat{\mathscr{A}})\Phi^n\right\|_{\hat{\mathscr{H}}} \geq \left|\Re e\langle(i\gamma_n I - \hat{\mathscr{A}})\Phi^n, \Phi^n\rangle_{\hat{\mathscr{H}}}\right|,$$

together with (4.3.68)–(4.3.70), we get

$$z^n \to 0, \quad \gamma_n y^n \to 0, \quad \text{and } y^n \to 0, \quad \text{in } L^2(0, 1), \tag{4.3.76}$$

as well as

$$\xi^n = z^n(0) = v^n(0) \to 0\,, \quad v^n(1) \to 0 \ \text{ in } \mathbb{C}. \tag{4.3.77}$$

In view of (4.3.73), we have

$$v^n(x) = v^n(0)\, e^{-i\tau\gamma_n x} + \tau \int_0^x e^{-i\tau\gamma_n(x-s)} K^n(s)\, ds.$$

Recalling that $K^n \to 0$ in $L^2(0, 1)$ (see (4.3.73)) and invoking (4.3.77), the latter yields

$$v^n \longrightarrow 0 \ \text{ in } \ L^2(0, 1). \tag{4.3.78}$$

Returning to (4.3.72), we get

$$\frac{(ay_x^n)_x}{\gamma_n} = \left(\frac{\sigma}{\gamma_n} + i\right) z^n - \frac{J^n}{\gamma_n},$$

which implies by virtue of (4.3.76)

$$\frac{(ay_x^n)_x}{\gamma_n} \longrightarrow 0 \ \text{ in } \ L^2(0, 1). \tag{4.3.79}$$

Now, since

$$(aw_x^n)(1) = \int_x^1 (ay_r^n)_r\, dr + ay_x^n,$$

we deduce from Young's inequality that

$$\left|(aw_x^n)(1)\right|^2 \le c\left(\|(ay_x^n)_x\|_2^2 + \|ay_x^n\|_2^2\right),$$

where $\|\cdot\|_2$ is the usual norm in $L^2(0, 1)$, and c is a positive constant (independent of n). In the sequel, we shall use, for convenience, the same letter c to represent a positive constant that is independent of n. Whereupon, the last estimate yields

$$\frac{(ay_x^n)(1)}{\gamma_n} \longrightarrow 0 \ \text{ in } \mathbb{C}, \tag{4.3.80}$$

by means of (4.3.69) and (4.3.79). Amalgamating (4.3.74) and (4.3.80), we have

$$\eta^n = z^n(1) \to 0, \quad \text{in } \mathbb{C}. \tag{4.3.81}$$

Evoking Gagliardo-Nirenberg interpolation inequality (see Theorem 2.1.4), we have

$$\left\| a y_x^n \right\|_2^2 \leq c \, \frac{\left\| (a y_x^n)_x \right\|_2}{|\gamma_n|} \left\| \gamma_n y^n \right\|_2 ,$$

for some positive constant c. Thereby,

$$\left\| a y_x^n \right\|_2 \longrightarrow 0 \quad \text{in} \quad L^2(0,1), \tag{4.3.82}$$

due to (4.3.79) and (4.3.76).

Lastly, the findings in (4.3.76)–(4.3.78), (4.3.81), and (4.3.82) contradict the property $\|\Phi^n\|_{\hat{\mathscr{H}}} = 1, \ \forall\, n \in \mathbb{N}$. Therefore, we managed to show that (2.3.16) and (2.3.17) hold, and thus the proof of Theorem 3.2.2 is achieved. □

Remark 4.3.2 An immediate consequence of Theorem 3.2.2 is the exponential convergence of the solutions of (4.3.43) in $\mathscr{H}$ to $(\eta, 0, 0, 0, 0)$ as $t \longrightarrow +\infty$, where η is given by (4.3.56).

4.3.4.1 Lack of Convergence When $\alpha \geq \beta$

This subsection is intended to provide an answer to the following question: What happens to the solutions of (4.3.43) in the case where (4.3.48) does not hold, that is, if $\alpha \geq \beta$? Obviously, the answer to this question could be provided once we show the existence of a delay τ, for which the semigroup $e^{t\hat{\mathscr{A}}}$ can be unstable in $\hat{\mathscr{H}}$. To proceed, suppose that $a = 1$, and then let us look for a positive real number γ and a nonzero function $g \in H^2(0,1)$ so that $y(x,t) = e^{\gamma t} g(x)$ is a solution of the system

$$\begin{cases} y_{tt}(x,t) - y_{xx}(x,t) + \sigma y_t(x,t) = 0, & 0 < x < 1, \ t > 0, \\ m w_{tt}(0,t) - y_x(0,t) = \alpha y_t(0, t-\tau) - \beta y_t(0,t), & t > 0, \\ M w_{tt}(1,t) + y_x(1,t) = 0, & t > 0. \end{cases} \tag{4.3.83}$$

Note that this system is governed on $\hat{\mathscr{H}}$ by the operator $\hat{\mathscr{A}}$ (see (4.3.61)). Then, we can claim $\|y\|_{L^2(0,1)} = e^{\gamma t} \|g\|_{L^2(0,1)} \rightarrow +\infty$, and so (4.3.83) has at least one unstable solution.

One can readily verify that $y(x,t) = e^{\gamma t} g(x)$ is a solution to (4.3.83) if g is a nontrivial solution to

$$\begin{cases} g_{xx} - \left(\gamma^2 + \sigma\gamma\right) g = 0, \\ g_x(0) + \gamma \left(\alpha e^{-\gamma\tau} - m\gamma - \beta\right) g(0) = 0, \\ g_x(1) + M\gamma^2 g(1) = 0. \end{cases} \tag{4.3.84}$$

Solving the differential equation of (4.3.84) yields $y(x) = c_1 e^{-\sqrt{\gamma^2+\sigma\gamma}x} + c_2 e^{\sqrt{\gamma^2+\sigma\gamma}}$, where c_1 and c_2 are constants that can be obtained by means of the boundary conditions in (4.3.84). Indeed, the latter implies that g is a nontrivial solution of (4.3.84) if and only if one finds $\gamma \neq 0$ solution of the following equation:

$$\left[\gamma\left(\alpha e^{-\gamma\tau}-\gamma m-\beta\right)-\sqrt{\gamma^2+\sigma\gamma}\right]\left[M\gamma^2+\sqrt{\gamma^2+\sigma\gamma}\right]e^{\sqrt{\gamma^2+\sigma\gamma}}$$
$$-\left[\gamma\left(\alpha e^{-\gamma\tau}-\gamma m-\beta\right)+\sqrt{\gamma^2+\sigma\gamma}\right]\left[M\gamma^2-\sqrt{\gamma^2+\sigma\gamma}\right]e^{-\sqrt{\gamma^2+\sigma\gamma}}=0. \quad (4.3.85)$$

Now let us take $\gamma=\sigma>0$ and $M=\sqrt{2}/\sigma$. Thereafter, (4.3.85) gives

$$\alpha e^{-\frac{\sqrt{2}}{M}\tau}-\beta-\frac{\sqrt{2}}{M}m-\sqrt{2}=0.$$

Lastly, solving the above equation, we get $\tau=\frac{M}{\sqrt{2}}\ln\left(\frac{\alpha M}{(\beta+\sqrt{2})M+\sqrt{2}m}\right)>0$ provided that $\alpha\geq\beta+\sqrt{2}\left(1+\frac{m}{M}\right)$. Bearing in mind these choices, we conclude that (4.3.85) is satisfied with $\gamma=\sigma>0$. Therefore, we have the optimality result of the condition (4.3.48):

Theorem 4.3.8 *If the assumption (4.3.48) is not satisfied, then there exists at least one time delay τ for which the convergence of solutions of the closed-loop system (4.3.43) does not hold.*

4.3.5 Numerical Examples

The aim of this subsection is to illustrate, via numerical examples, some of the convergence result obtained for the closed-loop system (4.1.4), (4.2.1), and (4.3.1). Specifically, the ultimate objective is to give prominence to the importance of the presence of the damping term for the convergence result. To do so, in order to avoid unnecessary complex computations and without loss of generality, we shall assume $a=1$. Then, a straightforward calculation shows that λ is an eigenvalue of the system if and only

$$\left[\lambda\left(\alpha e^{-\lambda\tau}-\lambda m-\beta\right)-\sqrt{\lambda^2+\sigma\lambda}\right]\left[M\lambda^2+\sqrt{\lambda^2+\sigma\lambda}\right]e^{\sqrt{\lambda^2+\sigma\lambda}}$$
$$-\left[\lambda\left(\alpha e^{-\lambda\tau}-\lambda m-\beta\right)+\sqrt{\lambda^2+\sigma\lambda}\right]\left[M\lambda^2-\sqrt{\lambda^2+\sigma\lambda}\right]e^{-\sqrt{\lambda^2+\sigma\lambda}}=0.$$

Next, we fix the delay $\tau=0.01$ and take $m=2,\ M=5$. Then, solving numerically the above equation by using Maple 13 via the function "fsolve," we are able to get the first 11 eigenvalues of the system in the following cases:

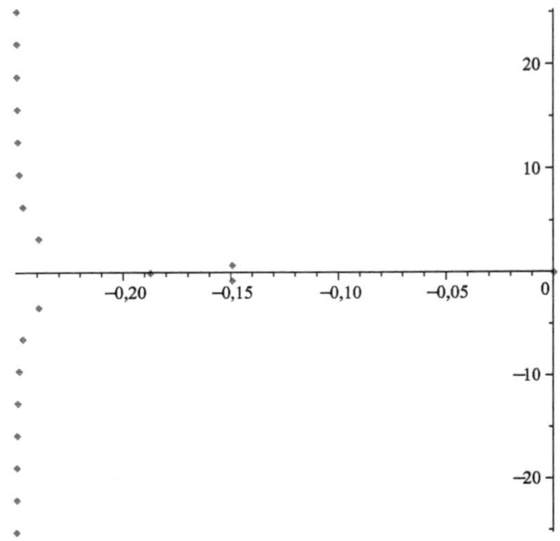

Fig. 4.4 $\mu = 0.1$, $\nu = 1$, and $\sigma = 0.5$

Case 1: Let $\alpha = 0.1,\ \beta = 1$, and $\sigma = 0.5$. In this case, we obtain, of course, one zero eigenvalue (see Fig. 4.4), while the others lie in the left-half plane. This ties with Theorem 3.2.2.

Case 2: Now, we take as in the previous case $\alpha = 0.1,\ \beta = 1$, but we increase the feedback gain $\sigma = 1$. As expected, we obtain the same spectrum configuration but considerably shifted to the left (see Fig. 4.5). This agrees with the outcome of Sect. 4.3.3, as it means that the damping term related to σ has a positive effect on the convergence of the system.

Case 3: Consider $\alpha = 0.1,\ \sigma = 0.5$, but we decrease $\beta = 0.5$. In such an event, the spectrum moves to the right as shown in Fig. 4.6. This explains the importance of the compensating term $\beta y_t(0,t)$ in 4.1.4 to eliminate the negative impact of the delayed term $\alpha y_t(0, t - \tau)$.

Case 4: In order to highlight the previous case, we keep decreasing the value of β to $\beta = 0.11$. It is noticed (see Fig. 4.7) that the first nonzero eigenvalue moves to the left, and this confirms the importance of $\beta y_t(0,t)$ as mentioned in Sect. 4.3.1. Moreover, this case highlights on one hand the relevance of the condition (4.3.48) and confirms on the other hand the result of Theorem 4.3.8.

4.3.6 Concluding Discussion

This section has addressed the problem of improving the convergence rate of solutions of an overhead crane system modeled by a hyperbolic PDE coupled with two ODEs and subject to the effect of a time delay in the boundary. First,

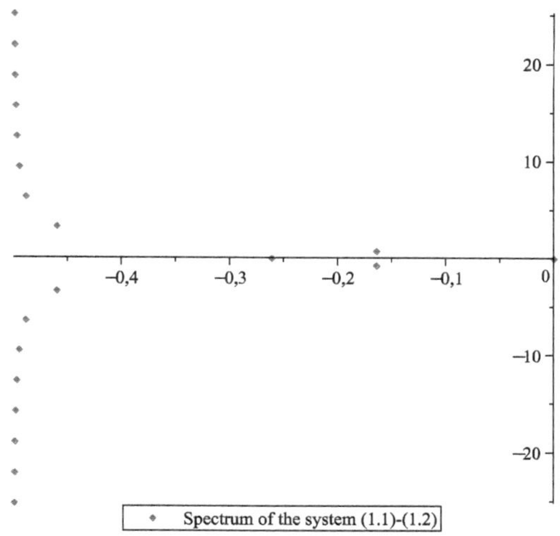

Fig. 4.5 $\mu = 0.1$, $\nu = 1$, and $\sigma = 1$

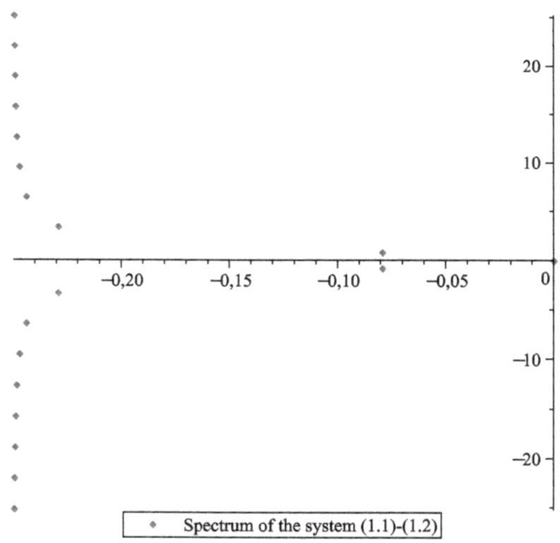

Fig. 4.6 $\mu = 0.1$, $\nu = 0.5$, and $\sigma = 0.5$

an interior damping control is proposed. Then, the system is shown to be well-posed in a functional space with an appropriate norm. Next, it is proved that the solutions converge to a steady state. Last but not least, we show that the convergence rate of solutions is exponential. This finding improves that of Sect. 4.2, where the convergence is only polynomial. This outcome is illustrated by some numerical examples.

One promising research problem is to determine, for practical reasons, the best estimate of the exponential convergence rate ω mentioned in Theorem 4.3.4 and Theorem 3.2.2. This will definitely provide the optimal convergence rate of solutions to the stationary position. Furthermore, in practice, the delay often depends

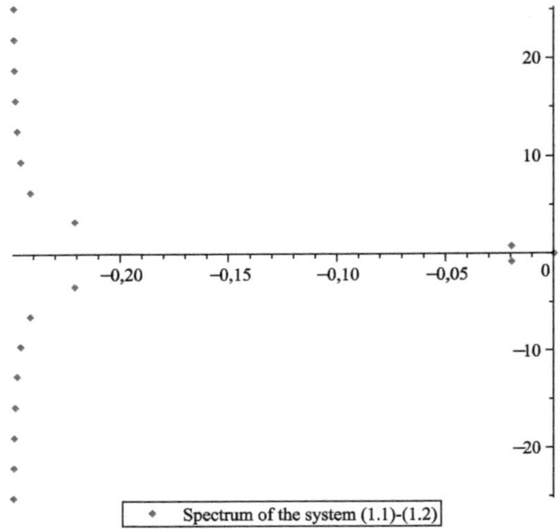

Fig. 4.7 $\mu = 0.1$, $\nu = 1$, and $\sigma = 0.1$

on time, and hence, it is legitimate to investigate the system when the length of the cable varies with time. Finally, noticing that the interior damping is global, one could wonder whether a localized interior damping is enough to stabilize the same system considered in this section. This will be considered in the next section.

4.4 Boundary Delay: Exponential Stabilization via a Localized Interior Control

This section is dedicated to studying the energy decay of the flexible crane structure with a local interior damping. We manage to show that the presence of a localized distributed damping is strong enough to obtain the exponentially decay of the energy of the system. This improves the result obtained in Sect. 4.3, where the damping was distributed over the entire spatial domain.

Our system is

$$\begin{cases} y_{tt}(x,t) - (ay_x)_x(x,t) + b(x)y_t(x,t) = 0, & 0 < x < 1,\ t > 0, \\ my_{tt}(0,t) - (ay_x)(0,t) = -\sigma y(0,t) + \alpha y_t(0,t-\tau) - \beta y_t(0,t), & t > 0, \\ My_{tt}(1,t) + (ay_x)(1,t) = 0, & t > 0, \\ y(x,0) = y_0(x),\ y_t(x,0) = y_1(x), & x \in (0,1), \\ y_t(0,t-\tau) = f(t-\tau), & t \in (0,\tau), \end{cases} \tag{4.4.1}$$

in which y is the cable displacement, $\alpha \geq 0$, $\beta > 0$, and $\tau > 0$ is the time delay. Furthermore, the space variable function $b(x) \in W^{1,\infty}(0,1)$ has local support on

(0, 1) and is strictly positive on some subinterval of (0, 1) so that $\int_0^1 b(x)\,dx > 0$. Finally, recall that $a(x)$ is the modulus of tension of the cable that satisfies (4.1.5).

As mentioned earlier, the exponential convergence result obtained in Sect. 4.3 is established as long as a global interior damping acts on the system, that is, when $b(x) = b$, $\forall x \in [0, 1]$, where b is a positive constant. The main contribution of the current section is to weaken the "global" condition of the distributed damping and show that the exponential result can still be achieved as long as the distributed function $b(x)$ is positive in a subinterval of [0, 1], despite the presence of the boundary *time delay* and the *dynamic* feature of both boundaries.

4.4.1 Well-Posedness of the System (4.4.1)

Let K be a positive constant to be determined, and let us define the energy associated to (4.4.1) as follows:

$$E(t) = \frac{1}{2}\left[\int_0^1 \left[y_t^2(x,t) + a(x)y_x^2(x,t) + K\tau y_t^2(0, t - x\tau)\right] dx + \sigma y^2(0,t) + m y_t^2(0,t) + M y_t^2(1,t)\right]. \tag{4.4.2}$$

Differentiating E formally, using (4.4.1), and integrating by parts, we obtain after a straightforward computation

$$E'(t) \le -\int_0^1 b(x)y_t^2(x,t)\,dx + \left(\frac{K}{2} + \frac{\alpha}{2} - \beta\right) y_t^2(0,t) + \frac{1}{2}(\alpha - K)\, y_t^2(0, t-\tau).$$

Clearly, the energy $E(t)$ is decreasing if we assume that

$$\alpha < \beta, \qquad \alpha \le K \le 2\beta - \alpha. \tag{4.4.3}$$

Subsequently, for $x \in (0, 1)$, let $u(x,t) = y_t(0, t - x\tau)$ [149, 184], and consider the state variable $\Phi = (y, z, u, \xi, \eta)$, where $z(\cdot, t) = y_t(\cdot, t)$, $\xi(t) = y_t(0, t)$ and $\eta(t) = y_t(1, t)$. Next, the state space of our system is

$$\mathscr{H} = H^1(0,1) \times \left(L^2(0,1)\right)^2 \times \mathbb{R}^2,$$

equipped with the following real inner product

$$\langle (y, z, u, \xi, \eta), (\tilde{y}, \tilde{z}, \tilde{u}, \tilde{\xi}, \tilde{\eta}) \rangle_{\mathscr{H}} = \int_0^1 (a y_x \tilde{y}_x + z\tilde{z})\,dx + K\tau \int_0^1 u\tilde{u}\,dx + \sigma y(0)\tilde{y}(0) + m\xi\tilde{\xi} + M\eta\tilde{\eta}, \tag{4.4.4}$$

where $K > 0$ satisfies (4.4.3). One can easily prove that the state space $\mathscr{H}$, endowed with the inner product (4.4.4), is a Hilbert space provided that (4.1.5) and (4.4.3) are fulfilled.

Now, the system (4.4.1) can be written as follows:

$$\begin{cases} \Phi_t(t) = \mathscr{A}\Phi(t), \\ \Phi(0) = \Phi_0, \end{cases} \tag{4.4.5}$$

in which $\mathscr{A}$ is a linear operator defined by

$$\begin{aligned} &\mathscr{D}(\mathscr{A}) = \left\{(y, z, u, \xi, \eta) \in \mathscr{H};\ y \in H^2(0,1),\quad z, u \in H^1(0,1),\right.\\ &\quad \left.\xi = u(0) = z(0),\ \eta = z(1)\right\},\\ &\mathscr{A}(y, z, u, \xi, \eta) = \left(z, (ay_x)_x - b(x)z, -\frac{u_x}{\tau}, \frac{1}{m}\left[-\sigma y(0) + (ay_x)(0)\right.\right.\\ &\left.\left. -\beta\xi + \alpha u(1)\right], -\frac{(ay_x)(1)}{M} \right), \end{aligned} \tag{4.4.6}$$

whereas $\Phi_0 = (y_0, y_1, f(-\tau\cdot), \xi_0, \eta_0)$.

The well-posedness result is stated below.

Theorem 4.4.1 *Assume that (4.1.5) and (4.4.3) hold. Then, the operator $\mathscr{A}$ defined by (4.4.6) is densely defined in $\mathscr{H}$ and generates on $\mathscr{H}$ a C_0-semigroup of contractions $e^{t\mathscr{A}}$. Consequently, if $\Phi_0 \in \mathscr{D}(\mathscr{A})$, then the solution Φ is strong and belongs to $C([0,\infty); \mathscr{D}(\mathscr{A}) \cap C^1([0,\infty); \mathscr{H})$. However, the system (4.4.5) has a unique mild solution $\Phi \in C([0,\infty); \mathscr{H})$ if $\Phi_0 \in \mathscr{H}$. Finally, the spectrum of $\mathscr{A}$ only consists of isolated eigenvalues of finite algebraic multiplicity.*

Proof Given $\Phi \in \mathscr{D}(\mathscr{A})$ and recalling (4.4.4) as well as (4.4.6), we obtain after several integrations by parts

$$\langle \mathscr{A}\Phi, \Phi)\rangle_{\mathscr{H}} = -\int_0^1 b(x)z^2\,dx + \alpha\xi u(1) - \frac{K}{2}u^2(1) + \frac{K}{2}u^2(0) - \beta\xi^2.$$

Invoking Young's inequality as well as (4.4.6), we deduce that

$$\langle \mathscr{A}\Phi, \Phi)\rangle_{\mathscr{H}} \le -\int_0^1 b(x)z^2\,dx + \left[-\beta + \frac{K+\alpha}{2}\right]\xi^2 + \frac{\alpha - K}{2}u^2(1). \tag{4.4.7}$$

Thenceforth, the operator $\mathscr{A}$ is dissipative in light of (4.4.3).

On the other hand, similarly to the previous sections, one can show by means of Lax-Milgram theorem (see Theorem 2.1.6) that $\lambda I - \mathscr{A}$ is onto $\mathscr{H}$, for $\lambda > 0$. In turn, Sobolev embedding [40] implies that $(\lambda I - \mathscr{A})^{-1}$ is compact, and hence, the spectrum of $\mathscr{A}$ consists of isolated eigenvalues of finite algebraic multiplicity only (see Remark 2.1.2). In turn, semigroups theory [158] permits to vindicate the other assertions of our proposition. □

4.4.2 *Exponential Stability of the System (4.4.1)*

The ultimate outcome is to establish the exponential stability of the solutions of the system (4.4.1). The proof depends on an essential way on the resolvent method stated in Theorem 2.3.8 (see also [118, 163]).

Our main result is:

Theorem 4.4.2 *Assume that (4.1.5) and (4.4.3) hold. Then, the solutions of (4.4.1) are exponentially stable in $\mathscr{H}$, that is, the energy $E(t)$ given in (4.4.2) exponentially decays toward zero.*

Proof The proof consists in showing that the two conditions of Theorem 2.3.8 hold.

First, we will show, by contradiction, that the operator $\mathscr{A}$ satisfies the first condition in Theorem 2.3.8. Suppose that there is $\gamma \in \mathbb{R}$ such that $i\gamma$ is an eigenvalue for $\mathscr{A}$, and hence one can find a corresponding eigenfunction $\Phi = (y, z, u, \xi, \eta) \in D(\mathscr{A})$. Consequently, we have

$$\mathscr{A}\Phi = i\gamma\Phi. \tag{4.4.8}$$

Our immediate aim is to prove that this equation has only $\Phi = 0$ as a solution, which contradicts the definition of an eigenfunction. First, the Eq. (4.4.8) is equivalent to considering the following system:

$$z = i\gamma y, \tag{4.4.9}$$

$$(a(x)y_x)_x - b(x)z = i\gamma z, \tag{4.4.10}$$

$$u_x = -i\gamma\tau u, \tag{4.4.11}$$

$$(ay_x)(0) - \sigma y(0) - \beta\xi + \alpha u(1) = i\gamma m\xi, \tag{4.4.12}$$

$$(ay_x)(1) = -i\gamma M\eta. \tag{4.4.13}$$

Second, we will consider two cases:

- Case $\gamma = 0$: In this case, by (4.4.9), we have $z = 0$, which gives $\xi = \eta = 0$ since $z(0) = \xi$ and $z(1) = \eta$. Therefore, (4.4.10) and (4.4.13) yield $y = k_1 \in \mathbb{R}$. Finally, using (4.4.11) and (4.4.12), we find that $u = 0$ and $y = 0$. Hence, (y, z, u, ξ, η) must be the trivial solution of (4.4.8), which is the desired result.
- Case $\gamma \neq 0$: Taking the $L^2(0, 1)$-inner product with Φ of both sides of (4.4.8) and using (4.4.7), we immediately obtain

$$0 = \Re e\left(\langle \mathscr{A}\Phi, \Phi\rangle_{\mathscr{H}}\right) \leq -\int_0^1 b(x)z^2\,dx + \left(-\beta + \frac{K+\alpha}{2}\right)\xi^2$$
$$+ \frac{\alpha - K}{2}u^2(1) \leq 0. \tag{4.4.14}$$

Whereupon, $\xi = u(0) = z(0) = y(0) = 0$, $u(1) = 0$, and $\int_0^1 b(x)z^2\,dx = 0$. The latter clearly implies that $b(x)z = 0$, for all $x \in [0, 1]$ as $\|b(x)z\|^2 \leq \|b\|_{L^\infty(0,1)} \int_0^1 b(x)z^2\,dx = 0$. Recalling that $u(0) = 0$ and using (4.4.11), we obtain $u = 0$. Combining these facts with (4.4.9)–(4.4.13), we deduce that y satisfies the boundary value problem:

$$\begin{cases} (ay_x)_x + \gamma^2 y = 0, \\ y(0) = y_x(0) = 0. \end{cases}$$

Next, let

$$\mu(x) = \exp\left(\int_x^1 \left|\frac{a_x}{a}\right|(s)\,ds\right)\left(-1 - \int_x^1 \exp\left(-\int_s^1 \left|\frac{a_x}{a}\right|(r)\,dr\right) ds\right), \quad \forall x \in [0, 1], \tag{4.4.15}$$

which satisfies $\mu(1) = -1$ and $\mu_x(x) \geq 1$. Moreover, we have

$$(a\mu_x - |a_x|\mu)(x) \geq a(x), \;\; \mu(x) < 0, \;\; \forall x \in [0, 1].$$

Multiplying the differential equation of y by $\mu\overline{y_x}$ and integrating by parts, one can readily verify that our boundary value problem has only one trivial solution $y = 0$. Thereby, going back to (4.4.9) and (4.4.13), we deduce that the only solution of (4.4.8) is the trivial one.

Now, our concern is to show that the resolvent operator of $\mathscr{A}$ satisfies the second condition of Theorem 2.3.8 by means of a contradiction argument. Indeed, if such a condition is not true, then there exist a sequence of real numbers γ_n and a sequence of vectors $\Phi_n = (y_n, z_n, u_n, \xi_n, \eta_n) \in D(\mathscr{A})$ with $\|\Phi_n\|_{\mathscr{H}} = 1$ such that

$$(i\gamma_n I - \mathscr{A})\Phi_n = \mathscr{F}_n = (f_n, g_n, v_n, p_n, q_n) \to 0, \quad \text{as } n \to \infty \;\; \text{in } \mathscr{H}. \tag{4.4.16}$$

Our goal is to obtain, from (4.4.16), a contradiction. For sake of clarity, we shall proceed through several steps.

Step 1: First, we note that (4.4.16) may be recast as follows:

$$i\gamma_n y_n - z_n = f_n \to 0, \quad \text{in} \quad H^1(0, 1), \tag{4.4.17}$$

$$i\gamma_n z_n - (a(x)(y_n)_x)_x + b(x)z_n = g_n \to 0, \quad \text{in} \quad L^2(0, 1), \tag{4.4.18}$$

$$i\gamma_n u_n + \frac{(u_n)_x}{\tau} = v_n \to 0, \quad \text{in} \quad L^2(0, 1), \tag{4.4.19}$$

$$i\gamma_n \xi_n - \frac{1}{m}\left[(a(y_n)_x)(0) - \sigma y_n(0) - \beta\xi_n + \alpha u_n(1)\right] = p_n \to 0, \quad \text{in} \quad \mathbb{C}, \tag{4.4.20}$$

$$i\gamma_n \eta_n + \frac{(a(y_n)_x)(1)}{M} = q_n \to 0, \quad \text{in} \quad \mathbb{C}. \tag{4.4.21}$$

Taking the inner product, in $\mathscr{H}$, of both sides of (4.4.16) with the bounded element Φ_n and invoking (4.4.7), we get

$$\int_0^1 b(x)z_n^2\,dx \to 0, \tag{4.4.22}$$

$$u_n(0) = z_n(0) = \xi_n \to 0, \quad \text{and} \quad u_n(1) \to 0. \tag{4.4.23}$$

Solving (4.4.19) and using (4.4.23) yield

$$u_n \to 0 \quad \text{in} \quad L^2(0,1). \tag{4.4.24}$$

In light of (4.4.17) and the fact that $\|\Phi_n\|_{\mathscr{H}} = 1$, we reach

$$y_n = \frac{1}{i\gamma_n}(z_n + f_n) \to 0, \quad \text{in} \ \ L^2(0,1).$$

Now, since $\|\mathscr{F}_n\|_{\mathscr{H}} \to 0$, it follows that

$$f_n(0) \to 0. \tag{4.4.25}$$

Then, by (4.4.23) and (4.4.17), we have

$$i\gamma_n y_n(0) = z_n(0) + f_n(0) \to 0 \quad \text{and} \quad y_n(0) \to 0. \tag{4.4.26}$$

Reporting (4.4.23) and (4.4.26) in (4.4.20), we find

$$\frac{1}{i\gamma_n}(a(y_n)_x)(0) = o(1). \tag{4.4.27}$$

Next, taking the inner product of (4.4.18) with $a(x)(y_n)_x$ in $L^2(0,1)$, we have

$$i\gamma_n \int_0^1 z_n a(x)(\overline{y}_n)_x\,dx - \int_0^1 (a(x)(y_n)_x)_x a(\overline{y}_n)_x\,dx + \int_0^1 b(x)z_n a(x)(\overline{y}_n)_x\,dx$$
$$= \int_0^1 g_n a(x)(\overline{y}_n)_x\,dx.$$

Integrating by parts the second integral in the above identity and using the fact that $\|\Phi_n\|_{\mathscr{H}} = 1$, $g_n \to 0$ in $L^2(0,1)$ and (4.4.22), we get

$$\frac{1}{\gamma_n^2}\left(|(a(y_n)_x)(1)|^2 - |(a(y_n)_x)(0)|^2\right) = o(1).$$

This, with (4.4.27), leads to claim that

$$\frac{(a(y_n)_x)(1)}{\gamma_n} = o(1).$$

Whereupon, (4.4.21) gives

$$\eta_n \to 0. \tag{4.4.28}$$

Thanks to (4.4.25), we have

$$f_n(1) = \int_0^1 (f_n)_x(x)\,dx + f_n(0) \to 0,$$

and hence, (4.4.17) yields

$$i\gamma_n y_n(1) = z_n(1) + f_n(1) = \eta_n + f_n(1) \to 0 \quad \text{and} \quad y_n(1) \to 0. \tag{4.4.29}$$

Step 2: Now, reporting (4.4.17) in (4.4.18) and multiplying the resulting equation by $\rho(x)\overline{(y_n)_x}$, where ρ is a function in $C^1([0, 1], \mathbb{R})$, we obtain

$$\int_0^1 (-\gamma_n^2 y_n - (a(x)(y_n)_x)_x)\rho(x)\overline{(y_n)_x}\,dx + \int_0^1 b(x)z_n\rho(x)\overline{(y_n)_x}\,dx$$
$$= \int_0^1 (g_n + i\gamma_n f_n)\rho(x)\overline{(y_n)_x}\,dx. \tag{4.4.30}$$

One can readily prove that the right-hand side of (4.4.30) converges to zero by simple integration by parts and by using the facts that f_n and g_n (resp. $f_n(0)$, $f_n(1)$, $\gamma_n y_n(0)$, and $\gamma_n y_n(1)$) converge to zero in $H^1(0, 1)$ and $L^2(0, 1)$ (resp. $\mathbb{C}$).

Bearing in mind these findings and integrating by parts the left-hand side of (4.4.30), we get

$$\begin{aligned}
&2\Re e\left\{\int_0^1 -\gamma_n^2 y_n\rho(x)\overline{(y_n)_x}\,dx\right\}\\
&= \int_0^1 \gamma_n^2\rho_x(x)|y_n(x)|^2\,dx + \left[-\gamma_n^2\rho(1)|y_n(1)|^2 + \gamma_n^2\rho(0)|y_n(0)|^2\right]\\
&= \int_0^1 \gamma_n^2\rho_x|y_n(x)|^2\,dx + o(1),\\
&2\Re e\left\{-\int_0^1 (a(x)(y_n)_x)_x\rho(x)\overline{(y_n)_x}\,dx\right\}\\
&= \int_0^1 (a\rho_x - a_x\rho)(x)|(y_n)_x|^2\,dx\\
&- \left(a(1)\rho(1)|(y_n)_x(1)|^2 - a(0)\rho(0)|(y_n)_x(0)|^2\right).
\end{aligned}$$

In view of (4.4.22) and the boundedness of $(y_n)_x$, we have

$$\int_0^1 b(x) z_n \rho(x) \overline{(y_n)_x}\, dx = o(1).$$

Amalgamating these last outcomes, we can claim that for any $\rho \in C^1([0,1], \mathbb{R})$, we have

$$\int_0^1 (a\rho_x - a_x \rho)(x) |(y_n)_x|^2\, dx + \int_0^1 \gamma_n^2 \rho_x(x) |y_n|^2\, dx - a(1)\rho(1)|(y_n)_x(1)|^2 \\ + a(0)\rho(0)|(y_n)_x(0)|^2 = o(1). \tag{4.4.31}$$

Step 3: First, let us take in (4.4.31)

$$\rho(x) = \exp\left(\int_0^x \left|\frac{a_x}{a}\right|(s)\, ds\right) \int_0^x \exp\left(\int_s^0 \left|\frac{a_x}{a}\right|(r)\, dr\right) ds, \ \forall x \in [0,1],$$

so that $\rho(0) = 0$, and for all $x \in [0,1]$, we have

$$\rho_x(x) \geq 1, \ (a\rho_x - |a_x|\rho)(x) \geq a(x), \ \rho(x) \geq 0.$$

This, together with the boundedness property $\|\Phi_n\|_{\mathscr{H}} = 1$, implies that $(y_n)_x(1)$ must be bounded in $\mathbb{C}$ and so is $(y_n)_x(0)$. Second, take the inner product of (4.4.18) with $(a\phi)_x y_n$ in $L^2(0,1)$, where ϕ is a smooth function. Finally, we integrate by parts, recall that g_n tends to 0 (see (4.4.18)), and use (4.4.22), (4.4.26), (4.4.29), and (4.4.19) as well as the boundedness of $(y_n)_x(0)$ and $(y_n)_x(1)$, and we get

$$\int_0^1 (a\phi)_x(x) a(x) |(y_n)_x|^2\, dx - \int_0^1 (a\phi)_x(x) |\gamma_n y_n|^2\, dx = o(1). \tag{4.4.32}$$

Step 4: Taking ρ in (4.4.31) so that

$$(\rho a_x - a\rho_x)(x) = (a(a\phi)_x)(x)$$

and using (4.4.32), we obtain

$$\left(\frac{\rho}{a}\right)(1)|(a(y_n)_x)(1)|^2 - \left(\frac{\rho}{a}\right)(0))|(a(y_n)_x)(0)|^2 \\ - \int_0^1 \left(\rho_x + \left(\frac{\rho}{a}\right)_x a\right)(x)|\gamma_n y_n|^2\, dx = o(1). \tag{4.4.33}$$

Subsequently, we choose

$$\rho(x) = \frac{\sqrt{a(x)}}{2} \int_0^x \frac{b(r)}{\sqrt{a(r)}} dr,$$

and hence,

$$\rho_x(x) + \left(\frac{\rho}{a}\right)_x (x) a(x) = b(x)$$

as well as $\rho(0) = 0$. Thereby, (4.4.33) becomes

$$\left(\frac{\rho}{a}\right)(1) |(a(y_n)_x)|^2(1) - \int_0^1 b(x) |\gamma_n y_n|^2 \, dx = o(1). \tag{4.4.34}$$

It suffices now to recall (4.4.22), which combined with (4.4.17) gives

$$\int_0^1 b(x) |\gamma_n y_n|^2 \, dx \to 0.$$

Inserting the latter into (4.4.34), we conclude that

$$(y_n)_x(1) = o(1). \tag{4.4.35}$$

Step 5: Taking the inner product of (4.4.17) (resp. (4.4.18)) with z_n (resp. y_n) in $L^2(0, 1)$ and then summing up both equations, we get after a simple integration by parts

$$\int_0^1 a(x) |(y_n)_x|^2 \, dx - \int_0^1 |z_n|^2 \, dx$$
$$= \Re e \left\{ \int_0^1 f_n \overline{z_n} \, dx + \int_0^1 g_n \overline{y_n} \, dx - \int_0^1 b(x) z_n \overline{y_n} \, dx \right\} = o(1).$$

In light of (4.4.23), (4.4.24), and (4.4.28), and recalling that $\|\Phi_n\|_{\mathcal{H}} = 1$, the above estimate yields

$$\int_0^1 |z_n|^2 \, dx \to 1/2, \quad \int_0^1 a(x) |(y_n)_x|^2 \, dx \to 1/2. \tag{4.4.36}$$

Lastly, it suffices to pick up ρ as defined in step 3, insert it into (4.4.31), and use (4.4.35) to obtain that

$$\int_0^1 a(x) |(y_n)_x|^2 \, dx = \int_0^1 |\gamma_n y_n|^2 \, dx = o(1).$$

This obviously contradicts (4.4.36), and whence the proof of Theorem 4.4.2 is achieved. □

Remark 4.4.1 The proof of Theorem 4.4.2 is considerably simpler when the modulus of tension of the cable $a(x)$ is constant. In this case, it suffices to take $\rho(x) = x$ and $\rho(x) = 1 - x$ in (4.4.31) to obtain

$$|(y_n)_x|^2(0) \to 1/a, \ |(y_n)_x|^2(1) \to 1/a.$$

Next, taking once again

$$\rho(x) = \int_0^x b(r)\,dr$$

in (4.4.31), it follows that

$$a\int_0^1 b(x)|(y_n)_x|^2\,dx + \int_0^1 b(x)|\gamma_n y_n|^2\,dx \to \int_0^1 b(r)\,dr > 0. \tag{4.4.37}$$

Finally, taking the inner product of (4.4.17) (resp. (4.4.18)) with $\gamma_n y_n$ (resp. by_n) in $L^2(0, 1)$, we get

$$\int_0^1 b(x)|\gamma_n y_n|^2\,dx \to 0$$

and

$$a\int_0^1 b(x)|(y_n)_x|^2\,dx + \int_0^1 b(x)|\gamma_n y_n|^2\,dx \to 0,$$

which contradicts (4.4.37).

Last but not least, the next result provides an idea on the case where the "smallness" condition (4.4.3) is violated, i.e., $\alpha \geq \beta$. We have:

Theorem 4.4.3 *Assume that (4.4.3) does not hold. Then, there exists at least one time delay τ for which the solution of (4.4.1) is unstable in $\mathscr{H}$.*

Proof The proof mimics that of a similar result obtained in Sect. 4.2.5, and hence, most of the details are omitted.

First, we restrict ourselves to the case where the constant coefficient $a(x) = a_0 > 0$ in order to avoid unnecessary complex computations. Next, assuming that $\alpha \geq \beta$, the objective is to seek a solution of (4.4.1) in the form: $y(x, t) = e^{\lambda t} f(x)$, with $\lambda > 0$, and f is a nonzero function of $H^2(0, 1)$. One can check after a direct computation that such a desired solution exists if and only if

$$\left(m\lambda^2 - \lambda\alpha e^{-\lambda\tau} + \beta\lambda + \sigma + \sqrt{(\lambda^2+\lambda)a_0}\right)\left(M\lambda^2 + \sqrt{(\lambda^2+\lambda)a_0}\right)e^{\sqrt{\frac{\lambda^2+\lambda}{a_0}}}$$

$$-\left(m\lambda^2 - \lambda\alpha e^{-\lambda\tau} + \beta\lambda + \sigma - \sqrt{(\lambda^2+\lambda)a_0}\right)\left(M\lambda^2 - \sqrt{(\lambda^2+\lambda)a_0}\right)e^{-\sqrt{\frac{\lambda^2+\lambda}{a_0}}} = 0. \tag{4.4.38}$$

Let us choose $M = 1$, $m = 0.7$, $a_0 = 0.5$, $\sigma = 0.2$, and $\alpha = 2.1 > \beta = 0.1$. Then, one can check that for $\tau = \ln(21/20)$, the characteristic equation (4.4.38) admits a positive solution $\lambda = 1$. This obviously implies that the corresponding solution $y(x,t) = e^t f(x)$ of (4.4.1) cannot be stable in $\mathcal{H}$. □

4.4.3 Conclusion

In this section, a localized damping is shown to be sufficient to exponentially stabilize the crane cable with delayed and dynamical boundary conditions.

Determining the optimal decay rate is of great interest in practice, and hence, this point deserves to be studied. Again, the problem 4.4.1 with time-dependent length remains interesting and is therefore definitely worth investigating.

4.5 Boundary Delay: Exponential Stabilization via Kelvin-Voigt Damping Control

This section deals with the stabilization problem of the delayed crane system by means of a higher-order boundary control. To be more precise, in contrast to previous sections, the delayed and nondelayed boundary terms are of different types. In order to handle this situation, the so-called Kelvin-Voigt damping is also invoked.

To start, the following stabilization problem is considered:

$$\begin{cases} y_{tt}(x,t) - (ay_x)_x(x,t) + \mathcal{F}(x,t) = 0, \ 0 < x < 1, \\ my_{tt}(0,t) - (ay_x)(0,t) = \mathcal{U}_1(t), \qquad t > 0, \\ My_{tt}(1,t) + (ay_x)(1,t) = \mathcal{U}_2(t), \qquad t > 0, \end{cases} \tag{4.5.1}$$

in which $\mathcal{F}(x,t)$ is the external controlling force (interior control) that drives the platform along the rail. Furthermore, $\mathcal{U}_1(t)$ is the boundary delayed control, while $\mathcal{U}_2(t)$ will play the role of the nondelayed control.

First, let us recall the results that are relevant to this section. In fact, we know from [148] that the exponential stability of the system (4.5.1) can be achieved via the action of a boundary control of *higher order* (angular velocity), that is,

$$\mathscr{U}_1(t) = -py(0,t) - pqy_t(0,t) + qy_{xt}(0,t), \quad p,\ q > 0,$$

or

$$\mathscr{U}_1(t) = -by(0,t) - (c+bd)y_t(0,t) + dy_{xt}(0,t),$$

where the positive constants b, c, and d obey suitable conditions. Therefore, it is legitimate to explore the case where a delay occurs in the boundary velocity term, that is, $y_t(0, t - \tau)$ is present, while the higher-order term $y_{xt}(0,t)$ acts as the input "contrasting" control. Inspired by this remark, the present section places primary emphasis on providing an answer to such a problem. Specifically, the main contribution is to show that whenever the following boundary delayed control is exerted

$$\mathscr{U}_1(t) = \alpha y_t(0, t-\tau) + \beta(ay_{xt})(0,t),\ \alpha > 0,\ \beta > 0, \tag{4.5.2}$$

and despite the presence of the delay term $y_t(0, t - \tau)$ ($\tau > 0$ is the time delay), the solutions of the closed-loop system (see (4.5.1) and (4.5.2)–(4.5.3)) can be *exponentially* driven to a stationary state. The latter is explicitly determined in terms of the initial conditions (see Theorem 4.5.2). In turn, such a desirable result is achieved, provided that two other feedback controls $\mathscr{F}(x,t)$ and $\mathscr{U}_2(t)$ are designed:

$$\begin{cases} \mathscr{F}(x,t) = -\beta(ay_{xt})_x(x,t), & 0 < x < 1, t > 0, \\ \mathscr{U}_2(t) = -\beta(ay_{xt})(1,t) - \gamma y_t(1,t), & t > 0, \end{cases} \tag{4.5.3}$$

where $\gamma > 0$. Note that (a priori) the above interior and boundary controls $\mathscr{F}$ and $\mathscr{U}_2$ are required because of the presence of higher-order angular velocity and also the dynamical feature of both ends of the cable. It is also worth noting that another interesting outcome of the current section is to show that a natural "smallness" condition of the delayed term is optimal and hence necessary to achieve the convergence of solutions. Indeed, we shall provide an example where the "destabilizing" effect of the delay term $y_t(0, t-\tau)$ may not be compensated through the action of the proposed feedback control if the "smallness" condition does not hold (see [83, 111, 149, 184] for similar results on different types of systems). Let us mention that the presence of a delay unexpectedly stabilizes unstable systems [35, 113] and in some cases does not affect the stability [112], whereas it may create periodic solutions [90, 91]. Furthermore, a spectral analysis of the system is also conducted. Accordingly, we manage to provide the asymptotic expansion of both eigenvalues and eigenfunctions of the system. Then, based on the spectral properties, the regularity of the solution of the system is further discussed. Last but not least, it is important to point out that we shall take into consideration the dynamical terms in the system (due to the movements of both masses m and M), whereas no position term arises in the closed-loop system. This means that the standard energy used in the literature is merely a seminorm, which impedes the utilization of classical Poincaré's or Wirtinger's inequality [115]. To ascertain this claim, it is a simple task to check that the standard energy of the closed-loop system (4.5.1) and (4.5.2)–(4.5.3)

$$E_0(t) = \frac{1}{2}\left\{\int_0^1 \left[y_t^2(x,t) + a(x)y_x^2(x,t)\right] dx + my_t^2(0,t) + My_t^2(1,t)\right.$$
$$\left. + K\tau \int_0^1 y_t^2(0, t - x\tau)\, dx\right\} \tag{4.5.4}$$

vanishes along nontrivial solutions of the system.

Let us briefly present an overview of this section. In Sect. 4.5.1, we show the well-posedness of the problem by means of semigroups theory. LaSalle's principle is applied in Sect. 4.5.2 to prove the convergence of solutions of the closed-loop system to a stationary state that depends on the initial data. In Sect. 4.5.3, the resolvent method is invoked to prove that the convergence of solutions to the equilibrium position is indeed exponential. When (4.5.8) does not hold, we show instability results in Sect. 4.5.4. Section 4.5.5 is devoted to the spectral analysis of the system and the regularity of its semigroup. In Sect. 4.5.6, the theoretical outcomes are validated through numerical simulations. Lastly, we end this section with a concluding discussion.

4.5.1 Preliminaries and Well-Posedness of the Problem

Differentiating formally (4.5.4) along the solutions of (4.5.1) and (4.5.2)–(4.5.3) and integrating by parts produce after a straightforward computation

$$E_0'(t) = -\beta \int_0^1 a(x) y_{xt}^2(x,t)\, dx - \gamma y_t^2(1,t) + \alpha y_t(0,t) y_t(0, t-\tau)$$
$$- \frac{K}{2}\left[y_t^2(0, t-\tau) - y_t^2(0,t)\right]. \tag{4.5.5}$$

Thereafter, invoking (4.1.5) as well as Young's inequality, one can readily verify the estimate

$$y_t^2(0,t) \le 2y_t^2(1,t) + \frac{2}{a_0}\int_0^1 a(x) y_{xt}^2(x,t)\, dx. \tag{4.5.6}$$

Consequently, we deduce from (4.5.5), after using once again Young's inequality, that

$$E_0'(t) \le (K + \alpha\mu - \gamma)y_t^2(1,t) + \frac{1}{2}(-K + \frac{\alpha}{\mu})y_t^2(0, t-\tau)$$
$$+ \left[\frac{1}{a_0}(\alpha\mu + K) - \beta\right]\int_0^1 a(x)y_{xt}^2(x,t)dx, \tag{4.5.7}$$

for $\mu > 0$. Thereby, the energy is decreasing once we fix $\mu > 0$, and then assume the "smallness" conditions:

$$\begin{cases} \alpha < (1 + 1/\mu)^{-1}\gamma, \\ \alpha < (1 + 1/\mu)^{-1}a_0\beta, \end{cases}$$

while K is chosen so that

$$\begin{cases} \alpha\mu^{-1} < K < \gamma - \alpha\mu, \\ \alpha\mu{-}1 < K < a_0\beta - \alpha\mu. \end{cases}$$

For sake of simplicity, we shall choose, for instance $\mu = 1$, and assume in the sequel that

$$2\alpha < \min\{\gamma, a_0\beta\}, \tag{4.5.8}$$

whereas K satisfies

$$\alpha \le K \le \min\{\gamma - \alpha, a_0\beta - \alpha\}. \tag{4.5.9}$$

Bearing in mind that $E_0(t)$ vanishes along constant solutions of the closed-loop system, one should seek another invariant functional to fix this issue. To this end, we introduce

$$E_1(t) = \frac{1}{2}\Bigg[\int_0^1 \{y_t(x,t) + \alpha\tau y_t(0, t - x\tau)\}\,dx + my_t(0,t) + My_t(1,t)$$
$$+\gamma y(1,t) - \alpha y(0,t)\Bigg]^2 \tag{4.5.10}$$

and define the system's energy by

$$\mathscr{E}(t) = E_0(t) + E_1(t), \tag{4.5.11}$$

where $E_0(t)$ is given in (4.5.4). Arguing as before, we obtain after a careful calculation that $E_1'(t) = 0$, for any $t \geq 0$, and along all solutions of the closed-loop system (4.5.1) and (4.5.2)–(4.5.3). This, together with (4.5.7)–(4.5.9), implies that

$$\mathscr{E}'(t) = E_0'(t) \leq 0,$$

for all $t \geq 0$.

Accordingly, this discussion motivates us to consider the state space

$$\mathscr{H} = H^1(0,1) \times L^2(0,1) \times L^2(0,1) \times \mathbb{R}^2,$$

equipped with the following real inner product

$$\begin{aligned}
&\langle (y, z, u, \xi, \eta), (\tilde{y}, \tilde{z}, \tilde{u}, \tilde{\xi}, \tilde{\eta}) \rangle_{\mathscr{H}} = \int_0^1 (a y_x \tilde{y}_x + z\tilde{z})\, dx \\
&+ K\tau \textstyle\int_0^1 u\tilde{u}\, dx + m\xi\tilde{\xi} + M\eta\tilde{\eta} \\
&+ \varsigma \left[\int_0^1 z\, dx + \tau\alpha \int_0^1 u\, dx + m\xi + M\eta + \gamma y(1) - \alpha y(0) \right] \\
&\times \left[\int_0^1 \tilde{z}\, dx + \tau\alpha \int_0^1 \tilde{u}\, dx + m\tilde{\xi} + M\tilde{\eta} + \gamma \tilde{y}(1) - \alpha \tilde{y}(0) \right],
\end{aligned} \tag{4.5.12}$$

where $K > 0$ satisfies (4.5.9) and ς is a positive constant sufficiently small so that the norm induced by (4.5.12) is equivalent to the usual norm of $\mathscr{H}$. The proof of this claim can be obtained by invoking (4.1.5), (4.5.8)–(4.5.9), and by proceeding as in Proposition 4.2.1 but with slight changes born out of necessity.

Subsequently, we introduce as usual the auxiliary state variables $u(x,t) = y_t(0, t - x\tau)$ (see [149, 184]). Then, let $z(x,t) = y_t(x,t)$ $\xi(t) = y_t(0,t)$ and $\eta(t) = y_t(1,t)$. Thereby, the closed-loop system (4.5.1) and (4.5.2)–(4.5.3) can be formulated as a differential equation in $\mathscr{H}$

$$\begin{cases} \varphi_t(t) = \mathscr{A}\varphi(t), \\ \varphi(0) = \varphi_0, \end{cases} \tag{4.5.13}$$

where $\varphi = (y, z, u, \xi, \eta)$, $\varphi_0 = (y_0, y_1, f(-\tau\cdot), \xi_0, \eta_0)$, and $\mathscr{A}$ is a linear operator defined as follows:

$$\begin{aligned}
&\mathscr{D}(\mathscr{A}) = \big\{ (y, z, u, \xi, \eta) \in \mathscr{H};\ a y_x + \beta a z_x \in H^1(0,1),\ u \in H^1(0,1), \\
&\quad \xi = u(0) = z(0),\ \eta = z(1) \big\}, \\
&\mathscr{A}(y, z, u, \xi, \eta) = \Bigg(z, (a y_x)_x + \beta (a z_x)_x, -\frac{u_x}{\tau}, \frac{1}{m} [(a y_x)(0) + \beta (a z_x)(0) \\
&\quad + \alpha u(1)], -\frac{1}{M} [(a y_x)(1) + \beta (a z_x)(1) + \gamma z(1)] \Bigg),
\end{aligned} \tag{4.5.14}$$

for all $(y, z, u, \xi, \eta) \in \mathscr{D}(\mathscr{A})$. The following result states, inter alia, that the problem is well-posed in the sense of semigroups theory.

Theorem 4.5.1 *Under the conditions (4.1.5), (4.5.8), and (4.5.9), the operator $\mathscr{A}$ defined by (4.5.14) is densely defined in $\mathscr{H}$ and generates a C_0-semigroup of contractions $e^{t\mathscr{A}}$ on $\mathscr{H}$. Herewith, for each initial data $\varphi_0 \in \mathscr{D}(\mathscr{A})$, there exists a unique strong solution $\varphi \in C([0,\infty); \mathscr{D}(\mathscr{A}) \cap C^1([0,\infty); \mathscr{H})$ of the Cauchy problem (4.5.13). In turn, in the event that $\varphi_0 \in \mathscr{H}$, the system (4.5.13) possesses a unique mild solution $\varphi \in C([0,\infty); \mathscr{H})$. Lastly, the spectrum of $\mathscr{A}$ only consists of isolated eigenvalues of finite algebraic multiplicity.*

Proof Let $\varphi = (y, z, u, \xi, \eta) \in \mathscr{D}(\mathscr{A})$. By virtue of (4.5.12) and (4.5.14), one can obtain after integrating by parts and using Young's inequality the estimate

$$\langle \mathscr{A}\varphi, \varphi \rangle \leq -\beta \int_0^1 a z_x^2 \, dx + \frac{K+\alpha}{2} z(0)^2 - \gamma z(1)^2 + \frac{\alpha - K}{2} u^2(1).$$

In light of (4.5.6), we deduce that

$$\langle \mathscr{A}\varphi, \varphi)\rangle \leq \left(\frac{1}{a_0}[\alpha + K] - \beta \right) \int_0^1 a z_x^2 \, dx + (K + \alpha - \gamma) z(1)^2 + \frac{\alpha - K}{2} u^2(1). \tag{4.5.15}$$

Thenceforth, $\mathscr{A}$ is dissipative thanks to (4.5.8)–(4.5.9).

The remaining task is to show that the range of $\lambda I - \mathscr{A}$ is $\mathscr{H}$, for $\lambda > 0$. To this end, let (f, g, v, p, q) in $\mathscr{H}$, and let us seek (y, z, u, ξ, η) in $\mathscr{D}(\mathscr{A})$ so that $(\lambda I - \mathscr{A})(y, z, u, \xi, \eta) = (f, g, v, p, q)$. The latter gives rise to the following system:

$$\begin{cases} z = \lambda y - f, \ \lambda z - (a y_x)_x - \beta (a z_x)_x = g, \\ u_x + \lambda \tau u = \tau v, \\ \lambda m \xi - (a y_x)(0) - \alpha u(1) - \beta (a z_x)(0) = mp, \\ \lambda M \eta + (a y_x)(1) + \beta (a z_x)(1) + \gamma z(1) = Mq, \\ \xi = u(0) = z(0), \ \ \eta = z(1). \end{cases}$$

The equation of u yields

$$u(x) = e^{-\tau\lambda x} z(0) + \tau \int_0^x e^{-\tau\lambda(x-s)} v(s) \, ds,$$

and hence,

$$u(1) = e^{-\tau\lambda} z(0) + \tau \int_0^1 e^{-\tau\lambda(1-s)} v(s) \, ds.$$

Letting $w := y + \beta z$ and recalling that $z = \lambda y - f$, we have

$$y = \frac{1}{1+\beta\lambda}(w + \beta f)$$

and

$$z = \frac{1}{1+\beta\lambda}(\lambda w - f).$$

Amalgamating these facts, it amounts to showing that the following elliptic problem

$$\begin{cases} \lambda^2 w - (1+\beta\lambda)(aw_x)_x = \lambda f + (1+\beta\lambda)g, \\ \lambda\left[m\lambda - \alpha e^{-\tau\lambda}\right] w(0) - (1+\beta\lambda)(aw_x)(0) = \tau\alpha(1+\beta\lambda)\displaystyle\int_0^1 e^{-\tau\lambda(1-s)}v(s)\,ds \\ \qquad\qquad\qquad\qquad + (m\lambda - \alpha e^{-\tau\lambda})f(0) + (1+\beta\lambda)mp, \\ \lambda(\gamma + M\lambda)w(1) + (1+\beta\lambda)(ay_x)(1) = (\lambda M + \gamma)f(1) + M(1+\beta\lambda)q \end{cases} \tag{4.5.16}$$

has a unique solution in $H^2(0,1)$. To proceed, multiplying the first equation in (4.5.16) by $\psi \in H^1(0,1)$ and integrating by parts, we get the weak formulation

$$\begin{aligned} &\int_0^1 [\lambda^2 \psi w + (1+\beta\lambda)aw_x\psi_x]\,dx + \lambda[m\lambda - \alpha e^{-\tau\lambda}]w(0)\psi(0) \\ &\quad + \lambda(\gamma + M\lambda)w(1)\psi(1) \\ &\quad = \int_0^1 [\lambda f + (\gamma + M\lambda)g]\,\psi\,dx + \Big[(\gamma + M\lambda)mp + (m\lambda - \alpha e^{-\tau\lambda})f(0) \\ &\quad + \tau\alpha(\gamma + M\lambda)\int_0^1 e^{-\tau\lambda(1-s)}v(s)\,ds\Big]\psi(0) + [Mq(\gamma + M\lambda) + \lambda M f(1)]\,\psi(1). \end{aligned} \tag{4.5.17}$$

One can verify that the left-hand side (resp. right-hand side) in (4.5.17) is a continuous and coercive bilinear form on $H^1(0,1) \times H^1(0,1)$ (resp. a continuous form on $H^1(0,1)$), and it follows from Lax-Milgram theorem (see Theorem 2.1.6) that there exists a unique solution $w \in H^1(0,1)$ of (4.5.17) for $\lambda > 0$. By a standard argument used for solving elliptic linear equations, one can show that $w \in H^2(0,1)$ and then recover the boundary conditions in (4.5.17). This establishes the maximality of $\mathscr{A}$, and hence, $(\lambda I - \mathscr{A})^{-1}$ exists and maps $\mathscr{H}$ into $\mathscr{D}(\mathscr{A})$, for $\lambda > 0$. In turn, Sobolev embedding [4] leads to claim that $(\lambda I - \mathscr{A})^{-1}$ is compact, and the spectrum of $\mathscr{A}$ consists of isolated eigenvalues of finite algebraic multiplicity only (see Remark 2.1.2). Lastly, invoking semigroups theory [158], one can conclude the other claims of Theorem 4.5.1. □

4.5.2 Long-Time Behavior of Solutions

We first point out that the energy $\mathscr{E}(t)$ (see (4.5.11)) of the closed-loop system is conservative along any nonzero constant solution since $\lambda = 0$ is an eigenvalue. This motivates us to determine exactly the state at which the solutions converge. In fact, we have:

Theorem 4.5.2 *Assume that the conditions (4.1.5) and (4.5.8)–(4.5.9) are satisfied. Then, for any initial data* $\varphi_0 = (y_0, y_1, f, \xi_0, \eta_0) \in \mathscr{H}$*, the solution* $\varphi(t) = (y, y_t, y_t(0, t - x\tau), y_t(0, t), y_t(1, t))$ *of the closed-loop system goes to* $(\chi, 0, 0, 0, 0)$ *in* $\mathscr{H}$*, as* $t \to +\infty$*, where*

$$\chi = \frac{1}{\gamma - \alpha}\left[\int_0^1 (y_1(x) + \alpha\tau f(-\tau x))\, dx + \gamma y_0(1) - \alpha y_0(0) + m\xi_0 + M\eta_0\right]. \tag{4.5.18}$$

Proof Given $\varphi_0 \in \mathscr{D}(\mathscr{A})$, consider the corresponding unique solution $\varphi(t) = e^{t\mathscr{A}}\varphi_0$ of (4.5.13). The set of solutions trajectories $\{\varphi(t)\}_{t\geq 0}$ is bounded for the graph norm. Hence, this set must precompact by virtue of the compactness of the operator $(\lambda I - \mathscr{A})^{-1}$, for $\lambda > 0$. This implies that one can use LaSalle's principle (see Theorem 2.3.11) to claim that the ω-limit set

$$\omega(\varphi_0) = \Big\{\psi \in \mathscr{H};\ \text{there exists a sequence } t_n \to \infty$$
$$\text{as } n \to \infty \text{ such that } \psi = \lim_{n\to\infty} e^{t_n\mathscr{A}}\varphi_0\Big\}$$

is non-empty, compact, invariant under the semigroup $e^{t\mathscr{A}}$. Furthermore, the solution $e^{t\mathscr{A}}\varphi_0$ satisfies the asymptotic property: $e^{t\mathscr{A}}\varphi_0 \longrightarrow \omega(\varphi_0)$, as $t \to \infty$. Thereafter, it remains to determine the ω-limit set $\omega(\Phi_0)$. To do so, let

$$\tilde{\varphi}_0 = (\tilde{y}_0, \tilde{y}_1, \tilde{f}, \tilde{\xi}, \tilde{\eta}) \in \omega(\varphi_0) \subset D(\mathscr{A}).$$

Subsequently, consider the unique strong solution $\tilde{\varphi}(t) = (\tilde{y}(t), \tilde{y}_t(t), \tilde{u}(t), \tilde{\xi}(t), \tilde{\eta}(t))$ of (4.5.13). As we know from [114, Theorem 2.1.3 p. 18], we have

$$\frac{\mathrm{d}}{\mathrm{d}t}\|\tilde{\varphi}(t)\|_{\mathscr{H}} = 0.$$

Then, using (4.5.15), it follows that $\|\tilde{z}_x\|_{L^2(0,1)} = 0$, $\tilde{\eta} = \tilde{z}(1, t) = 0$, and $\tilde{u}(1) = 0$. Thus, $\tilde{z} = \tilde{y}_t$ is constant, which also implies the same for $\tilde{y}$ by means of (4.5.13). Thereby, there exists a real constant χ such that $\omega(\varphi_0) = \{(\chi, 0, 0, 0, 0)\}$. The ultimate outcome is to find χ. This can be done by noting that if we consider an initial condition $\varphi_0 = (y_0, y_1, f, \xi_0, \eta_0) \in \mathscr{D}(\mathscr{A})$, then the energy $E_1(t)$ (see

(4.5.10)) along the corresponding solution of (4.5.13) satisfies $E_1'(t) = 0$, for any $t \geq 0$, and hence,

$$\int_0^1 (y_t(x,t) + \alpha\tau y_t(0, t - x\tau))\,dx + \gamma y(1,t) - \alpha y(0,t) + m y_t(0,t) + M y_t(1,t)$$
$$= \int_0^1 (y_1(x) + \alpha\tau f(-x\tau))\,dx \; + m\xi_0 + M\eta_0 + \gamma y_0(1) - \alpha y_0(0). \tag{4.5.19}$$

On the other hand, if $(\chi, 0, 0, 0, 0) \in \omega(\varphi_0)$, then

$$\varphi(t_n) = (y(t_n), y_t(t_n), u(t_n), \xi(t_n), \eta(t_n)) = e^{t_n\mathscr{A}}\varphi_0 \to (\chi, 0, 0, 0, 0),$$
$$\text{for some } t_n \to \infty, \text{ as } n \to \infty. \tag{4.5.20}$$

It suffices to take $t = t_n$ in (4.5.19), let $n \to \infty$, and use (4.5.20) to obtain χ as given in (4.5.18). Lastly, the result we have obtained so far will remain valid for $\varphi_0 \in \mathscr{H}$ due to the density of $\mathscr{D}(\mathscr{A})$ in $\mathscr{H}$ and the contraction of the semigroup. □

4.5.3 *Exponential Convergence Rate for the Closed-Loop System (4.5.1) and (4.5.2)–(4.5.3)*

The aim of this subsection is to go one step further by providing the convergence rate of the solutions of the system (4.5.13). Indeed, the convergence result obtained in the previous subsection will be shown to be of exponential type. The main ingredient of the proof is the frequency-domain method (see Theorem 2.3.8).

Before stating our result, let us define a subspace of $\mathscr{H}$ of codimension 1 as follows:

$$\hat{\mathscr{H}} = \left\{(y, z, u, \xi, \eta) \in \mathscr{H};\; \int_0^1 (z(x) + \alpha\tau u(x))\,dx + \gamma y(1) \right.$$
$$\left. -\alpha y(0) + m\xi + M\eta = 0\right\}.$$

Then, we introduce a new operator $\hat{\mathscr{A}}$

$$\begin{aligned} &\hat{\mathscr{A}} : \mathscr{D}(\hat{\mathscr{A}}) := \mathscr{D}(\mathscr{A}) \cap \hat{\mathscr{H}} \subset \hat{\mathscr{H}} \to \hat{\mathscr{H}} \\ &\hat{\mathscr{A}}\varphi = \mathscr{A}\varphi, \; \forall \varphi \in \mathscr{D}(\hat{\mathscr{A}}). \end{aligned} \tag{4.5.21}$$

According to the previous subsection, the operator $\hat{\mathscr{A}}$, defined by (4.5.21), clearly has the same properties as the original operator $\mathscr{A}$.

The first main result of this section is:

Theorem 4.5.3 *Assume that (4.1.5), (4.5.8), and (4.5.9) hold. Then, the* C_0*-semigroup of contractions* $e^{t\hat{\mathscr{A}}}$ *is exponentially stable. Thereby, the solutions of the closed-loop system (4.5.13) exponentially converge in* $\mathscr{H}$ *to* $(\chi, 0, 0, 0, 0)$ *as* $t \longrightarrow +\infty$*, where* χ *is given by (4.5.18).*

Proof In the first lieu, we are going to check (2.3.16), i.e.,

$$\rho(\hat{\mathscr{A}}) \supset \{i\nu;\ \nu \in \mathbb{R}\} \equiv i\mathbb{R}.$$

To do so, let $\nu \in \mathbb{R}$, and let us solve the eigenvalue problem

$$\hat{\mathscr{A}}\varphi = i\nu\varphi, \tag{4.5.22}$$

where $\varphi = (y, z, u, \xi, \eta) \in \mathscr{D}(\hat{\mathscr{A}})$. If $\nu = 0$, then (4.5.22) gives $z = (a\,y_x)_x = u_x = 0$ and $(ay_x)(0) + \alpha u(1) = (ay_x)(1) = 0$. This implies that $u = 0$ and $y = 0$ by means of the fact that

$$\int_0^1 (z(x) + \alpha\tau u(x))\, dx + \gamma y(1) - \alpha y(0) + m\xi + M\eta = 0.$$

Whereupon

$$\nu = 0 \notin \sigma(\hat{\mathscr{A}}).$$

Now, if $\nu \neq 0$, then (4.5.22) and (4.5.15) permit to claim that $z_x = 0$ in $L^2(0, 1)$ and $z(1) = u(1) = 0$. Thus $z = 0$, and (4.5.22) leading to (y, z, u, ξ, η) is the trivial solution, and once again $i\nu$ cannot be an eigenvalue of $\hat{\mathscr{A}}$.

The ultimate outcome is to verify (2.3.17), that is,

$$\limsup_{|\nu|\to\infty} \|(i\nu I - \mathscr{L})^{-1}\|_{\mathscr{L}(\mathscr{H})} < \infty.$$

Suppose that this is not the case. This implies, by Banach-Steinhaus theorem (see Theorem 2.1.7), that there exist a sequence of real numbers $\nu_n \to +\infty$ and a sequence of vectors $\varphi^n = (y^n, z^n, u^n, \xi^n, \eta^n) \in \mathscr{D}(\hat{\mathscr{A}})$ with $\|\varphi^n\|_{\hat{\mathscr{H}}} = 1$ such that

$$\|(i\nu_n I - \hat{\mathscr{A}})\varphi^n\|_{\hat{\mathscr{H}}} \to 0, \quad \text{as} \quad n \to \infty, \tag{4.5.23}$$

which implies that as $n \to \infty$:

$$i\nu_n y^n - z^n \to 0, \quad \text{in} \ \ H^1(0, 1); \tag{4.5.24}$$

$$i\nu_n z^n - (ay_x^n)_x - \beta(az_x^n)_x \to 0, \ \text{in } L^2(0, 1); \tag{4.5.25}$$

$$i\tau\nu_n u^n + u_x^n \equiv V^n \to 0, \quad \text{in } L^2(0,1); \tag{4.5.26}$$
$$im\nu_n\xi^n - \left[(ay_x^n)(0) + \alpha u^n(1) + \beta(az_x^n)(0)\right] \to 0, \text{ in } \mathbb{C};$$
$$iM\nu_n\eta^n + \left[(ay_x^n)(1) + \gamma z^n(1) + \beta(az_x^n)(1)\right] \to 0, \text{ in } \mathbb{C}.$$

In view of (4.5.15), (4.5.23), and the inequality

$$\|(i\nu_n I - \hat{\mathscr{A}})\varphi^n t\|_{\hat{\mathscr{H}}} \geq |\Re e\langle(i\nu_n I - \hat{\mathscr{A}})\varphi^n, \varphi^n\rangle_{\hat{\mathscr{H}}}|,$$

we get

$$z_x^n \to 0 \quad \text{in } L^2(0,1), \tag{4.5.27}$$

and

$$z^n(1) = \eta^n \to 0\,, \quad u^n(1) \to 0 \quad \text{in } \mathbb{C}. \tag{4.5.28}$$

By virtue of (4.5.26), we have

$$u^n(x) = u^n(1)\, e^{-i\tau\nu_n(x-1)} + \int_1^x e^{-i\tau\nu_n(x-s)} V^n(s)\, ds. \tag{4.5.29}$$

Combining (4.5.26), (4.5.28), and (4.5.29), we have

$$u^n \longrightarrow 0 \text{ in } L^2(0,1). \tag{4.5.30}$$

Amalgamating (4.5.27) and (4.5.28), we deduce that $z^n(0) = \xi^n \to 0$, in $\mathbb{C}$, and hence, $z^n \to 0$, in $L^2(0,1)$. This, together with (3.2.30), implies that

$$y^n \to 0 \quad \text{in } L^2(0,1). \tag{4.5.31}$$

Lastly, going back to (3.2.30), we have

$$y_x^n = \frac{1}{i\nu_n} z_x^n + \frac{1}{i\nu_n} f_x^n.$$

Since $f_x^n \to 0$ in $L^2(0,1)$ (see (3.2.30)) and using (4.5.27), the last identity gives

$$y_x^n \to 0 \quad \text{in } L^2(0,1). \tag{4.5.32}$$

Combining the results in (4.5.28) and (4.5.30)–(4.5.32), we can claim that $\varphi^n \longrightarrow 0$ in $\hat{\mathscr{H}}$, which contradicts the fact that $\|\varphi^n\|_{\hat{\mathscr{H}}} = 1, \ \forall\, n \in \mathbb{N}$. Therewith, the proof of Theorem 4.5.3 is achieved. □

Remark 4.5.1 It follows from Theorem 4.5.3 that the proposed feedback law ensures the exponential convergence of solutions of the closed-loop system to a well-defined steady state. Clearly, this equilibrium state becomes null provided that the initial condition satisfies the extra condition:

$$\int_0^1 (y_1(x) + \alpha\tau f(-\tau x))\, dx + \gamma y_0(1) - \alpha y_0(0) + m\xi_0 + M\eta_0 = 0.$$

4.5.4 *Lack of Convergence When (4.5.8) Does Not Hold*

In the sequel, we shall verify that the solutions of the closed-loop system (4.5.1) and (4.5.2)–(4.5.3) do not converge to their steady state if the condition (4.5.8) is not satisfied. This outcome shows that such a condition is optimal. To this end, we show that if (4.5.8) does not hold, then there exists a time delay τ, for which the semigroup $e^{t\hat{\mathscr{A}}}$ is unstable in $\hat{\mathscr{H}}$.

In fact, we have the second main result:

Theorem 4.5.4 *Assume that (4.5.8) does not hold. Then, there exists a delay $\tau > 0$ for which the solutions of the closed-loop system (4.5.13) do not converge to their stationary state.*

Proof Let $a = 1$ for sake of simplicity. The task ahead is to seek a positive number λ and a nonzero $g \in H^2(0, 1)$ so that $y(x, t) = e^{\lambda t} g(x)$ is a solution of the Cauchy problem on $\hat{\mathscr{H}}$, associated to the operator $\hat{\mathscr{A}}$ (see (4.5.21)). In such an event, we have: $\|y\|_{L^2(0,1)} = e^{\lambda t} \|g\|_{L^2(0,1)} \to +\infty$, and thereby the solution is unstable. This, in turn, leads to the result stated in our theorem.

Now, it amounts to finding a solution $y(x, t) = e^{\lambda t} g(x)$ of the boundary value problem:

$$\begin{cases} y_{tt} - y_{xx} - \beta y_{txx} = 0, \\ m y_{tt}(0, t) - y_x(0, t) = \alpha y_t(0, t - \tau) + \beta y_{xt}(0, t), \\ M y_{tt}(1, t) + y_x(1, t) = -\gamma y_t(1, t) - \beta y_{xt}(1, t). \end{cases} \tag{4.5.33}$$

The latter is equivalent to looking for a nontrivial solution g of the following problem:

$$\begin{cases} \lambda^2 g - (1 + \beta\lambda)\, g_{xx} = 0, \\ \lambda \left(m\lambda - \alpha e^{-\lambda\tau}\right) g(0) - (1 + \beta\lambda)\, g_x(0) = 0, \\ (1 + \beta\lambda)\, g_x(1) + \lambda\, (M\lambda + \gamma)\, g(1) = 0. \end{cases} \tag{4.5.34}$$

Solving the differential equation of (4.5.34) yields

$$g(x) = K_1 e^{-\frac{\lambda}{\sqrt{1+\beta\lambda}} x} + K_2 e^{\frac{\lambda}{\sqrt{1+\beta\lambda}} x},$$

where K_1 and K_2 are constants to be determined by the boundary conditions in (4.5.34), that is,

$$\left(\lambda m - \alpha e^{-\lambda\tau} + \sqrt{1+\beta\lambda}\right)\left(M\lambda + \gamma + \sqrt{1+\beta\lambda}\right)e^{\frac{\lambda}{\sqrt{1+\beta\lambda}}}$$
$$-\left(\lambda m - \alpha e^{-\lambda\tau} - \sqrt{1+\beta\lambda}\right)\left(M\lambda + \gamma - \sqrt{1+\beta\lambda}\right)e^{-\frac{\lambda}{\sqrt{1+\beta\lambda}}} = 0. \quad (4.5.35)$$

Let us choose $M = 1$, $\gamma > 1$, and then $\beta > 2\gamma + \sqrt{\gamma^2 - 1}$. This permits to claim that

$$\lambda^* := \frac{\beta - 2\gamma \pm \sqrt{\beta^2 - 4\gamma\beta + 4}}{2}$$

is a positive real number satisfying $\lambda + \gamma = \sqrt{1+\beta\lambda}$. Combining these facts with (4.5.35), we obtain

$$\alpha e^{-\lambda^*\tau} = m\lambda + \sqrt{1+\beta\lambda^*}.$$

Since (4.5.8) is not fulfilled, we have

$$2\alpha \geq \min\{\gamma, a_0\beta\} = \gamma,$$

and the last equation yields

$$\tau = \frac{1}{\lambda^*}\ln\left(\frac{\alpha}{m\lambda + \sqrt{1+\beta\lambda^*}}\right),$$

which is a positive delay as long as m is sufficiently small. □

4.5.5 Spectral Analysis and Regularity Results

This section is devoted to the asymptotic distribution of the spectrum for system (4.5.1) when (4.5.8) is satisfied. Exploring this outcome, the differentiability of its solution is also discussed.

First, recall, thanks to Theorem 4.5.1, that the spectrum consists only of eigenvalues of the operator $\mathscr{A}$.

Thereafter, let us consider the eigenvalue problem: $(\lambda I - \mathscr{A})\varphi = 0$, where $\varphi = (\widehat{y}, \widehat{z}, \widehat{u}, \widehat{\xi}, \widehat{\eta}) \in \mathscr{D}(\mathscr{A})$, i.e.,

$$\begin{cases} \lambda\widehat{y} - \widehat{z} = 0, \ \lambda\widehat{z} - (a\widehat{y}_x)_x - \beta(a\widehat{z}_x)_x = 0, \\ \widehat{u}_x + \lambda\tau\widehat{u} = 0, \\ \lambda m\widehat{\xi} - (a\widehat{y}_x)(0) - \alpha\widehat{u}(1) - \beta(a\widehat{z}_x)(0) = 0, \\ \lambda M\widehat{\eta} + (a\widehat{y}_x)(1) + \beta(a\widehat{z}_x)(1) + \gamma\widehat{z}(1) = 0, \\ \widehat{\xi} = \widehat{u}(0) = \widehat{z}(0), \ \widehat{\eta} = \widehat{z}(1). \end{cases} \tag{4.5.36}$$

In order to avoid unnecessary complex computations, we only consider the case where a is constant. Then, the third equation in (4.5.36) yields $\widehat{u}(x) = \widehat{u}(0)e^{-\lambda\tau x}$. This, together with the first equation of (4.5.36), leads to the following problem:

$$\begin{cases} \lambda^2\widehat{y} - a(1+\beta\lambda)\widehat{y}_{xx} = 0, \\ a(1+\beta\lambda)\widehat{y}_x(0) = (m\lambda^2 - \lambda e^{-\tau\lambda}\alpha)\widehat{y}(0), \\ a(1+\beta\lambda)\widehat{y}_x(1) = -(M\lambda^2 + \gamma\lambda)\widehat{y}(1). \end{cases} \tag{4.5.37}$$

Obviously, $\lambda = 0$ is an eigenvalue, and its corresponding eigenvector is $(1, 0, 0, 0, 0)$. Whence, we assume in the sequel that $\lambda \neq 0$. Set

$$\rho^2 = \frac{\lambda^2}{a(1+\beta\lambda)}. \tag{4.5.38}$$

By the first equation in (4.5.37), we get $\widehat{y}(x) = c_1 e^{\rho x} + c_2 e^{-\rho x}$, where $c_j, \ = 1, 2$ are constants to be determined. Substituting $\widehat{y}$ in the two boundary conditions of (4.5.37), we claim that

$$[a(1+\beta\lambda)\rho - (m\lambda^2 - \lambda e^{-\tau\lambda}\alpha)]c_1 + [-a(1+\beta\lambda)\rho - (m\lambda^2 - \lambda e^{-\tau\lambda}\alpha)]c_2 = 0,$$

$$[a(1+\beta\lambda)\rho e^{\rho} + \lambda(M\lambda+\gamma)e^{\rho}]c_1 + [-a(1+\beta\lambda)\rho e^{-\rho} + \lambda(M\lambda+\gamma)e^{-\rho}]c_2 = 0.$$

Thus, λ is an eigenvalue if and only if

$$\Delta(\lambda) = \det\begin{bmatrix} a(1+\beta\lambda)\rho - (m\lambda^2 - \lambda e^{-\tau\lambda}\alpha) & -a(1+\beta\lambda)\rho - (m\lambda^2 - \lambda e^{-\tau\lambda}\alpha) \\ a(1+\beta\lambda)\rho e^{\rho} + \lambda(M\lambda+\gamma)e^{\rho} & -a(1+\beta\lambda)\rho e^{-\rho} + \lambda(M\lambda+\gamma)e^{-\rho} \end{bmatrix} = 0,$$

that is,

$$\begin{aligned} \Delta(\lambda) &= e^{-\rho}(a(1+\beta\lambda)\rho - (m\lambda^2 - \lambda e^{-\tau\lambda}\alpha))(-a(1+\beta\lambda)\rho + \lambda(M\lambda+\gamma)) \\ &\quad + e^{\rho}(a(1+\beta\lambda)\rho + (m\lambda^2 - \lambda e^{-\tau\lambda}\alpha))(a(1+\beta\lambda)\rho + \lambda(M\lambda+\gamma)) = 0. \end{aligned} \tag{4.5.39}$$

Recall that, in light of (4.5.8), all the eigenvalues are located on the left-hand side of the complex plane due to the dissipativity of the system operator. Moreover, they are symmetric with respect to the real axis. Thus, by (4.5.38), it is sufficient to consider

$\arg\rho \in [\frac{\pi}{4}, \frac{\pi}{2}]$. We shall study the distribution of the eigenvalue by discussing the following two cases:

Case 1 $\arg\rho \to \frac{\pi}{4}$, *as* $|\rho| \to \infty$.

In this case, we have $e^{\rho-\tau\lambda} > e^{\rho}$ and $e^{-\rho},\ e^{-\rho-\tau\lambda} \to 0$ as $|\rho| \to \infty$. Thus, by (4.5.39), we obtain

$$e^{\rho}(a(1+\beta\lambda)\rho + m\lambda^2)(a(1+\beta\lambda)\rho + \lambda(M\lambda+\gamma)) - \lambda e^{\rho-\tau\lambda}\alpha(a(1+\beta\lambda)\rho + \lambda(M\lambda+\gamma)) = o(e^{-\rho-\tau\lambda}). \tag{4.5.40}$$

Thereby, we get

$$(a\rho + a\beta\lambda\rho + m\lambda^2)e^{\tau\lambda} - \alpha\lambda = o(1),$$

and hence,

$$\lambda e^{\tau\lambda} = \frac{\alpha}{m}[1 + O(\lambda^{-\frac{1}{2}})].$$

A straightforward computation gives

$$\lambda_{1n} = \frac{1}{\tau}\left[\ln\left(\frac{\alpha}{m}\right) - \ln\omega_n\right] + i(\omega_n + \pi) + O(n^{-\frac{1}{2}}), \tag{4.5.41}$$

where $\omega_n = \frac{2n-\frac{3}{2}}{\tau}$, for $n \in \mathbb{N}$, and

$$\rho_n = \frac{1}{\sqrt{a\beta}}e^{i\frac{\pi}{4}}\sqrt{\omega_n}\left(1 + i\frac{\ln\omega_n}{2\tau\omega_n}\right) + O(n^{-\frac{1}{2}}).$$

Moreover, the corresponding eigenvector is given by

$$\varphi_{1,n} = \left(\frac{1}{\lambda_{1,n}}\frac{\ln\omega_n}{\omega_n^2}e^{-\rho_n x}, \frac{\ln\omega_n}{\omega_n^2}e^{-\rho_n x}, \frac{\ln\omega_n}{\omega_n^2}e^{-\tau\lambda_{1,n}x}, \frac{\ln\omega_n}{\omega_n^2}, \frac{\ln\omega_n}{\omega_n^2}e^{-\rho_n}\right), \quad n \in \mathbb{N}.$$

Case 2 $\arg\rho \in (\frac{\pi}{4}, \frac{\pi}{2}]$, *as* $|\rho| \to \infty$.

In such an event, we have

$$e^{\rho-\tau\lambda} > e^{-\rho-\tau\lambda} > e^{\rho} > e^{-\rho}$$

and $e^{-\rho} \to 0$, as $|\rho| \to \infty$. Thereby, we have

$$e^{-\rho}\lambda\alpha[-a(1+\beta\lambda)\rho + \lambda(M\lambda+\gamma)] - e^{\rho}\lambda\alpha[a(1+\beta\lambda)\rho + \lambda(M\lambda+\gamma)] = O(e^{\rho+\tau\lambda})$$

and

$$e^{2\rho} = 1 + \frac{-\frac{a\rho}{\lambda^2} - \frac{a\beta\rho}{\lambda} + \frac{\gamma}{\lambda}}{M} + o(\lambda^{-2}).$$

The latter yields

$$\rho_n = i\left(n\pi + \frac{1}{2Mn\pi}\right) - \frac{\gamma}{2a\beta n^2\pi^2} + O(n^{-3}),\ n \in \mathbb{N},$$

and hence, the corresponding eigenvalue is

$$\lambda_{2,n} = -a\beta\left(n^2\pi^2 + \frac{1}{M}\right) + \frac{1}{\beta} - i\frac{\gamma}{n\pi} + O(n^{-2}),\ n \in \mathbb{N}. \tag{4.5.42}$$

Besides, we can calculate the corresponding eigenvector and get

$$\varphi_{2,n} = \left(\frac{1}{\lambda_n}\phi_n(x), \phi_n(x), O(n^{-1}), O(e^{-\tau a\beta n^2\pi^2}), O(n^{-1})\right),$$

where

$$\phi_n = \sin\left(n\pi + \frac{1}{2Mn\pi}\right)x + O(e^{-\tau a\beta n^2\pi^2}),\ n \in \mathbb{N}.$$

Summarizing the above findings, we have:

Proposition 4.5.1 *Let $\mathscr{A}$ be defined as before. Then the spectrum of $\mathscr{A}$ has two branches. Specifically, we have*

$$\sigma(\mathscr{A}) = \{\lambda_{1,n},\ \lambda_{2,n},\ \overline{\lambda_{1,n}},\ \overline{\lambda_{2,n}}\} \cup \{0\},$$

where $\lambda_{1,n}$ and $\lambda_{2,n}$ are respectively given by (4.5.41) and (4.5.42).

Now, based on the above spectral analysis of the operator $\mathscr{A}$, we can proceed to consider the regularity of the solution to the system. To do this, let us first note that $Im(z)$ will denote the imaginary part of a complex number z. Then, recall the following classical result on the characterization of differential semigroups [158].

Theorem 4.5.5 *Let $S(t)$ be a C_0-semigroup and A be its infinitesimal generator. If $\|S(t)\| \le Me^{\omega t}$ for some constants M, ω, then $S(t)$ is differentiable for $t > t_0$, where t_0 is some positive constant, if and only if the following two conditions hold.*

C_1. There exist real constants k and b, $C > 0$ such that

$$\rho(A) \supset \Theta_{k,b} = \{\lambda \in \mathbb{C} : Re\lambda \ge k - b\ln|Im\lambda|\};$$

C_2. $\|R(\lambda, A)\| \le C|Im\lambda|$, for $\lambda \in \Theta_{k,b}$, $\Re\lambda \le 0$.

We have the following result:

Theorem 4.5.6 *Let $\mathscr{A}$ be the system operator defined by (4.5.14) on $\mathscr{H}$. Then, the semigroup $e^{t\mathscr{A}}$ generated by the system operator $\mathscr{A}$ is differentiable for $t > t_0$, where t_0 is some positive constant.*

Proof According to the expressions of the spectrum of $\mathscr{A}$ given in Proposition 4.5.1, there exist some $k > \frac{1}{\tau}\ln(\frac{\alpha}{m})$ and $b < \frac{1}{3\tau}$ such that

$$\rho(A) \supset \Theta_{k,b} = \{\lambda \in \mathbb{C} : Re\lambda \geq k - b\ln|Im\lambda|\}.$$

Thus, the condition C_1 of Theorem 4.5.5 holds.

Now, we shall focus on verifying the condition C_2. If this were not true, then Banach-Steinhaus theorem (see Theorem 2.1.7) implies that there exist at least one sequence $\lambda_n = -\mu_n + i\omega_n \in \Theta_{k,b}$ with $\mu_n > 0$, $\omega_n \to \infty$, $\omega_n > 1$ and $\{U_n\}_{n=1}^{\infty} \subset \mathscr{D}(\mathscr{A})$ with $U_n = (y_n, z_n, u_n, \xi_n, \eta_n)$ and $\|U_n\|_{\mathscr{H}} = 1$ such that

$$\lim_{n\to\infty} \omega_n\|(\lambda_n I - \mathscr{A})U_n\|_{\mathscr{H}} = 0, \quad as\ n \to \infty,$$

that is,

$$\omega_n(\lambda_n y_n - z_n) \to 0, \text{ in } H^1(0,1); \tag{4.5.43}$$

$$\omega_n(\lambda_n z_n - a y_{n,xx} - \beta a z_{n,xx}) \to 0, \text{ in } L^2(0,1); \tag{4.5.44}$$

$$\omega_n\left(\lambda_n u_n + \frac{u_{n,x}}{\tau}\right) \equiv F_3^n \to 0, \text{ in } L^2(0,1); \tag{4.5.45}$$

$$\omega_n\left(\lambda_n\xi_n - \frac{1}{m}(a y_{n,x}(0) + \beta a z_{n,x}(0) + \alpha u_n(1))\right) \to 0, \text{ in } \mathbb{C}; \tag{4.5.46}$$

$$\omega_n\left(\lambda_n\eta_n + \frac{1}{M}(a y_{n,x}(1) + \beta a z_{n,x}(1) + \gamma z_n(1))\right) \to 0, \text{ in } \mathbb{C}. \tag{4.5.47}$$

Note that $\omega_n \Re e((\lambda_n I - \mathscr{A})U_n, U_n)_{\mathscr{H}} \to 0$. Due to the dissipativity of $\mathscr{A}$ (see (4.5.15)), we have for sufficiently large n:

$$\begin{aligned} &- \omega_n\left((\frac{1}{a_0}[\alpha + K] - \beta)\int_0^1 a z_{n,x}^2\,dx + (K + \alpha - \gamma)\,z_n(1)^2 + \frac{\alpha - K}{2}u_n^2(1)\right)\\ &\leq \omega_n\mu_n\|U_n\|_{\mathscr{H}}^2. \end{aligned}$$

Thus, we have: As $n \to \infty$,

$$- (\ln\omega_n)^{-1}((\frac{1}{a_0}[\alpha + K] - \beta)\int_0^1 a z_{n,x}^2\,dx + (K + \alpha - \gamma)\,z_n(1)^2$$

$$+ \frac{\alpha - K}{2} {u_n}^2(1)) \le \frac{\mu_n}{\ln \omega_n}.$$

Since $\lambda_n = -\mu_n + i\omega_n \in \Theta_{k,b}$, it follows that $\frac{\mu_n}{\ln \omega_n} < b$, and so there exists some constant $C > 0$ such that

$$(\ln \omega_n)^{-\frac{1}{2}} \|z_{n,x}\|_{L^2} \le C, \tag{4.5.48}$$

$$(\ln \omega_n)^{-\frac{1}{2}} |z_n(1)|, \ (\ln \omega_n)^{-\frac{1}{2}} |u_n(1)| \le C. \tag{4.5.49}$$

We can also get

$$\omega_n (\ln \omega_n)^{-\frac{1}{2}} \|y_{n,x}\|_{L^2} \le C,$$

by virtue of (4.5.43) and (4.5.48). Since $\omega_n (\ln \omega_n)^{-\frac{1}{2}} \to \infty, \ as\ n \to \infty$, we obtain

$$\|y_{n,x}\|_{L^2} \le C\omega_n^{-1} \ln \omega_n \to 0, \quad n \to \infty. \tag{4.5.50}$$

Dividing (4.5.44) by ω_n and taking the L^2 inner product with $\frac{z_n}{\ln \omega_n}$, it follows that

$$\left\langle \frac{\lambda_n}{\ln \omega_n} z_n, z_n \right\rangle - \frac{1}{\ln \omega_n} \langle a y_{n,xx} + \beta a z_{n,xx}, z_n \rangle \to 0.$$

A direct calculation yields

$$\left\langle \frac{\lambda_n}{\ln \omega_n} z_n, z_n \right\rangle - \frac{1}{\ln \omega_n} \langle a y_{n,x} + \beta a z_{n,x}, z_n \rangle |_0^1 + \frac{1}{\ln \omega_n} \langle a y_{n,x} + \beta a z_{n,x}, z_{n,x} \rangle \to 0. \tag{4.5.51}$$

On one hand, it follows from (4.5.48) the boundedness of

$$\frac{1}{\ln \omega_n} |\langle a y_{n,x} + \beta a z_{n,x}, z_{n,x} \rangle|.$$

Recall here that we adopted the standard notation: $f(n) = \text{o}(1)$ means that $\lim_{n\to\infty}(f(n)/K) \to 0$ for any positive constant K, and hence, $\lim_{n\to\infty} f(n) = 0$. On the other hand, by virtue of (4.5.46) and (4.5.47), we deduce that the second term in (4.5.51) satisfies

$$\frac{1}{\ln \omega_n} \langle a y_{n,x} + \beta a z_{n,x}, z_n \rangle |_0^1 + \left\langle \lambda_n \eta_n, \frac{M\eta_n}{\ln \omega_n} \right\rangle$$
$$+ \left\langle \gamma \eta_n, \frac{\eta_n}{\ln \omega_n} \right\rangle + \left\langle \lambda_n \xi_n, \frac{m\xi_n}{\ln \omega_n} \right\rangle - \left\langle \alpha u_n(1), \frac{\xi_n}{\ln \omega_n} \right\rangle = o(1). \tag{4.5.52}$$

Substituting (4.5.52) into (4.5.51), we obtain

$$\begin{aligned}&\left\langle\tfrac{\lambda_n}{\ln\omega_n}z_n, z_n\right\rangle+\left\langle\lambda_n\eta_n, \tfrac{M\eta_n}{\ln\omega_n}\right\rangle+\left\langle\lambda_n\xi_n, \tfrac{m\xi_n}{\ln\omega_n}\right\rangle+\left\langle\gamma\eta_n, \tfrac{\eta_n}{\ln\omega_n}\right\rangle\\&-\left\langle\alpha u_n(1), \tfrac{\xi_n}{\ln\omega_n}\right\rangle+\frac{1}{\ln\omega_n}\left\langle\beta a z_{n,x}, z_{n,x}\right\rangle\to 0.\end{aligned}\tag{4.5.53}$$

Using Gagliardo-Nirenberg inequality (see Theorem 2.1.4), we get

$$\left\|\frac{z_n}{(\ln\omega_n)^{1/4}}\right\|_{L^\infty}\le d_1\left\|\frac{z_{n,x}}{(\ln\omega_n)^{1/2}}\right\|^{1/2}\|z_n\|^{1/2}+d_2\left\|\frac{z_n}{(\ln\omega_n)^{1/4}}\right\|.$$

Due to (4.5.48) and the fact that $z_n\in L^2(0,1)$, we deduce the boundedness of $\left\|\frac{z_n}{(\ln\omega_n)^{1/4}}\right\|_{L^\infty}$, $\left|\frac{z_n(0)}{(\ln\omega_n)^{1/4}}\right|$, and $\left|\frac{z_n(1)}{(\ln\omega_n)^{1/4}}\right|$. Then by (4.5.49), we have

$$\left\langle\alpha u_n(1), \frac{m z_n(0)}{\ln\omega_n}\right\rangle=\left\langle\alpha\frac{u_n(1)}{(\ln\omega_n)^{\frac{3}{4}}}, \frac{m z_n(0)}{(\ln\omega_n)^{\frac{1}{4}}}\right\rangle\to 0.$$

Now, considering the imaginary part of (4.5.53), we obtain

$$\left\langle\frac{\omega_n}{\ln\omega_n}z_n, z_n\right\rangle+\left\langle\omega_n\eta_n, \frac{M\eta_n}{\ln\omega_n}\right\rangle+\left\langle\omega_n\xi_n, \frac{m\xi_n}{\ln\omega_n}\right\rangle\to 0,$$

which implies that as $n\to\infty$,

$$\left\|\frac{\omega_n^{\frac{1}{2}}}{(\ln\omega_n)^{\frac{1}{2}}}z_n\right\|, \left|\frac{\omega_n^{\frac{1}{2}}}{(\ln\omega_n)^{\frac{1}{2}}}\eta_n\right|, \left|\frac{\omega_n^{\frac{1}{2}}}{(\ln\omega_n)^{\frac{1}{2}}}\xi_n\right|\to 0,\tag{4.5.54}$$

and hence, $\left|\frac{\omega_n^{\frac{1}{2}}}{(\ln\omega_n)^{\frac{1}{2}}}u_n(0)\right|=\left|\frac{\omega_n^{\frac{1}{2}}}{(\ln\omega_n)^{\frac{1}{2}}}\xi_n\right|\to 0$. In turn, by (4.5.45), we get

$$u_n(x)=e^{-\lambda\tau x}u_n(0)+\tau\int_0^x e^{-\lambda\tau(x-s)}\frac{F_3^n(s)}{\omega_n}ds.$$

Note from the beginning of the proof that $\frac{\mu_n}{\ln\omega_n}<b<\frac{1}{3\tau}$. Thus, as $n\to\infty$, we have

$$\left\|e^{-\lambda\tau(x-s)}\frac{F_3^n(s)}{\omega_n}\right\|=|e^{\mu_n\tau}|\left\|\frac{F_3^n(s)}{\omega_n}\right\|\le\omega_n^{\frac{1}{3}}\left\|\frac{F_3^n(s)}{\omega_n}\right\|\to 0.$$

Furthermore, it follows from (4.5.54) that

$$\|e^{-\lambda\tau x}u_n(0)\|=|e^{\mu_n\tau}u_n(0)|\le\omega_n^{\frac{1}{3}}|u_n(0)|\to 0, n\to\infty.$$

Thereby, we obtain that $\|u_n\|_{L^2(0,1)} \to 0$, as $n \to \infty$. Hence, by (4.5.50), (4.5.51), (4.5.54), we have proved that $\|U_n\|_{\mathscr{H}} = \|(y_n, z_n, u_n, \xi_n, \eta_n)\|_{\mathscr{H}} \to 0$, as $n \to \infty$, which contradicts the fact that $\|U_n\|_{\mathscr{H}} = 1$. Thus, the condition C_2 holds, and then by Theorem 4.5.5, we conclude that there exists some constant $t_0 > 0$ such that the semigroup $e^{t\mathscr{A}}$ generated by the system operator $\mathscr{A}$ is differentiable for $t > t_0$. The proof is completed. □

Remark 4.5.2 As a consequence of the spectral mapping theorem for differential semigroups, the spectrum-determined-growth condition (see Definition 2.3.7) always holds, that is, the sharp convergence rate of the solution is given by the supremum of the real part of the spectrum.

4.5.6 Numerical Simulations

This subsection is consecrated to numerical simulations for the distributions of the spectrum as well as the dynamical behavior of the system under consideration.

Simulation Part I: Spectrum of the Closed-Loop System
Let us set the parameters of the system:

$$a = a_0 = 3,\ \beta = 10,\ m = 40, M = 20,\ \alpha = 2,\ \gamma = 30,\ \tau = 1. \tag{4.5.55}$$

Clearly, (4.5.8) is fulfilled. Substituting (4.5.55) into (4.5.39) and calculating its zeros, we obtain Fig. 4.8. As the spectrum of $\mathscr{A}$ is symmetrically located with respect to the real axis, the spectrum is depicted only in the upper-half complex plane.

We present the spectrum of $\mathscr{A}$ in two different figures so as to clearly show the spectral distribution. Indeed, from these two figures, we can recognize the two branches of the spectrum, which is consistent with the distributions of the spectrum obtained in Proposition 4.5.1.

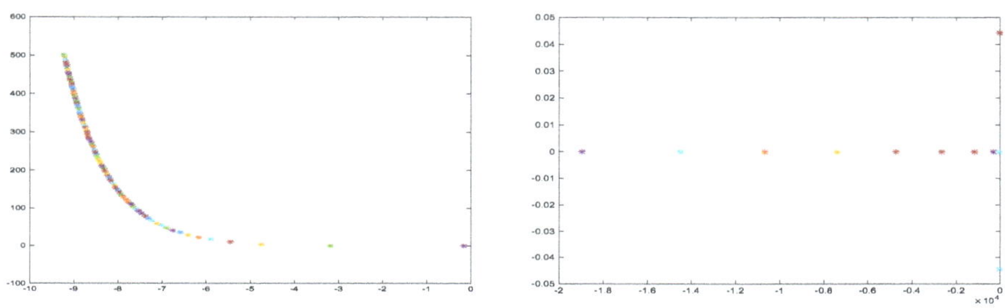

Fig. 4.8 Spectrum of $\mathscr{A}$

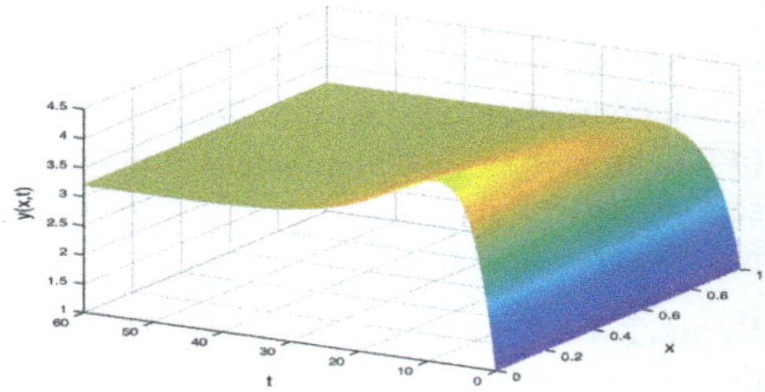

Fig. 4.9 $y(x, t)$

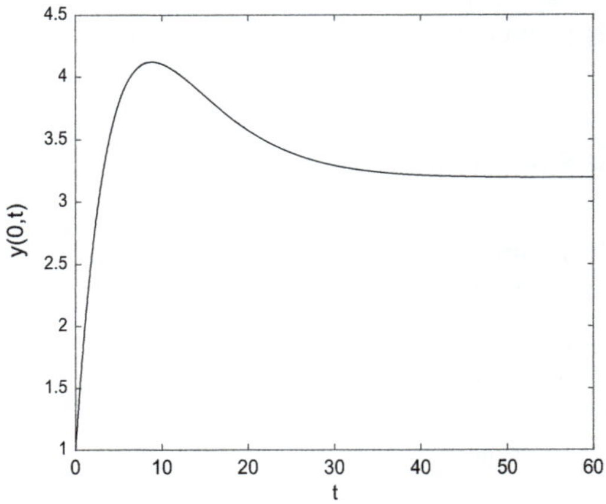

Fig. 4.10 $y(0, t)$

Simulation Part II: Dynamical Behavior of the Closed-Loop System

Firstly, recall (4.5.55) and then consider the initial conditions $y(x, 0) = y_t(x, 0) = 1$, whereas the initial condition for the delayed term is: $y_t(0, t - \tau) = 1,\ 0 \leq t < \tau$. In order to visualize the dynamical behavior of the system, we employ the Chebyshev spectral method in space ($N = 40$), and the backward Euler method in time ($dt = 0.001$), and program via Matlab R2014b [178]. The behavior of $y(x, t)$ in the time interval $[0, 60]$ is presented in Fig. 4.9.

By virtue of Theorem 4.5.2 and (4.5.55), one obtains $\chi = 3.25$, and hence, the solution must converge to $(3.25, 0, 0, 0, 0)$, which means that $y(x, t)$ goes to the state 3.25 as t is sufficiently large. From Figs. 4.9, 4.10, and 4.11, we can see that the solution $y(x, t)$ and the components $y(0, t)$ and $y(1, t)$ tend to the desired value, which agrees with Theorem 4.5.2.

In turn, if we set $\beta = 5$, then we obtain Figs. 4.12, 4.13, and 4.14. From these figures, we notice that the solution $y(x, t)$ is still convergent to the same value as in Fig. 4.9. This is reasonable, since it is known from (4.5.18) that the equilibrium state $(\chi, 0, 0, 0, 0)$ is irrelevant to the choice of β, although the dynamical behavior of the solution depends on β as depicted in Figs. 4.9 and 4.12.

However, if we modify the values of α and β so that the "smallness" condition (4.5.8) is not fulfilled, for example, $\alpha = 16,\ \beta = 5$, we get Figs. 4.15, 4.16,

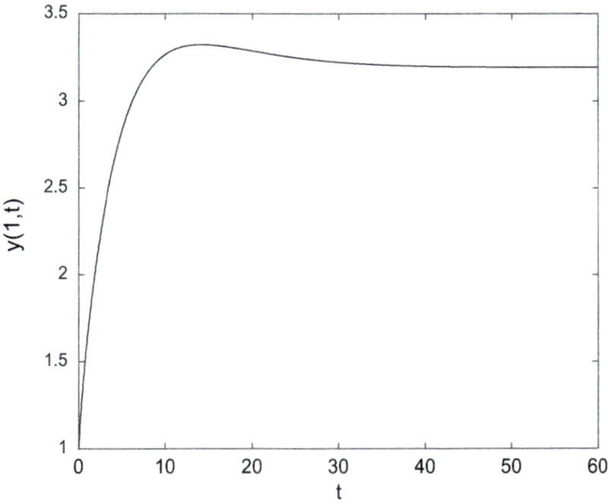

Fig. 4.11 $y(1, t)$

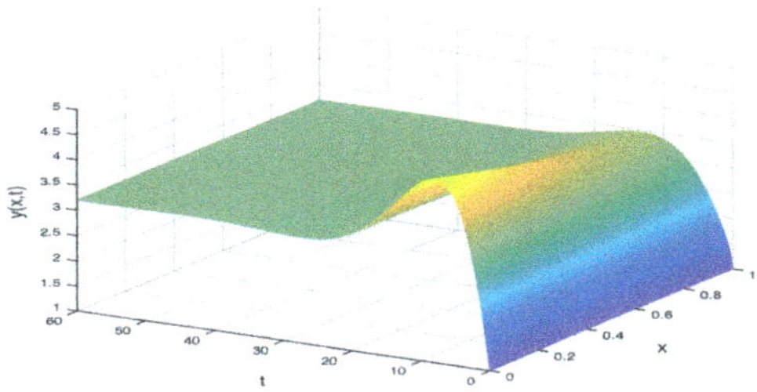

Fig. 4.12 $y(x, t)$

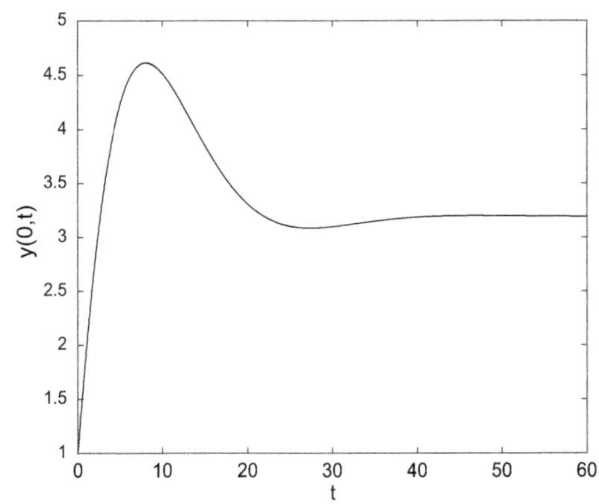

Fig. 4.13 $y(0, t)$

Fig. 4.14 $y(1, t)$

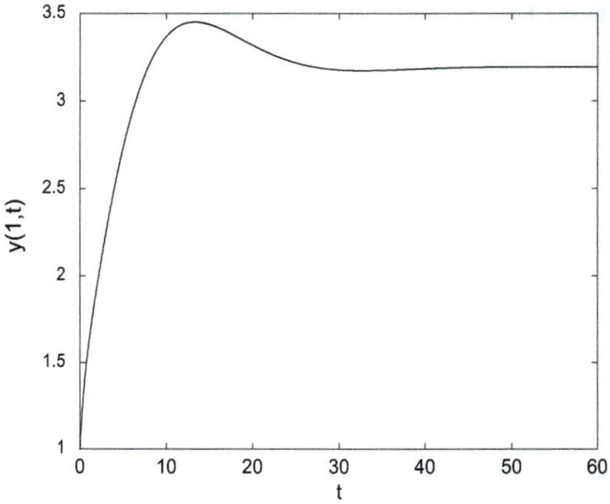

Fig. 4.15 $y(x, t)$

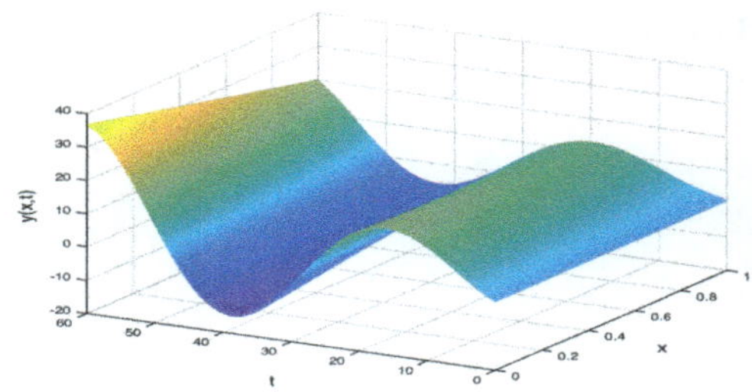

Fig. 4.16 $y(0, t)$

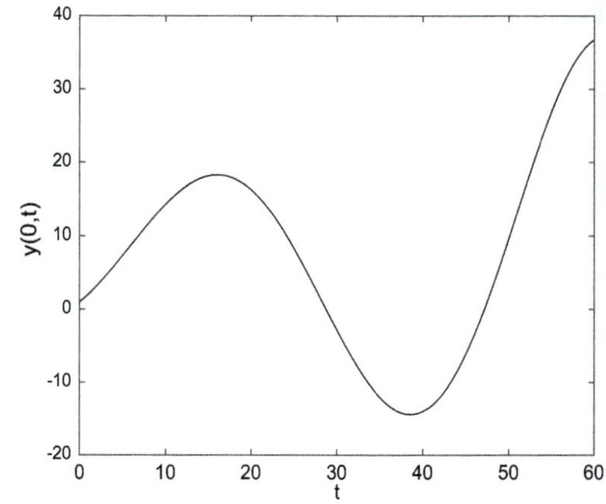

Fig. 4.17 $y(1, t)$

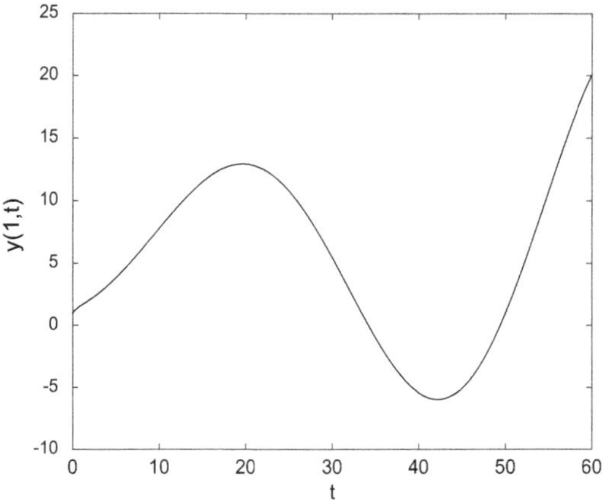

and 4.17. In this case, we observe that the solution does not converge to a stationary state. In fact, the values of $y(x, t)$ become larger as t increases. This is also consistent with Theorem 4.5.4.

4.5.7 Concluding Discussion

In this section, the overhead crane model is considered by assuming that the input control at the top end of the cable consists of an angular velocity and a delayed velocity. In order to neutralize the effect of the delay, a feedback control law is designed. Then, it is shown that the solutions of the closed-loop system exponentially converge to a stationary state, under a "smallness" condition of the delayed term. Moreover, the asymptotic distribution of the spectrum and the asymptotic expansion of the eigenfunctions are provided. This leads to the differentiability of the semigroup generated by the system operator. Last but not least, it is shown that the spectrum-determined-growth condition always holds, and hence, the sharp convergence rate of the solutions is given by the supremum of the real part of the spectrum.

As the reader may have noticed, the convergence outcome is reached via the action of two controls (one control is interior (of type Kevin-Voigt), while the other one acts on the right-end boundary). These two controls are needed to contrast the effect of the left-end boundary delay. Clearly, this situation is too costly in practice, and hence, it would be interesting to reduce the number of contrasting controls. This problem is addressed in the next section. In turn, the time-dependent delay and/or cable length case should be investigated.

4.6 Boundary Delay: Exponential Stabilization via a Single Boundary Control

Motivated by the study of Sect. 4.5, the main concern here is to weaken the stability conditions. Indeed, the interior control of high-order $(y_{xt})_x(x,t)$ considered in Sect. 4.5 is viewed as a very strong damping [141]. Furthermore, the presence of another boundary control at the lower end of the cable shows how "costly" could be the implementation of the feedback law proposed in Sect. 4.5.

4.6.1 *Introduction*

In order to highlight the inducement of this section, the system that we study is

$$\begin{cases} y_{tt}(x,t) - (ay_x)_x\,(x,t) = 0, & x \in I,\ t > 0, \\ my_{tt}(0,t) - (ay_x)\,(0,t) = -\sigma y(0,t) + \nu y_t(0,t-\tau) - \zeta y_t(0,t) & \\ \qquad\qquad +\omega(ay_{xt})(0,t), & t > 0, \\ My_{tt}(1,t) + (ay_x)\,(1,t) = 0, & t > 0, \\ y(x,0) = y_0(x),\ y_t(x,0) = y_1(x), & x \in I, \\ y_t(0,t-\tau) = f(t-\tau), & t \in (0,\tau), \end{cases} \tag{4.6.1}$$

in which $I = (0,1)$, the parameters $\nu,\ \zeta,\ \omega,\ \sigma$ are positive feedback gain, and $\tau > 0$ is the time delay. Finally, f is the initial condition corresponding to the presence of the delay τ. The major purpose of the present section is twofold: First, we aim to show that the presence of the Kelvin-Voigt damping $(y_{xt})_x(x,t)$ and the boundary control at $x = 1$ in Sect. 4.5 are unnecessary to achieve the exponential stability of the crane system. Second, contrary to the previous sections, an estimate of the exponential decay is desired in the current section.

Obviously, our results will be obtained under an expected smallness condition on the parameter ν of the delayed term (see [149, 184] for similar situations). This outcome considerably improves that of Sect. 4.5 since neither the interior control $(ay_{xt})_x(x,t)$ nor the boundary control $(ay_{xt})(1,t)$ and $y_t(1,t)$ are required. In other words, the action of $(ay_{xt})(0,t)$ is strong enough to get the desirable result as the exponential stability. It is also worth mentioning that if $\omega = 0$, that is, the term $(ay_{xt})(0,t)$ is excluded, then the system may be unstable (see Sect. 4.2), and accordingly, the presence of $(ay_{xt})(0,t)$ is unavoidable if one hopes to obtain the stability of the system.

Of course, our findings remain valid if $\sigma = 0$ except that the exponential stability result takes another form, namely, the exponential convergence of solutions to a steady state. In this case, it suffices to argue as in the previous sections.

The remainder of this section is organized as follows: In Sect. 4.6.2, we show that the system (4.6.1) is well-posed in the sense of semigroups theory of linear operators. Section 4.6.3 is devoted to the proof of the exponential stability of the system (4.6.1). The main ingredient of the proof is the combination of the frequency-domain method and the multipliers technique. In Sect. 4.6.4, we provide an estimate of the exponential rate by means of an appropriate Lyapunov functional. Numerical examples are presented in Sect. 4.6.5 through an extensive investigation of the spectrum of the system. Lastly, a conclusion is given in the last section.

4.6.2 Well-Posedness of the Problem

First, we consider the following auxiliary variable:

$$\xi(t) = my_t(0,t) + \omega\left[\sigma y(0,t) - (ay_x)(0,t)\right]. \tag{4.6.2}$$

Thereafter, assuming from now on that $\zeta = \delta + \sigma\omega$, for some positive constant δ and

$$m > \delta\omega, \tag{4.6.3}$$

we define the energy corresponding to the system (4.6.1) as follows:

$$\begin{aligned} E(t) = \frac{1}{2}\Bigg[\int_0^1 \left[y_t^2(x,t) + a(x)y_x^2(x,t) + K\tau y_t^2(0,t-x\tau)\right] dx + \sigma y^2(0,t) \\ + \frac{1}{m-\delta\omega}\xi^2(t) + My_t^2(1,t)\Bigg], \end{aligned} \tag{4.6.4}$$

where K is a positive constant to be determined. Clearly, (4.1.5) and (4.6.3) permit to claim that the energy $E(t)$ defined above is indeed positive. Next, multiplying the first equation in (4.6.1) by $y_t(x,t)$ and formally integrating with respect to x, we get

$$\begin{aligned} \frac{1}{2}\frac{d}{dt}\left(\int_0^1 \left[y_t^2(x,t) + a(x)y_x^2(x,t)\right] dx\right) + (ay_x)(0,t)y_t(0,t) \\ - (ay_x)(1,t)y_t(1,t) = 0. \end{aligned} \tag{4.6.5}$$

Using (4.6.2) and the boundary condition at $x = 0$, we have

$$(ay_x)(0,t) = \xi'(t) + \sigma y(0,t) - \nu y_t(0,t-\tau) + \delta y_t(0,t). \tag{4.6.6}$$

Then, by (4.6.5)–(4.6.6) and the boundary condition at $x = 1$, we find

$$\begin{aligned} \frac{1}{2}\frac{d}{dt}\left(\int_0^1 \left[y_t^2(x,t) + a(x)y_x^2(x,t)\right] dx + \sigma y^2(0,t) + My^2(1,t)\right) \\ = -\xi'(t)y_t(0,t) + \nu y_t(0,t-\tau)y_t(0,t) - \delta y_t^2(0,t). \end{aligned} \tag{4.6.7}$$

On the other hand, it is easy to check that

$$\frac{K\tau}{2}\frac{d}{dt}\int_0^1 y_t^2(0,t-x\tau)\,dx = \frac{K}{2}y_t^2(0,t) - \frac{K}{2}y_t^2(0,t-\tau), \tag{4.6.8}$$

whereas (4.6.2) and the boundary conditions yield

$$\xi'(t) = -\frac{1}{\omega}\xi(t) + \frac{m-\delta\omega}{\omega}y_t(0,t) + \nu y_t(0,t-\tau). \tag{4.6.9}$$

Then, (4.6.7), (4.6.8), and (4.6.9) give

$$\begin{aligned} E'(t) = {} & \frac{K}{2}y_t^2(0,t) - \frac{K}{2}y_t^2(0,t-\tau) - \delta y_t^2(0,t) - \frac{\omega}{m-\delta\omega}\xi'^2(t) \\ & + \nu y_t(0,t-\tau)y_t(0,t) + \frac{\nu\omega}{m-\delta\omega}y_t(0,t-\tau)\xi'(t). \end{aligned} \tag{4.6.10}$$

Applying Young's inequality, the latter becomes

$$\begin{aligned} E'(t) \le {} & \left(\frac{K}{2} + \frac{\nu}{4\epsilon_1} - \delta\right) y_t^2(0,t) + \left(\nu\epsilon_1 + \frac{\nu\omega\epsilon_2}{m-\delta\omega} - \frac{K}{2}\right) y_t^2(0,t-\tau) \\ & + \left(\frac{\nu}{4\epsilon_2} - 1\right)\frac{\omega}{m-\delta\omega}\xi'^2(t), \end{aligned} \tag{4.6.11}$$

for any positive constants ϵ_1 and ϵ_2. Considering, for sake of simplicity, the particular case $\epsilon_1 = \epsilon_2 = \frac{1}{2}$, we get

$$\begin{aligned} E'(t) \le {} & \left(\frac{K+\nu}{2} - \delta\right) y_t^2(0,t) + \frac{1}{2}\left(\nu + \frac{\nu\omega}{m-\delta\omega} - K\right) y_t^2(0,t-\tau) \\ & + \left(\frac{\nu}{2} - 1\right)\frac{\omega}{m-\delta\omega}\xi'^2(t). \end{aligned} \tag{4.6.12}$$

In order to make the energy $E(t)$ decrease, we shall assume that

$$\nu < \min\left\{2, \frac{\delta}{1 + \frac{\omega}{2(m-\delta\omega)}}\right\}, \tag{4.6.13}$$

and then choose K such that

$$\nu\left(1 + \frac{\omega}{m-\delta\omega}\right) \le K \le 2\delta - \nu, \tag{4.6.14}$$

which is possible in light of (4.6.13). Lastly, under (4.6.3), (4.6.13), and (4.6.14), the energy $E(t)$ defined by (4.6.4) is decreasing.

Remark 4.6.1 It is important to point out that (4.6.13) yields $\nu < \delta$, which in turn implies $\nu < \zeta$ since $\zeta = \delta + \sigma\omega > \delta$. Whereupon, we retrieve the usual "smallness" condition $\nu < \zeta$ for the system (4.6.1) as known in the literature for many physical systems [149, 184].

Now, let us introduce the standard auxiliary variable [149, 184]

$$u(x,t) = y_t(0, t - x\tau), \tag{4.6.15}$$

which implies that the system (4.6.1) becomes

$$\begin{cases} y_{tt}(x,t) - (ay_x)_x(x,t) = 0, & x \in I,\ t > 0, \\ \tau u_t(x,t) + u_x(x,t) = 0, & x \in I,\ t > 0, \\ my_{tt}(0,t) - (ay_x)(0,t) = -\sigma y(0,t) + \nu u(1,t) & \\ \quad -(\delta + \sigma\omega)u(0,t) + \omega y_{xt}(0,t), & t > 0, \\ My_{tt}(1,t) + (ay_x)(1,t) = 0, & t > 0, \\ y(x,0) = y_0(x),\ y_t(x,0) = y_1(x), & x \in I, \\ y_t(0, t-\tau) = f(t-\tau), & t \in (0,\tau). \end{cases} \tag{4.6.16}$$

Subsequently, let $\Phi = (y, z, u, \xi, \eta)$, where $z(.,t) = y_t(.,t)$ and $\eta(t) = y_t(1,t)$. Then, define the state space $\mathscr{H}$ by

$$\mathscr{H} = H^1(I) \times L^2(I) \times L^2(I) \times \mathbb{R}^2, \tag{4.6.17}$$

equipped with the following real inner product:

$$\begin{aligned} \langle (y,z,u,\xi,\eta), (\tilde{y},\tilde{z},\tilde{u},\tilde{\xi},\tilde{\eta}) \rangle_{\mathscr{H}} = & \int_0^1 (ay_x\tilde{y}_x + z\tilde{z})\, dx \\ & + K\tau \int_0^1 u\tilde{u}\, dx + \sigma y(0)\tilde{y}(0) + \frac{1}{m - \delta\omega}\xi\tilde{\xi} \\ & + M\eta\tilde{\eta}, \end{aligned} \tag{4.6.18}$$

in which $K > 0$ satisfies the condition (4.6.14). Assuming that (4.1.5), (4.6.13), and (4.6.14), we can easily check that $\mathscr{H}$ endowed with the inner product (4.6.18) is a Hilbert space.

Whereupon, we rewrite the problem (4.6.1) (see also (4.6.16)) as follows:

$$\begin{cases} \Phi_t(t) = \mathscr{A}\Phi(t), \\ \Phi(0) = \Phi_0, \end{cases} \tag{4.6.19}$$

in which $\mathscr{A}$ is a linear operator defined by

$$\mathscr{D}(\mathscr{A}) = \left\{ \begin{array}{r} (y,z,u,\xi,\eta) \in \mathscr{H};\ y \in H^2(I),\ z, u \in H^1(I),\ \xi = mu(0) + \omega[\sigma y(0) - (ay_x)(0)], \\ u(0) = z(0),\ \eta = z(1) \end{array} \right\}, \tag{4.6.20}$$

$$\mathscr{A}(y, z, u, \xi, \eta) = \left(z, (ay_x)_x, -\frac{u_x}{\tau}, -\frac{1}{\omega}\xi + \frac{m - \delta\omega}{\omega} u(0) + \nu u(1), -\frac{(ay_x)(1)}{M} \right), \tag{4.6.21}$$

with $\Phi = (y, z, u, \xi, \eta)$ and $\Phi_0 = (y_0, y_1, f(-\tau\cdot), \xi_0, \eta_0)$.

The well-posedness result is stated below.

Theorem 4.6.1 *Suppose that (4.1.5), (4.6.3), and (4.6.13) hold. Then the operator $\mathscr{A}$ defined by (4.6.20)–(4.6.21) is densely defined in $\mathscr{H}$ and generates on $\mathscr{H}$ a C_0-semigroup of contractions $e^{t\mathscr{A}}$. Thereby, if $\Phi_0 \in \mathscr{D}(\mathscr{A})$, then the solution Φ of (4.6.19) is called a strong solution and belongs to $C((0,\infty); \mathscr{D})(\mathscr{A}) \cap C^1((0,\infty); \mathscr{H})$. In particular, we have*

$$\begin{gathered} y \in C^2\left((0,\infty); L^2(I)\right) \cap C^1\left((0,\infty); H^1(I)\right) \cap C\left((0,\infty); H^2(I)\right); \\ u \in C^1\left((0,\infty); L^2(I)\right) \cap C\left((0,\infty); H^1(I)\right); \\ \xi,\ \eta \in C^1\left((0,\infty); \mathbb{R}\right). \end{gathered}$$

However, if the initial condition Φ_0 belongs to $\mathscr{H}$, then the system (4.6.19) has a unique mild solution $\Phi \in C((0,\infty); \mathscr{H})$. In this case, we have

$$\begin{gathered} y \in C^1\left((0,\infty); L^2(I)\right) \cap C\left((0,\infty); H^1(I)\right); \\ u \in C\left((0,\infty); L^2(I)\right); \\ \xi,\ \eta \in C\left((0,\infty); \mathbb{R}\right). \end{gathered}$$

Finally, the spectrum of $\mathscr{A}$ consists of isolated eigenvalues of finite algebraic multiplicity only (see Remark 2.1.2).

Proof Let $\Phi = (y, z, u, \xi, \eta) \in \mathscr{D}(\mathscr{A})$. Consider the following change of variable:

$$\rho := -\frac{1}{\omega}\xi + \frac{m - \delta\omega}{\omega} u(0) + \nu u(1) = -\sigma y(0) + \nu u(1) - \delta u(0) + (ay_x)(0). \tag{4.6.22}$$

In light of (4.6.18) and (4.6.20)–(4.6.22), we obtain after several integrations by parts

$$\begin{aligned} \langle \mathscr{A}\Phi, \Phi)\rangle_{\mathscr{H}} &= [\sigma y(0) - (ay_x)(0)]z(0) - \frac{K}{2}[u^2(1) - u^2(0)] + \frac{\rho}{m - \delta\omega}\xi \\ &= \frac{\xi - mu(0)}{\omega} z(0) - \frac{K}{2}[u^2(1) - u^2(0)] + \frac{\rho}{m - \delta\omega}\xi \\ &= -\delta u^2(0) + \nu u(0)u(1) - \frac{\omega}{m - \delta\omega}\rho^2 + \frac{\nu\omega}{m - \delta\omega}\rho u(1) \\ &\quad - \frac{K}{2}[u^2(1) - u^2(0)]. \end{aligned}$$

Subsequently, using Young's inequality, we deduce that

$$\langle \mathscr{A}\Phi, \Phi) \rangle_{\mathscr{H}} \leq \left(\frac{\nu}{2}+\frac{K}{2}-\delta\right) u^2(0) + \left(\frac{\nu}{2}-\frac{K}{2}+\frac{\nu\omega}{2(m-\delta\omega)}\right) u^2(1) + \left(\frac{\nu}{2}-1\right)\frac{\omega}{m-\delta\omega}\rho^2. \tag{4.6.23}$$

Herewith, the operator $\mathscr{A}$ is dissipative by virtue of the assumptions of the theorem.

Next, we prove the maximality of $\mathscr{A}$, that is, given $(f, g, h, p, q) \in \mathscr{H}$, we look for an element $(y, z, u, \xi, \eta) \in \mathscr{D}(\mathscr{A})$ such that $(\lambda I - \mathscr{A})(y, z, u, \xi, \eta) = (f, g, h, p, q)$ for $\lambda > 0$. One can readily verify that it amounts to finding the solution of the following system:

$$\begin{cases} \lambda^2 y - (ay_x)_x = g + \lambda f, \\ u_x + \lambda\tau u = \tau h, \\ \omega\left[\sigma(\lambda\omega+1)+\lambda(\lambda m+\delta)\right] y(0) - \omega(\lambda\omega+1)(ay_x)(0) \\ -\nu\omega u(1) = \omega p + \omega(m\lambda+\delta) f(0), \\ \lambda^2 M y(1) + (ay_x)(1) = Mq + \lambda M f(1), \\ z = \lambda y - f, \\ \xi = mu(0) + \omega[\sigma y(0) - (ay_x)(0)], \; u(0) = z(0) = \lambda y(0) - f(0), \\ \eta = z(1) = \lambda y(1) - f(1). \end{cases} \tag{4.6.24}$$

We start by solving the second equation in (4.6.24). We obtain

$$u(x) = e^{-\tau\lambda x}(\lambda y(0) - f(0)) + \tau\int_0^x e^{-\tau\lambda(x-s)} h(s)\, ds, \tag{4.6.25}$$

and hence,

$$u(1) = e^{-\tau\lambda}(\lambda y(0) - f(0)) + \tau\int_0^1 e^{-\tau\lambda(1-s)} h(s)\, ds. \tag{4.6.26}$$

In view of (4.6.24)–(4.6.26), we need to seek $y \in H^2(I)$ such that

$$\begin{cases} \lambda^2 y - (ay_x)_x = g + \lambda f, \\ u_x + \lambda\tau u = \tau h, \\ \omega\left[\sigma(\lambda\omega+1)+\lambda(\lambda m+\delta) - \nu\lambda e^{-\tau\lambda}\right] y(0) - \omega(\lambda\omega+1)(ay_x)(0) \\ = p\omega + \left[\omega(m\lambda+\delta) - \nu\omega e^{-\tau\lambda}\right] f(0) + \nu\tau\omega\int_0^1 e^{-\tau\lambda(1-s)} h(s)\, ds, \\ \lambda^2 M y(1) + (ay_x)(1) = Mq + \lambda M f(1). \end{cases} \tag{4.6.27}$$

Multiplying the first equation in (4.6.27) by $\phi \in H^1(I)$ and integrating over $I = (0, 1)$, we obtain the variational form of the problem (4.6.27)

$$\mathscr{P}(y, \phi) = \mathscr{Q}(\phi),$$

where $\mathscr{P}$ is the real bilinear form in $H^1(I) \times H^1(I)$ given by

$$\mathscr{P}(y, \phi) = \int_0^1 (\lambda^2 y\phi + a y_x \phi_x)\, dx + \frac{1}{\omega\lambda + 1}\left[\sigma(\lambda\omega + 1) + \lambda(\lambda m + \delta) - \nu\lambda e^{-\tau\lambda}\right]$$

$$y(0)\phi(0) + \lambda^2 M y(1)\phi(1)$$

and $\mathscr{Q}$ is a linear form defined by $\mathscr{Q} : H^1(I) \longrightarrow \mathbb{R}$, with

$$\mathscr{Q}(\phi) = \int_0^1 (\lambda f + g)\phi\, dx$$

$$+\left[\frac{1}{\omega(\lambda\omega + 1)}\left(p\omega + \omega(m\lambda + \delta) - \nu\omega e^{-\tau\lambda}\right) f(0) + \frac{\nu\tau}{\lambda\omega + 1}\int_0^1 e^{-\tau\lambda\tau(1-s)} h(s)\, ds\right]\phi(0).$$

It is a simple task to check that $\mathscr{P}$ is a continuous coercive bilinear form, whereas $\mathscr{Q}$ is a continuous linear form as long as the assumptions of our theorem are satisfied. Thus, thanks to Lax-Milgram theorem (see Theorem 2.1.6), we can show the existence of a unique solution $y \in H^2(I)$ of (4.6.27) as long as $\lambda > 0$, which implies that the range of $\lambda I - \mathscr{A}$ is $\mathscr{H}$, for $\lambda > 0$. Then, in light of semigroup theory [158], the operator $\mathscr{A}$ is densely defined in $\mathscr{H}$ and generates on $\mathscr{H}$ a C_0-semigroup of contractions, while the solutions possess the regularity results stated in our theorem depending on the initial data Φ_0.

Moreover, Sobolev embedding [4] implies that $(\lambda I - \mathscr{A})^{-1}$ is compact, and hence, the spectrum of $\mathscr{A}$ consists of isolated eigenvalues of finite algebraic multiplicity only (see Remark 2.1.2). □

4.6.3 Exponential Stability

In this subsection, we will be concerned with the exponential stability of the solutions of the system (4.6.1). To this end, we will use Theorem 2.3.8.

The main result is given below:

Theorem 4.6.2 *Suppose that (4.1.5), (4.6.3), and (4.6.13) are fulfilled. Then, the system (4.6.1) (see also (4.6.16)) is exponentially stable in* $\mathscr{H}$*, i.e., there exist* $C > 0$ *and* $\varpi > 0$ *such that for all* $t > 0$ *we have the following the decay estimate:*

$$\left\| e^{t\mathscr{A}} \Phi \right\|_{\mathscr{H}} \leq Ce^{-\varpi t} \left\| \Phi \right\|_{\mathscr{H}}, \quad \forall \Phi \in \mathscr{H}.$$

Equivalently, the energy $E(t)$ *defined by (4.6.3) decays exponentially.*

Proof First, let us consider a sufficiently regular solution Φ so that our computations are valid (take, for instance, a solution Φ stemmed from an initial condition $\Phi \in \mathscr{D}(\mathscr{A}^2)$). Then, using a standard density argument, one can extend the result to $\mathscr{H}$. The proof is divided into two parts:

(I) The first part concerns (2.3.16), whose proof will be by contradiction. To proceed, suppose that there is $\lambda \in \mathbb{R}$ such that $i\lambda$ is an eigenvalue for $\mathscr{A}$. Let $\Phi = (y, z, u, \xi, \eta) \in D(\mathscr{A})$ the associated eigenvector. Then we have the following equation:

$$\mathscr{A}\Phi = i\lambda\Phi. \tag{4.6.28}$$

Our purpose is to prove that this equation has only a trivial solution $\Phi = 0$. Equation (4.6.28) is equivalent to considering the following system:

$$z = i\lambda y, \tag{4.6.29}$$

$$(ay_x)_x = i\lambda z, \tag{4.6.30}$$

$$-\frac{u_x}{\tau} = i\lambda u, \tag{4.6.31}$$

$$-\frac{1}{\omega}\xi + \frac{m - \delta\omega}{\omega}u(0) + \nu u(1) = i\lambda\xi, \tag{4.6.32}$$

$$-\frac{(ay_x)(1)}{M} = i\lambda\eta. \tag{4.6.33}$$

We shall distinguish two cases:

- First, assume that $\lambda = 0$. Then, (4.6.29) implies that $z = 0$, which gives $u(0) = \eta = 0$ due to (4.6.20). Using (4.6.30) and (4.6.31), we obtain $ay_x = k_1$ and $u = k_2$, where k_1 and k_2 are two real constants. Since (4.1.5), $u(0) = 0$, and (4.6.33), we deduce $k_1 = k_2 = 0$, and thus, $u = 0$ and $y = k_3$. Moreover, by (4.6.32), we have $\xi = 0$, and by (4.6.20), we conclude that $y(0) = 0$, and hence, $k_3 = 0$. Thus, (y, z, u, ξ, η) must be the trivial solution of (4.6.28), which is our desired result.

- Second, we treat the case $\lambda \neq 0$. Taking the inner product with Φ on both sides of (4.6.28), considering the real parts, and using (4.6.23), (4.6.13), and (4.6.14), we immediately find

$$0 = \Re e\left(\langle \mathscr{A}\Phi, \Phi \rangle\rangle_{\mathscr{H}}\right) \leq \tag{4.6.34}$$

$$\left(\frac{\nu}{2} + \frac{K}{2} - \delta\right) u^2(0) + \left(\frac{\nu}{2} - \frac{K}{2} + \frac{\nu\omega}{2(m-\delta\omega)}\right) u^2(1) + \left(\frac{\nu}{2} - 1\right) \frac{\omega}{m-\delta\omega}\rho^2 \leq 0,$$

which yields

$$u(0) = u(1) = \rho = 0. \tag{4.6.35}$$

Therefore, $z(0) = 0$ in view of (4.6.20). Moreover, by (4.6.32), we have $\xi = 0$. Solving (4.6.31) and using (4.6.35), we get $u = 0$. Moreover, since $z(0) = 0$, it follows from (4.6.29) that $y(0) = 0$, which implies, owing to (4.6.20), that $(ay_x)(0) = 0$. Then, combining these facts and reporting (4.6.29) in (4.6.30), we deduce that y satisfies the boundary value problem

$$\begin{cases} (ay_x)_x + \lambda^2 y = 0, \\ y(0) = y_x(0) = 0, \\ (ay_x)(1) = \lambda^2 M y(1). \end{cases} \tag{4.6.36}$$

Next, consider the following function (see [12, 74, 167] for similar cases):

$$\theta(x) = \exp\left(\int_x^1 \left|\frac{a_x}{a}\right|(s)\,ds\right)\left(-M - \int_x^1 \exp\left(-\int_s^1 \left|\frac{a_x}{a}\right|(r)\,dr\right) ds\right),$$
$$\forall x \in [0,1], \tag{4.6.37}$$

which satisfies

$$\theta(1) = -M \text{ and } \theta_x(x) \geq 1, \;\; (a\theta_x - |a_x|\theta)(x) \geq a(x), \;\; \theta(x) < 0, \;\; \forall x \in [0,1]. \tag{4.6.38}$$

Multiplying the PDE in (4.6.36) by $\theta\overline{y_x}$ and integrating by parts, one can readily verify that our boundary value problem has only the trivial solution $y = 0$. Finally, going back to (4.6.33), we get $\eta = 0$. Finally, we observe that the only solution of (4.6.28) is the null one.

Note that this first part of the proof implies that our system (4.6.1) (see also (4.6.16)) is strongly stable, i.e., $\lim_{t\to\infty} E(t) = 0$.

(II) Now, our concern is to show, by means of a contradiction argument, that the resolvent operator of $\mathscr{A}$ satisfies the second condition of Theorem 2.3.8. To this end, suppose such a requirement is not true. Then, by the Banach-Steinhaus theorem (see Theorem 2.1.7), there exist a sequence of real numbers $\lambda_n \to +\infty$ and a sequence of vectors $\Phi_n = (y_n, z_n, u_n, \xi_n, \eta_n) \in D(\mathscr{A})$ with $\|\Phi_n\|_{\mathscr{H}} = 1$ such that

$$(i\lambda_n I - \mathscr{A})\Phi_n = \mathscr{F}_n = (f_n, g_n, h_n, p_n, q_n) \to 0, \quad \text{as } n \to \infty \quad \text{in } \mathscr{H}. \tag{4.6.39}$$

In order to achieve the contradiction, it suffices to prove that

$$\lim_{t\to\infty} \Phi_n = 0 \text{ in } \mathscr{H}.$$

For sake of clarity, we shall proceed through several steps.
Step 1: First, we note that (4.6.39) may be recast as

$$i\lambda_n y_n - z_n := f_n \to 0, \quad \text{in} \quad H^1(I), \tag{4.6.40}$$

$$i\lambda_n z_n - (a(x)(y_n)_x)_x := g_n \to 0, \quad \text{in} \quad L^2(I), \tag{4.6.41}$$

$$i\lambda_n u_n + \frac{(u_n)_x}{\tau} := h_n \to 0, \quad \text{in} \quad L^2(I), \tag{4.6.42}$$

$$i\lambda_n \xi_n + \frac{1}{\omega}\xi_n - \frac{m-\delta\omega}{\omega} u_n(0) - \nu u_n(1) := p_n \to 0, \quad \text{in} \quad \mathbb{C}, \tag{4.6.43}$$

$$i\lambda_n \eta_n + \frac{(a(y_n)_x)(1)}{M} := q_n \to 0, \quad \text{in} \quad \mathbb{C}. \tag{4.6.44}$$

Taking the inner product, in $\mathscr{H}$, of both sides of (4.6.39) with the bounded element Φ_n and invoking (4.6.23), we get

$$u_n(0) \to 0, \quad u_n(1) \to 0 \quad \text{and} \quad \rho_n \to 0, \tag{4.6.45}$$

which implies since (4.6.22) that

$$\xi_n \to 0, \tag{4.6.46}$$

and since $\Phi_n \in D(\mathscr{A})$

$$z_n(0) \to 0 \text{ and } \sigma y_n(0) - (a(y_n)_x)(0) \to 0. \tag{4.6.47}$$

Now, using the expression of $\|\mathscr{F}_n\|_{\mathscr{H}}$, which converges to 0, we get

$$f_n(0) \to 0. \tag{4.6.48}$$

Thereby, since (4.6.40), we obtain

$$i\lambda_n y_n(0) \to 0 \text{ and thus } y_n(0) \to 0. \tag{4.6.49}$$

Whereupon, (4.6.47) gives

$$(y_n)_x(0) \to 0. \tag{4.6.50}$$

Solving (4.6.42), we find the following:

$$u_n(x) = u_n(0)e^{-i\tau\lambda_n x} + \tau \int_0^x e^{-i\tau\lambda_n(x-s)}h_n(s)\, ds, \tag{4.6.51}$$

which implies, using (4.6.45), that

$$u_n \to 0 \quad \text{in} \quad L^2(I). \tag{4.6.52}$$

In light of (4.6.40) and the fact that $\|\Phi_n\|_{\mathscr{H}} = 1$, we deduce

$$y_n = \frac{1}{i\lambda_n}(z_n + f_n) \to 0 \quad \text{in} \quad L^2(I). \tag{4.6.53}$$

Next, taking the inner product in $L^2(I)$ of (4.6.41) with $\frac{1}{\lambda_n^2}a(x)(\overline{y}_n)_x$, we have

$$\frac{i}{\lambda_n}\int_0^1 z_n a(x)(\overline{y}_n)_x\, dx - \frac{1}{\lambda_n^2}\int_0^1 (a(x)(y_n)_x)_x a(\overline{y}_n)_x\, dx = \frac{1}{\lambda_n^2}\int_0^1 g_n a(x)(\overline{y}_n)_x\, dx.$$

Integrating by parts the second integral in the above identity, using the boundedness of Φ_n in $\mathscr{H}$, and recalling (4.6.50) together with $g_n \to 0$ in $L^2(I)$, we obtain

$$\frac{(a(y_n)_x)(1)}{\lambda_n} = o(1). \tag{4.6.54}$$

Whereupon, (4.6.44) gives

$$\eta_n \to 0. \tag{4.6.55}$$

In light of (4.6.49) and keeping in mind that $f_n \to 0$ in $H^1(I)$, we have

$$f_n(1) = \int_0^1 (f_n)_x(x)\, dx + f_n(0) \to 0. \tag{4.6.56}$$

Therefore, thanks to (4.6.40), we obtain

$$i\lambda_n y_n(1) = \eta_n + f_n(1) \to 0 \quad \text{and} \quad y_n(1) \to 0. \tag{4.6.57}$$

Step 2: First, the combination of (4.6.40) and (4.6.41) yields

$$-\lambda_n^2 y_n - (a(y_n)_x)_x = i\lambda_n f_n + g_n.$$

Next, multiplying the latter by $\theta(\overline{y}_n)_x$, where θ is the function in $C^1([0,1],\mathbb{R})$ defined in (4.6.37), and integrating the resulting identity over (I), we get

$$\int_0^1 (-\lambda_n^2 y_n - (a(x)(y_n)_x)_x)\theta(x)\overline{(y_n)_x}\,dx = \int_0^1 (g_n + i\lambda_n f_n)\theta(x)\overline{(y_n)_x}\,dx. \tag{4.6.58}$$

By the previous step, we have that f_n and g_n (resp. $f_n(0)$, $f_n(1)$, $\lambda_n y_n(0)$, and $\lambda_n y_n(1)$) converge to zero in $H^1(I)$ and $L^2(I)$ (resp. $\mathbb{C}$), which gives, by a simple integration by parts, that it is easy to prove that the right-hand side of (4.6.58) converges to zero.

Now, we turn to the left-hand side of this last identity. By simply integrating parts for the two corresponding integrals, we get

$$\begin{aligned}
2\Re e\left\{\int_0^1 -\lambda_n^2 y_n\theta(x)\overline{(y_n)_x}\,dx\right\} &= \int_0^1 \lambda_n^2\theta_x(x)|y_n(x)|^2\,dx \\
&\quad - \lambda_n^2\theta(1)|y_n(1)|^2 + \lambda_n^2\theta(0)|y_n(0)|^2 \\
&= \int_0^1 \lambda_n^2\theta_x(x)|y_n(x)|^2\,dx + o(1), \\
2\Re e\left\{-\int_0^1 (a(x)(y_n)_x)_x\theta(x)\overline{(y_n)_x}\,dx\right\} &= \int_0^1 (a\theta_x - a_x\theta)(x)|(y_n)_x|^2\,dx \\
&\quad - a(1)\theta(1)|(y_n)_x(1)|^2 \\
&\quad + a(0)\theta(0)|(y_n)_x(0)|^2, \\
&= \int_0^1 (a\theta_x - a_x\theta)(x)|(y_n)_x|^2\,dx \\
&\quad - a(1)\theta(1)|(y_n)_x(1)|^2 + o(1).
\end{aligned}$$

Gathering all these facts from this step, we can deduce

$$\begin{aligned}
&\int_0^1 (a\theta_x - a_x\theta)(x)|(y_n)_x|^2\,dx + \int_0^1 \lambda_n^2\theta_x(x)|y_n(x)|^2\,dx \\
&\quad - a(1)\theta(1)|(y_n)_x(1)|^2 = o(1).
\end{aligned} \tag{4.6.59}$$

Reporting (4.6.38) in (4.6.59), we find

$$\int_0^1 a(x)|(y_n)_x|^2\,dx + \int_0^1 \lambda_n^2|y_n|^2\,dx + Ma(1)|(y_n)_x(1)|^2 = o(1). \tag{4.6.60}$$

Finally, putting in mind (4.1.5), it follows that

$$y_n \to 0 \quad \text{in} \quad H^1(I), \tag{4.6.61}$$

and

$$(y_n)_x(1) \to 0 \quad \text{in} \quad \mathbb{C}. \tag{4.6.62}$$

Step 3: It remains to show that z_n converges to zero in $L^2(0,1)$. To proceed, multiplying both sides of (4.6.41) by $\overline{y}_n$, integrating the result equation over I, and using (4.6.40), we obtain

$$-\int_0^1 z_n(w)\overline{z_n(x) + f_n(x)}\,dx - \int_0^1 (a(y_n)_x)_x(x)\overline{y}_n(x)\,dx = \int_0^1 g_n(x)\overline{y}_n(x)\,dx.$$

Thereafter, integrating by parts and using the facts that $\|\Phi_n\|_{\mathscr{H}} = 1$ and $f_n,\ g_n$ converge to zero in $L^2(I)$, we get

$$\int_0^1 a(x)|(y_n)_x(x)|\,dx - \int_0^1 |z_n(x)|^2\,dx \to 0, \tag{4.6.63}$$

which proves, owing to (4.6.61), that

$$z_n \to 0 \text{ in } L^2(I). \tag{4.6.64}$$

Finally, the contradiction that we sought is given by (4.6.46), (4.6.52), (4.6.55), (4.6.61), and (4.6.64). □

4.6.4 An Estimate of the Exponential Decay Rate

Clearly, Theorem 4.6.2 ensures the exponential convergence to zero of the solutions of the system (4.6.1) (see also (4.6.16)), but no estimate of the exponential rate ϖ is provided. Accordingly, in order to obtain such an estimate, one should opt for another way other than the resolvent method such as the multipliers method. To do so, we will use a Lyapunov functional along regular solutions Φ stemmed from an initial condition $\Phi \in \mathscr{D}(\mathscr{A}^2)$, and the result will be extended to $\mathscr{H}$ by a standard density argument. Then, inspired by the works [148, 149, 167], we define, along a regular solution of (4.6.1) (see also (4.6.16)), the functional:

$$\mathscr{E}(t) = E(t) + \varepsilon\mathscr{V}(t), \tag{4.6.65}$$

in which ε is an appropriate positive constant (see (4.6.83)) and $\mathscr{V}(t) = V_0(t) + V_1(t) + V_2(t)$, where

$$\begin{cases} V_0(t) = 2\int_0^1 \theta(x) y_t y_x \, dx, \\ V_1(t) = K\tau \int_0^1 e^{-\vartheta \tau x} y_t^2(0, t - x\tau)\, dx, \ K \text{ satisfies (4.6.14)} \\ \quad \text{and } \vartheta \text{ is an arbitrary positive constant,} \\ V_2(t) = C_2 y(0,t) \left(\int_0^1 y_t \, dx + M y_t(1,t) \right), \ C_2 \\ \quad \text{is a positive constant to be chosen.} \end{cases} \tag{4.6.66}$$

First, we claim the existence of positive constants $A_1 0$ and A_2 such that

$$A_1 E(t) \leq \mathscr{E}(t) \leq A_2 E(t), \tag{4.6.67}$$

provided that $\varepsilon > 0$ is chosen sufficiently small (see (4.6.83)). Indeed, applying Young's and Cauchy-Schwarz inequalities to $\mathscr{V}(t)$ leads to

$$|\mathscr{V}(t)| \leq C_* E(t), \tag{4.6.68}$$

where

$$C_* = \left(2 \left\| \frac{\theta}{\sqrt{a}} \right\|_{L^\infty(I)} + 2 + \frac{\sqrt{M}+1}{\sqrt{\sigma}} C_2 \right). \tag{4.6.69}$$

Accordingly, we obtain (4.6.67), where

$$A_1 = 1 - \varepsilon C_* \quad \text{and} \quad A_2 = 1 + \varepsilon C_*, \tag{4.6.70}$$

as long as $\varepsilon < 1/C_*$. Next, we are going to estimate the derivative of $\mathscr{E}(t)$. To proceed, we recall that (4.6.12) yields

$$E'(t) \leq -B_1 y_t^2(0,t) - B_2 y_t^2(0, t-\tau) - B_3 \xi'^2(t), \tag{4.6.71}$$

with

$$B_1 = \delta - \frac{K+\nu}{2}, \quad B_2 = \frac{1}{2}\left(K - \nu - \frac{\nu\omega}{m - \delta\omega} \right) \text{ and } B_3 = \left(1 - \frac{\nu}{2} \right) \frac{\omega}{m - \delta\omega}. \tag{4.6.72}$$

It is important to note that B_i, $i = 1, 2, 3$, are positive by virtue of (4.6.3), (4.6.13), and (4.6.14). Likewise, using (4.6.16) and (4.6.37)–(4.6.38), a straightforward computation shows that the first functional of (4.6.66) gives

$$\begin{aligned} V_0'(t) &= \theta(1)(ay_x^2)(1,t) - \theta(0)(ay_x^2)(0,t) - \int_0^1 a\left(\frac{\theta}{a}\right)_x ay_x^2\,dx \\ &\quad + \theta(1)y_t^2(1,t) - \theta(0)y_t^2(0,t) - \int_0^1 \theta_x y_t^2\,dx \\ &\le -\int_0^1 y_t^2\,dx - \int_0^1 ay_x^2\,dx - My_t^2(1,t) - \theta(0)y_t^2(0,t) - \theta(0)(ay_x^2)(0,t). \end{aligned} \tag{4.6.73}$$

Owing to (4.6.6), it follows that

$$\begin{aligned} (ay_x)^2(0,t) - 2\sigma(ay_x)(0,t)y(0,t) + \sigma^2 y^2(0,t) &= \xi'^2(t) \\ + \delta^2 y_t^2(0,t) + \nu^2 y_t^2(0,t-\tau) - 2\nu\delta y_t(0,t)y_t(0,t-\tau) &+ 2\delta\xi'(t)y_t(0,t) \\ &-2\nu\xi'(t)y_t(0,t-\tau). \end{aligned} \tag{4.6.74}$$

Once more, owing to Young's inequality, (4.6.74) yields

$$\begin{aligned} -\theta(0)(ay_x^2)(0,t) \le &-2\sigma\frac{\theta(0)}{a(0)}(ay_x)(0,t)y(0,t) + \sigma^2\frac{\theta(0)}{a(0)}y^2(0,t) \\ &-3\frac{\theta(0)}{a(0)}\left(\xi'^2(t) + \delta^2 y_t^2(0,t) + \nu^2 y_t^2(0,t-\tau)\right). \end{aligned} \tag{4.6.75}$$

Thereby, one can deduce from (4.6.73) and (4.6.75) the following estimate:

$$\begin{aligned} V_0'(t) \le &-\int_0^1 y_t^2\,dx - \int_0^1 ay_x^2\,dx - My_t^2(1,t) - \theta(0)y_t^2(0,t) \\ &-2\sigma\tfrac{\theta(0)}{a(0)}(ay_x)(0,t)y(0,t) \\ &+\sigma^2\frac{\theta(0)}{a(0)}y^2(0,t) - 3\frac{\theta(0)}{a(0)}\left(\xi'^2(t) + \delta^2 y_t^2(0,t) + \nu^2 y_t^2(0,t-\tau)\right). \end{aligned} \tag{4.6.76}$$

Secondly, considering the second functional of (4.6.66), integration by parts gives

$$\begin{aligned} V_1'(t) &= Ky_t^2(0,t) - Ke^{-\vartheta\tau}y_t^2(0,t-\tau) - K\vartheta\tau\int_0^1 e^{-\vartheta\tau x}y_t^2(0,t-x\tau)\,dx \\ &\le Ky_t^2(0,t) - Ke^{-\vartheta\tau}y_t^2(0,t-\tau) - K\vartheta e^{-\vartheta\tau}\tau\int_0^1 y_t^2(0,t-x\tau)\,dx. \end{aligned} \tag{4.6.77}$$

Thirdly, recalling the third functional of (4.6.66), it follows from (4.6.16) and Young's inequality that

$$\begin{aligned} V_2'(t) &= C_2 y_t(0,t)\left(\int_0^1 y_t\,dx + M y_t(1,t)\right) \\ &\quad + C_2 y(0,t)\left(\int_0^1 y_{tt}\,dx + M y_{tt}(1,t)\right) \\ &= C_2 y_t(0,t)\left(\int_0^1 y_t\,dx + M y_t(1,t)\right) - C_2 y(0,t)(ay_x)(0,t) \\ &\le \frac{1}{2}\int_0^1 y_t^2\,dx + \frac{M}{2} y_t^2(1,t) + \frac{C_2^2}{2}(M+1) y_t^2(0,t) - C_2 (ay_x)(0,t) y(0,t). \end{aligned} \tag{4.6.78}$$

Combining (4.6.76), (4.6.77), and (4.6.78), it follows that

$$\begin{aligned} \mathscr{V}'(t) &\le -\frac{1}{2}\int_0^1 y_t^2\,dx - \int_0^1 a y_x^2\,dx - \frac{M}{2} y_t^2(1,t) + \sigma^2 \frac{\theta(0)}{a(0)} y^2(0,t) \\ &\quad - K\vartheta e^{-\vartheta\tau}\tau \int_0^1 y_t^2(0,t-x\tau)\,dx \\ &\quad + \left(-2\sigma\frac{\theta(0)}{a(0)} - C_2\right)(ay_x)(0,t)y(0,t) - 3\frac{\theta(0)}{a(0)}\xi'^2(t) \\ &\quad + \left(-3\frac{\theta(0)}{a(0)}\nu^2 - Ke^{-\vartheta\tau}\right) y_t^2(0,t-\tau) \\ &\quad + \left(-\theta(0) - 3\frac{\theta(0)}{a(0)}\delta^2 + C_2^2\frac{M+1}{2} + K\right) y_t^2(0,t). \end{aligned} \tag{4.6.79}$$

In turn, by virtue of the relationship between ξ and ξ' (see (4.6.9)), we have

$$\begin{aligned} \frac{1}{2(m-\delta\omega)}\xi^2(t) &= \frac{m-\delta\omega}{2} y_t^2(0,t) + \frac{\omega^2\nu^2}{2(m-\delta\omega)} y_t^2(0,t-\tau) + \frac{\omega^2}{2(m-\delta\omega)}\xi'^2(t) \\ &\quad - \frac{\omega^2\nu}{m-\delta\omega}\xi'(t) y_t(0,t-\tau) + \omega\nu y_t(0,t) y_t(0,t-\tau) \\ &\quad - \omega y_t(0,t)\xi'(t) \\ &\le \left(\frac{m-\delta\omega}{2} + 1\right) y_t^2(0,t) + \omega^2\nu^2\left(\frac{1}{m-\delta\omega} + \frac{1}{2}\right) y_t^2(0,t-\tau) \\ &\quad + \omega^2\left(\frac{1}{m-\delta\omega} + \frac{1}{2}\right)\xi'^2(t). \end{aligned} \tag{4.6.80}$$

By choosing $C_2 = -2\sigma \dfrac{\theta(0)}{a(0)}$ in (4.6.79) together with (4.6.80), we get

$$\begin{aligned}\mathcal{V}'(t) \leq & -\frac{1}{2}\int_0^1 y_t^2\,dx - \int_0^1 a y_x^2\,dx - \frac{M}{2}y_t^2(1,t) + \sigma^2\frac{\theta(0)}{a(0)}y^2(0,t)\\ & -K\vartheta e^{-\vartheta\tau}\tau\int_0^1 y_t^2(0,t-x\tau)\,dx\\ & -\frac{1}{2(m-\delta\omega)}\xi^2(t) + K_1 y_t^2(0,t) + K_2 y_t^2(0,t-\tau) + K_3\xi'^2(t),\end{aligned} \tag{4.6.81}$$

where

$$\begin{aligned}K_1 = & -\theta(0) - 3\frac{\theta(0)}{a(0)}\delta^2 + C_2^2\frac{M+1}{2} + K + \frac{m-\delta\omega}{2} + 1,\ K_2 = -3\frac{\theta(0)}{a(0)}\nu^2\\ & - Ke^{-\vartheta\tau} + \omega^2\nu^2\left(\frac{1}{m-\delta\omega} + \frac{1}{2}\right)\end{aligned}$$

and

$$K_3 = -3\frac{\theta(0)}{a(0)} + \omega^2\left(\frac{1}{m-\delta\omega} + \frac{1}{2}\right).$$

Therefore, combining (4.6.71) and (4.6.81), we deduce that

$$\begin{aligned}\mathcal{E}'(t) = & E'(t) + \varepsilon\mathcal{V}'(t)(t)\\ \leq & \varepsilon\left\{-\frac{1}{2}\int_0^1 y_t^2\,dx - \int_0^1 a y_x^2\,dx - \frac{M}{2}y_t^2(1,t) + \sigma^2\frac{\theta(0)}{a(0)}y^2(0,t)\right.\\ & \left. - K\vartheta e^{-\vartheta\tau}\tau\int_0^1 y_t^2(0,t-x\tau)\,dx\right\}\\ & -\varepsilon\frac{1}{2(m-\delta\omega)}\xi^2(t) + (\varepsilon K_1 - B_1)y_t^2(0,t) + (\varepsilon K_2 - B_2)y_t^2(0,t-\tau)\\ & +(\varepsilon K_3 - B_3)\xi'^2(t).\end{aligned} \tag{4.6.82}$$

Lastly, choosing ε such that

$$\varepsilon \leq \min\left\{\frac{B_1}{K_1}, \frac{B_2}{K_2}, \frac{B_3}{K_3}, \frac{1}{C_*}\right\}, \tag{4.6.83}$$

and using (4.6.67), we obtain

$$\mathcal{E}'(t) \leq -c\varepsilon E(t) \leq -c\varepsilon A_2^{-1}\mathcal{E}(t), \tag{4.6.84}$$

where

$$c = \min\left\{1, 2\vartheta e^{-\vartheta\tau}, -2\sigma\frac{\theta(0)}{a(0)}\right\}. \tag{4.6.85}$$

Now, it suffices to solve the differential inequality (4.6.84) and use (4.6.67) once again to get

$$E(t) \leq A_2 A_1^{-1} e^{-\varpi t} E(0), \tag{4.6.86}$$

where $\varpi = c\varepsilon A_2^{-1}$.

To summarize, we have proved the second main result:

Theorem 4.6.3 *Under the assumptions (4.1.5), (4.6.3), and (4.6.13), the energy $E(t)$ defined by (4.6.3) of the system (4.6.1) (see also (4.6.16)) exponentially decays to zero with a decay rate $\varpi = A_2^{-1}\varepsilon c > 0$, where A_2, ε, and c satisfy respectively (4.6.70), (4.6.83), and (4.6.85).*

Remark 4.6.2 It is worth mentioning that in order to deal with the tracking problem, that is, to steer the cable to an equilibrium position y_e, we can redefine the energy $E(t)$ as follows:

$$\begin{aligned} E(t) = \frac{1}{2}\Bigg[\int_0^1 \left[y_t^2(x,t) + a(x)y_x^2(x,t) + K\tau y_t^2(0, t - x\tau)\right] dx \\ + \sigma(y(0,t) - y_e)^2 + \frac{1}{m - \delta\omega}\xi^2(t) + My_t^2(1,t)\Bigg]. \end{aligned}$$

Then, one can deduce from our Theorem 3 that $y(0,t)$ exponentially tends to y_e as t is sufficiently large. On the other hand, the reader has certainly noticed that our system (4.6.1) is not subject to disturbances or uncertainties. Otherwise, one can design an adaptive state feedback controller and use Lyapunov functional to guarantee the disturbance rejection of uncertainties and disturbances in (4.6.1) [143, 189].

4.6.5 Numerical Examples

We shall illustrate, in this subsection, our stability results proved in the previous subsections. To proceed, we will focus on the spectrum of the crane system (4.6.1) (see (4.6.16)) and discuss the impact of changing the problem parameters on the spectrum behavior. In particular, we shall highlight the importance of the smallness condition of the delayed term (4.6.13). For sake of simplicity, we take the tension of the cable $a(x) = 1$. This is in the fact the case for a nearly "perfectly" flexible cable, which can be interpreted as almost uniform stretching along the cable. Hence, $a(x)$

can be approximated by a positive constant, and then using scaling arguments, such a positive constant can be taken as 1.

Subsequently, a straightforward computation shows that λ is an eigenvalue of the system if and only it is a solution of the following characteristic equation:

$$\begin{aligned} & e^{2\lambda}\Big[-M(m+\omega)\lambda^3+\lambda^2\left(-(m+\omega)+M(\nu e^{-\lambda\tau}-\delta-\sigma\omega-1)\right) \\ & +\lambda\left(\nu e^{-\lambda\tau}-\delta-\sigma\omega-1-\sigma M\right)\Big] \\ & -\sigma e^{2\lambda}+M(m-\omega)\lambda^3+\lambda^2\left(\omega-m-M(\nu e^{-\lambda\tau}-\delta-\sigma\omega+1)\right) \\ & +\lambda\left(\nu e^{-\lambda\tau}-\delta-\sigma\omega+1+\sigma M\right)-\sigma=0. \end{aligned} \tag{4.6.87}$$

Thereafter, we numerically solve (4.6.87) by means of *MATLAB R2019a* via the function "fsolve" and hence obtain the first eigenvalues of the system. For sake of clarity of the presentation, this task will be conducted for an exhaustive list of the parameters.

4.6.5.1 System Spectrum When ν Varies

Let us fix $\delta = 1$, $\omega = 2$, $m = 3$, $M = 10$, $\tau = 0.01$, and $\sigma = 0.05$, which clearly shows that (4.6.3) holds. Then, we vary the values of the parameter ν of the delayed term $y_t(0, t-\tau)$.

Figure 4.18 depicts the eigenvalues of the system for $\nu = 0.1$, which satisfies the condition (4.6.13). Needless to mention that the two subfigures represent the same situation, but the only difference is that the second subfigure only shows one part of the spectrum (eigenvalues with small real parts). We observe from the first subfigure of Fig. 4.18 that the system spectrum possesses two branches, which both lie in the left-half plane (see, in particular, the first branch shown in the second part of Fig. 4.18). Hence, the system is stable. Likewise, the same remark remains

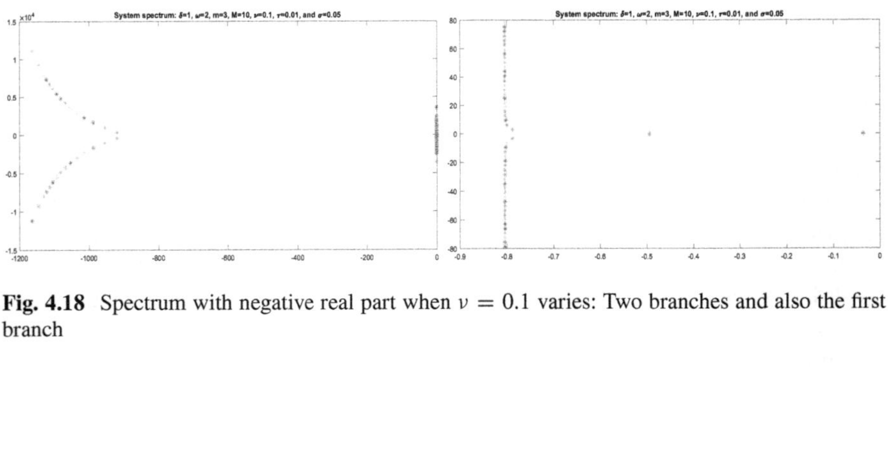

Fig. 4.18 Spectrum with negative real part when $\nu = 0.1$ varies: Two branches and also the first branch

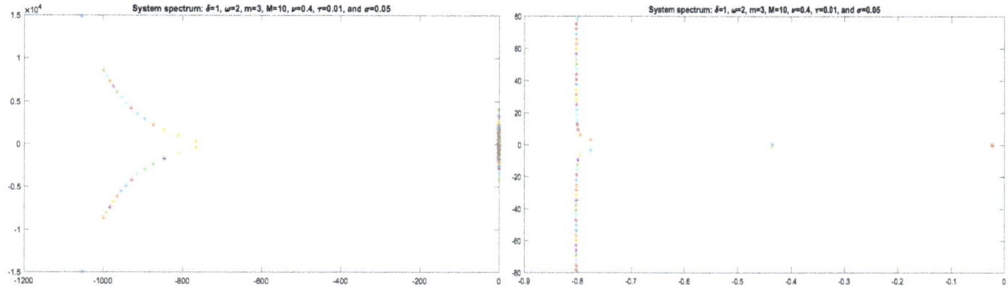

Fig. 4.19 Spectrum with negative real part when $\nu = 0.4$: Two branches and also the first branch

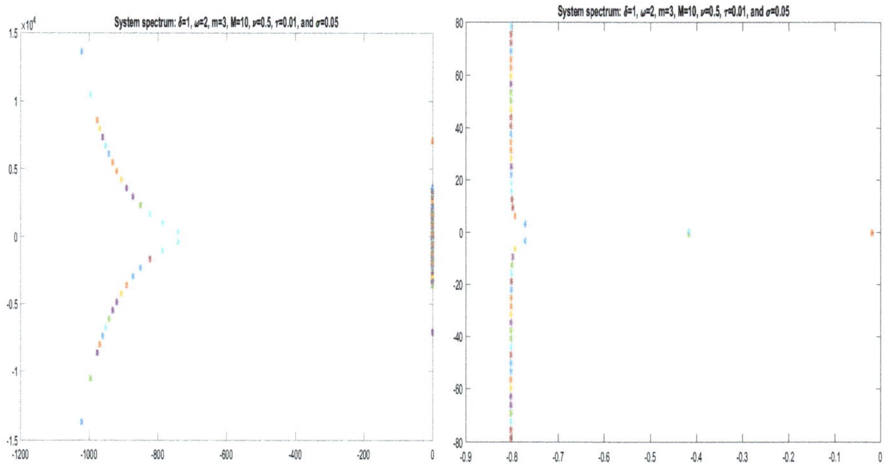

Fig. 4.20 System spectrum when ν varies: Eigenvalues with real part close to zero

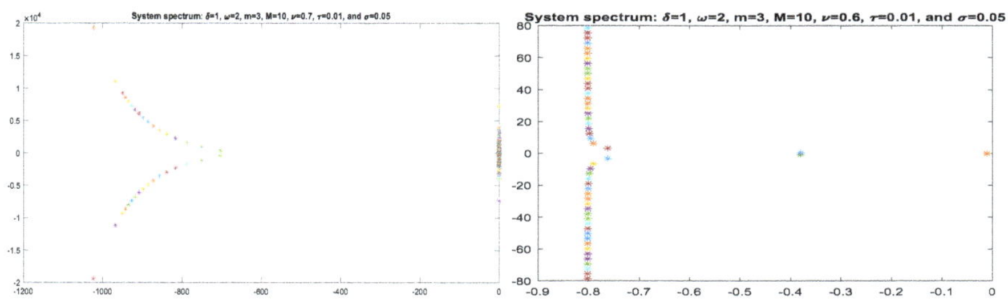

Fig. 4.21 System spectrum when ν varies: Eigenvalues with real part near zero

valid when $\nu = 0.4$ as shown in Fig. 4.19. This ties with Theorem 4.6.2 (see also Theorem 4.6.3) since (4.6.13) holds.

It is also apparent, from Figs. 4.20, 4.21, and 4.22, that as ν increases, the spectrum moves to the right and more importantly the first eigenvalues move to the right. Indeed, it is observed from the second part in each of Figs. 4.20 and 4.21 that the system has eigenvalues close to the origin. In turn, an eigenvalue on the

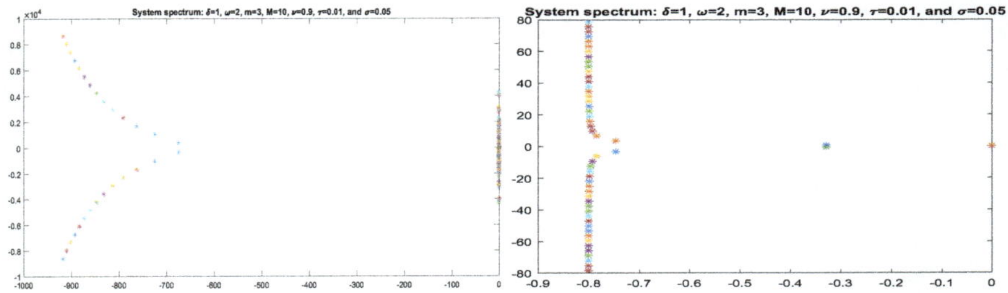

Fig. 4.22 System spectrum when ν varies: Eigenvalue on the imaginary axis

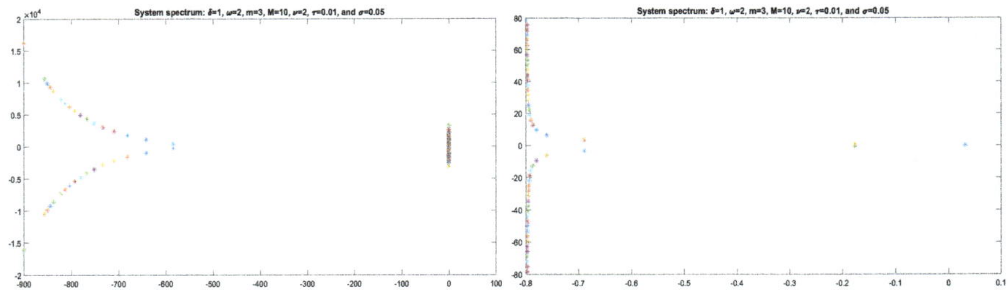

Fig. 4.23 System spectrum when ν varies: Eigenvalue with positive real part

imaginary axis appears in the second part of Fig. 4.22. We notice from Fig. 4.23 that the spectrum may have an eigenvalue with a positive real part. This is due to the fact that (4.6.13) is highly violated for such values of ν. Undoubtedly, this discussion agrees with Theorem 4.6.2 (see also Theorem 4.6.3).

4.6.5.2 The Time-Delay τ Effect on the Spectrum

In this part of the numerical simulations, we vary the values of the time delay τ, while we take $\delta = 1$, $\omega = 2$, $m = 3$, $M = 10$, $\nu = 0.1$ and $\sigma = 0.05$. Recalling that $\tau = 0.01$ in the previous case (see the first subfigure in Fig. 4.18), it is apparent from Fig. 4.24 that as we increase the values of τ, the whole spectrum moves to the right, although it is still in the left-half plane. In particular, the first eigenvalues move to the right, and consequently, the stability of the system is rather sensitive to the increase of the time delay τ.

4.6.5.3 System Spectrum When σ Varies

In Fig. 4.18, $\sigma = 0.05$. Now, if we increase the value of σ, we notice from Fig. 4.25 that the second branch of the system remains more or less comparable, but the first eigenvalues move in the opposite direction. Indeed, the first eigenvalue moves to

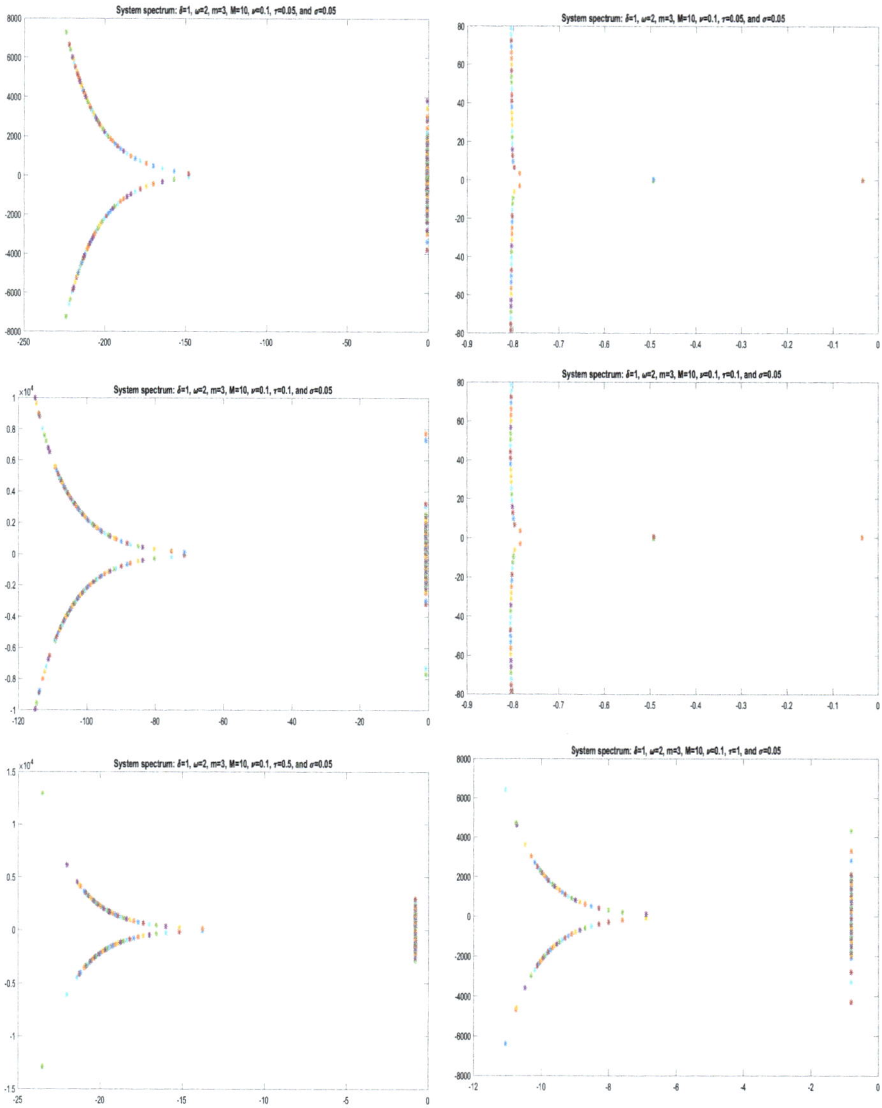

Fig. 4.24 System spectrum: time-delay effect

the right, whereas the second one moves to the left. This phenomenon may be interpreted by the fact that the parameter σ appears in two terms in the boundary condition of (4.6.16), namely, the displacement term $y(0, t)$ as well as the velocity term $y_t(0, t)$.

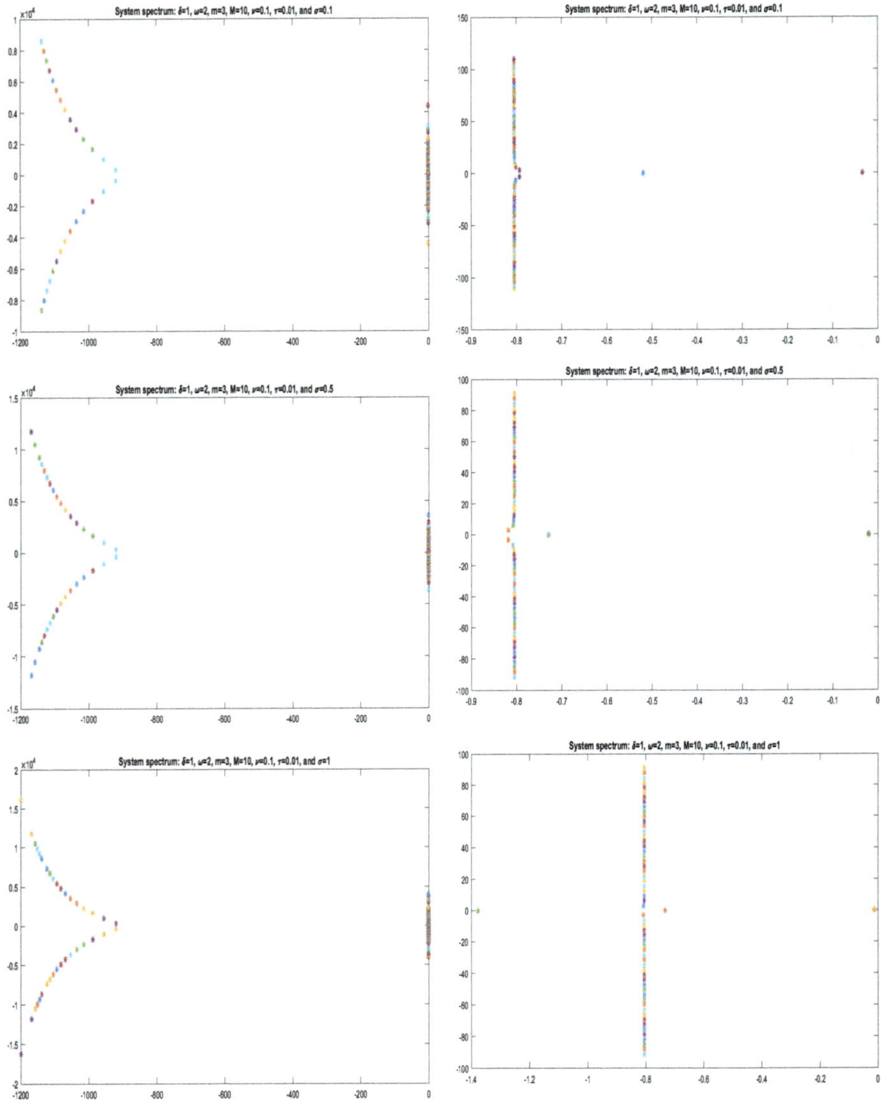

Fig. 4.25 System spectrum when σ varies

4.6.5.4 System Spectrum When δ Varies

The intention in this part is to numerically study the behavior of the system spectrum as the parameter δ changes values (see Fig. 4.18 for the case $\delta = 1$). Figure 4.26 indicates that when δ increases even by relatively small values, the first eigenvalues move to the left, and hence, the stability of the system should improve. This is

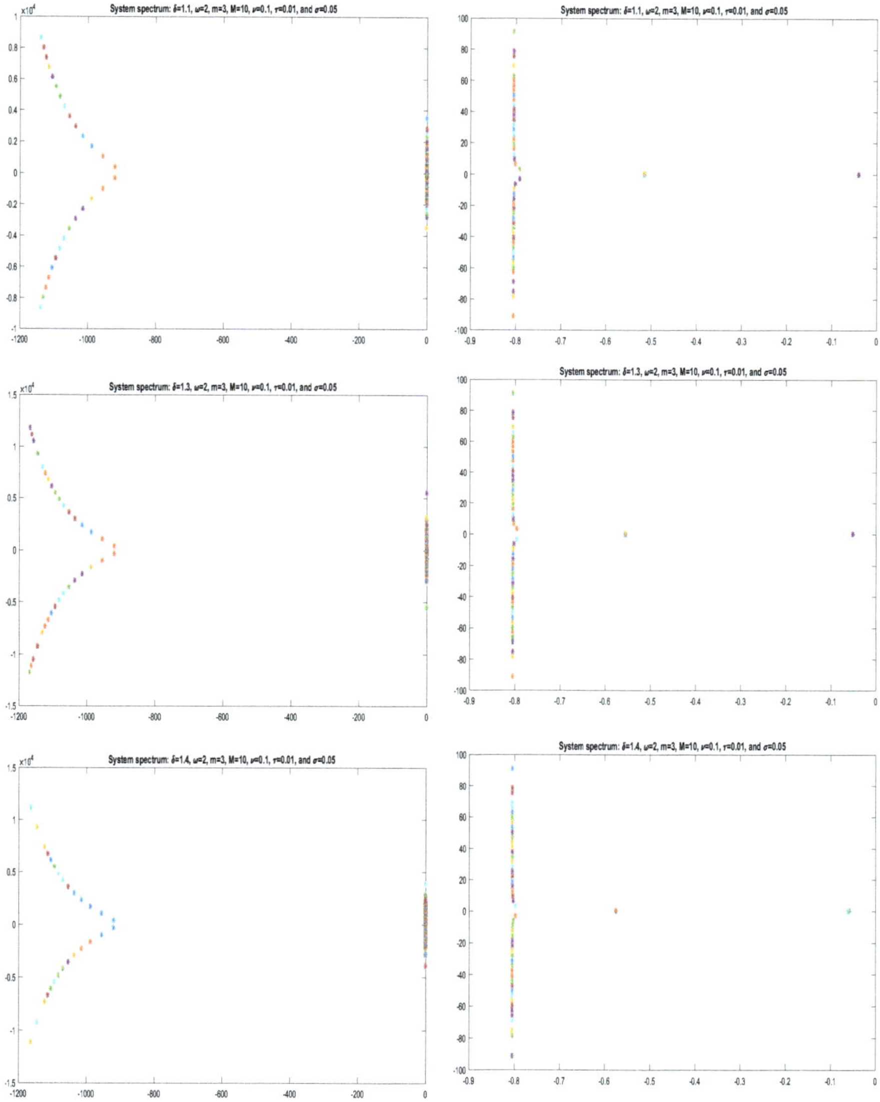

Fig. 4.26 System spectrum when δ varies

expected as δ appears in the nondelayed velocity term $y_t(0, t)$, which should play a positive role in stabilizing the system.

4.6.5.5 System Spectrum When ω Varies

Now, we aim to highlight the importance of the parameter ω in the behavior of the system spectrum and a fortiori in the stability property. In Fig. 4.18, we set $\omega = 2$. Next, varying the values of ω, we observe in Fig. 4.27 that the second branch moves to the right but more importantly the first branch of the spectrum, including the first eigenvalues, moves to the left, which means that the stability of the system is enhanced. Once again, such a change is expected in the first eigenvalues since the parameter ω appears in the angular velocity term $y_{xt}(0, t)$. In fact, the latter is crucial for the exponential stability of the system as discussed in the introduction. Not only that, recall that ω also appears in the velocity term $y_t(0, t)$.

4.6.5.6 System Spectrum When m and M Vary

This last part of our numerical simulations is concerned with the numerical study of the spectrum when the mass m of the cart of the crane system and the mass M of the payload change.

First, keeping in mind that $m = 3$ in Fig. 4.18, we notice from Fig. 4.28 that as we increase the mass m, the second branch of the spectrum moves to the left, while the first branch of the spectrum, including the first eigenvalues, moves to the right.

In turn, referring to Fig. 4.18 where the mass of the payload is $M = 10$, we remark from Fig. 4.29 that as M increases, the main two branches of the spectrum remain in their positions, whereas the first eigenvalues move in opposite directions. More precisely, the first eigenvalue moves to the right, while the second eigenvalue moves to the left. To sum up, we conclude that the decay rate becomes less satisfactory as M increases.

4.6.6 Concluding Remarks

In this section, the overhead crane system is considered by taking into account the dynamics of both ends of the cable, and the presence of a boundary delay. Despite this situation, a simple boundary control is designed to show, by means of the resolvent method, that the system is exponentially stable under reasonable conditions on the feedback gains. Moreover, an estimate of the decay is provided thanks to a suitable Lyapunov functional. As pointed out earlier, the novelty of the outcome, in contrast to the previous section, is about reducing the number

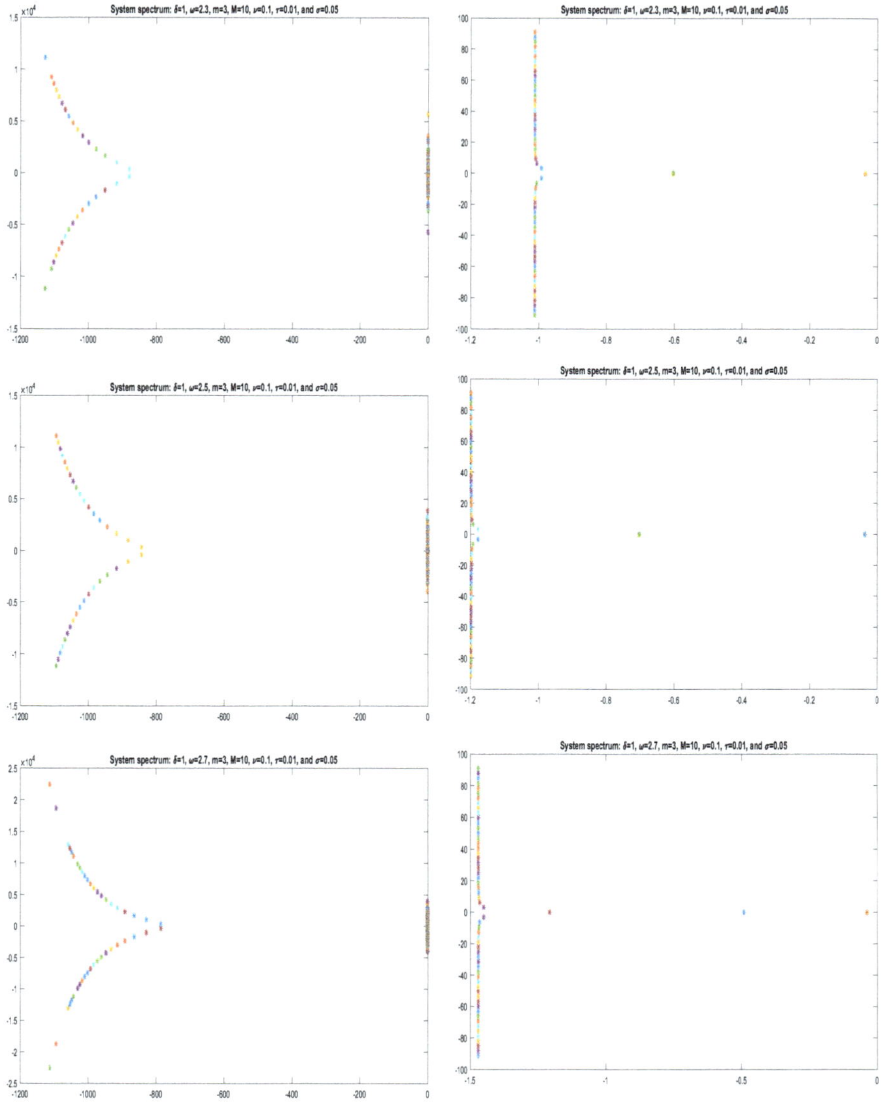

Fig. 4.27 System spectrum when ω varies

of controls to only one in order to obtain the exponential stability of the system. Although an estimate of the exponential decay rate is provided, the optimality of such a decay still subsists. Furthermore, one could consider the case of time-dependent length of the cable and investigate the stabilization problem with the same type of controls.

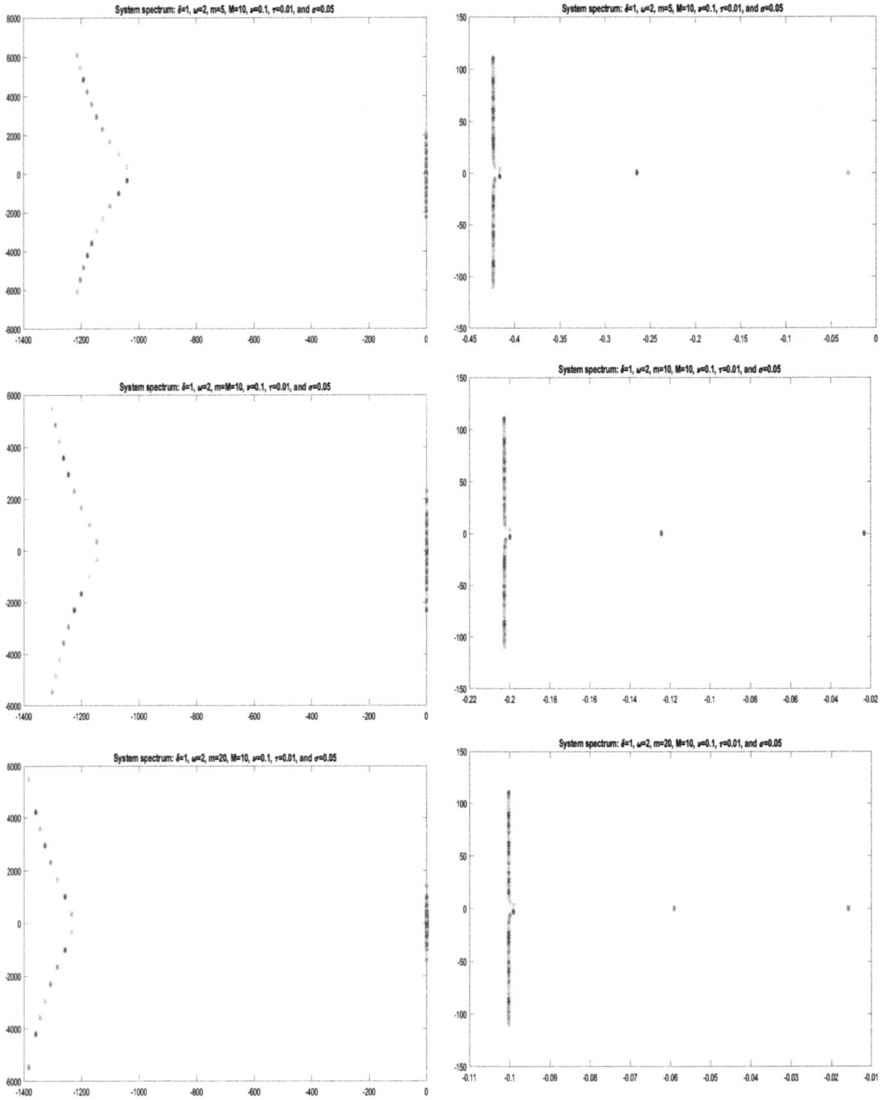

Fig. 4.28 System spectrum when the mass m of the cart varies

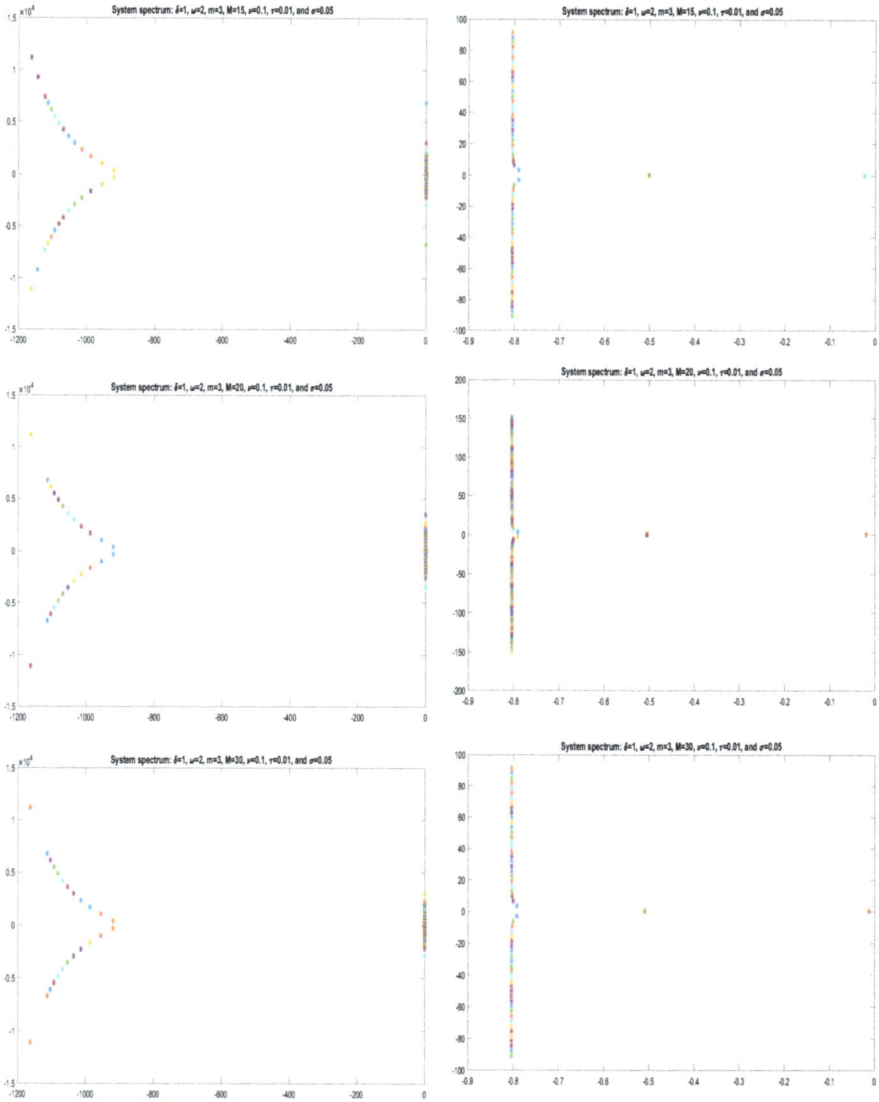

Fig. 4.29 System spectrum when the mass M of the payload varies

4.7 Boundary Time-Dependent Delay: Exponential Stabilization

In this section, we assume that a time-dependent delay occurs in one boundary of the crane system. As in the previous sections, the main concern is to stabilize the dynamics of the cable as well as the dynamical terms related to the cart and the load mass. To do so, we first prove that the problem is well-posed in the sense of semigroups theory provided that some conditions on the delay function are satisfied. Thereafter, an appropriate Lyapunov function is put forward, which leads to the exponential decay of the energy as well as to an estimate of the decay rate.

4.7.1 Introduction

Consider the system

$$\begin{cases} y_{tt}(x,t) - (ay_x)_x(x,t) = 0, & x \in I,\ t > 0, \\ my_{tt}(0,t) - (ay_x)(0,t) = -\beta y(0,t) + \zeta y_t(0,t-\tau(t)) - \varrho y_t(0,t) & \\ \qquad +\mu(ay_{xt})(0,t), & t > 0, \\ My_{tt}(1,t) + (ay_x)(1,t) = 0, & t > 0, \\ y(x,0) = y_0(x),\ y_t(x,0) = y_1(x), & x \in I, \\ y_t(0,t-\tau(0)) = f(t-\tau(0)), & t \in (0,\tau(0)), \end{cases} \tag{4.7.1}$$

in which $I = (0, 1)$, the parameters ζ, ϱ, μ and β are positive feedback gains, whereas $\tau(t)$ is the time-dependent delay. Also, note that the expression

$$-\beta y(0,t) + \zeta y_t(0,t-\tau(t)) - \varrho y_t(0,t) + \mu(ay_{xt})(0,t)$$

is viewed as a boundary *time-dependent delayed* control.

The main contribution of this section is to show that the system (4.7.1) is exponentially stable despite the presence of the time-dependent delay as long as the latter satisfies reasonable conditions.

In particular, we are able to prove that one can stabilize the cable as well as the dynamics of its both ends. Needless to note that such a result improves those obtained in the previous sections, where the delay was constant. Obviously, in practice, the delay involved in systems control is often time dependent. Furthermore, inspired by Nicaise et al. [155, 156], our stability result is obtained via the Lyapunov method, which is direct and also provides an estimate of the convergence rate.

Throughout the remainder of this section, the following assumptions are imposed on the time-dependent delay:

There exist constants $\kappa < 1$ and $\tau_1 > 0$, $\tau_2 > 0$ such that

$$\tau(\cdot) \in W^{2,\infty}([0,T]), \quad \forall T > 0; \tag{4.7.2}$$

$$0 < \tau_1 \leq \tau(t) \leq \tau_2, \quad \forall t > 0; \tag{4.7.3}$$

$$\dot{\tau}(t) \leq \kappa < 1, \quad \forall t > 0. \tag{4.7.4}$$

Furthermore, the tension of the cable $a(x)$ is assumed, as previously done, to satisfy 4.1.5.

This section is organized as follows. Section 4.7.2 is devoted to the well-posedness of the problem by using semigroups theory [72, 158]. In Sect. 4.7.3, we show the exponential stability of the system (4.7.1) by means of the energy method. Finally, a short discussion is provided in Sect. 4.7.4.

4.7.2 Well-Posedness of the Problem

Let us first introduce a new variable

$$\xi(t) := my_t(0, t) + \mu \left[\beta y(0, t) - (ay_x)(0, t)\right]. \tag{4.7.5}$$

Next, we consider the standard auxiliary variable $u(x, t) = y_t(0, t - x\tau(t))$, where $x \in (0, 1)$ [155, 156]. Thereby, (4.7.1) can be written as

$$\begin{cases} y_{tt}(x, t) - (ay_x)_x(x, t) = 0, & x \in I, \ t > 0, \\ \tau(t)u_t(x, t) + (1 - x\dot{\tau}(t))u_x(x, t) = 0, & x \in I, \ t > 0, \\ my_{tt}(0, t) - (ay_x)(0, t) = -\beta y(0, t) + \zeta u(1, t) & \\ -(\theta + \beta\mu)u(0, t) + \mu y_{xt}(0, t), & t > 0, \\ My_{tt}(1, t) + (ay_x)(1, t) = 0, & t > 0, \\ y(x, 0) = y_0(x), \ y_t(x, 0) = y_1(x), & x \in I, \\ u(x, 0) = f(-x\tau(0)), & t \in (0, \tau), \end{cases} \tag{4.7.6}$$

where we assumed, for convenience, that $\varrho = \theta + \beta\mu$, for some positive constant θ. Next, let $\Phi = (y, z, u, \xi, \eta)$, where $z(., t) = y_t(., t)$ and $\eta(t) = y_t(1, t)$, and suppose, from now on, that

$$m > \theta\mu. \tag{4.7.7}$$

This allows us to consider the state space

$$\mathscr{H} = H^1(I) \times L^2(I) \times L^2(I) \times \mathbb{R}^2, \tag{4.7.8}$$

equipped with the inner product

$$\langle (y, z, u, \xi, \eta), (\tilde{y}, \tilde{z}, \tilde{u}, \tilde{\xi}, \tilde{\eta}) \rangle_{\mathscr{H}} = \int_0^1 (ay_x\tilde{y}_x + z\tilde{z})\, dx + \int_0^1 u\tilde{u}\, dx + \beta y(0)\tilde{y}(0)$$

$$+\frac{1}{m-\theta\mu}\xi\tilde{\xi}+M\eta\tilde{\eta}.$$

One can readily check that $\mathscr{H}$ is a Hilbert space. Then, (4.7.6) may be brought to a first-order evolution equation

$$\begin{cases}\Phi_t(t)=\mathscr{A}(t)\Phi(t),\\ \Phi(0)=\Phi_0,\end{cases}\tag{4.7.9}$$

where the time-varying linear operator $\mathscr{A}(t)$ is defined by

$$\mathscr{A}(t)(y,z,u,\xi,\eta)\\ =\left(z,\,(ay_x)_x,\,\frac{x\dot{\tau}(t)-1}{\tau(t)}u_x,\,-\frac{1}{\mu}\xi+\frac{m-\theta\mu}{\mu}u(0)+\zeta u(1),\,-\frac{(ay_x)(1)}{M}\right),\tag{4.7.10}$$

and its domain is given by

$$\mathscr{D}(\mathscr{A}(t))=\left\{\begin{array}{r}(y,z,u,\xi,\eta)\in\mathscr{H};\ y\in H^2(I),\ z,u\in H^1(I),\ \xi=mu(0)\\ +\mu[\beta y(0)-(ay_x)(0)],\,u(0)=z(0),\ \eta=z(1)\end{array}\right\}.\tag{4.7.11}$$

This domain is clearly independent of time t as we have

$$\mathscr{D}(\mathscr{A}(t))=\mathscr{D}(\mathscr{A}(0)),\quad \text{for all}\, t\geq 0.\tag{4.7.12}$$

One can also readily check that $\mathscr{D}(\mathscr{A}(0))$ is dense in $\mathscr{H}$ (see [155, 156] for similar situations).

The first main outcome of this section concerns the well-posedness of our problem and is stated below:

Theorem 4.7.1 *Assume that 4.1.5, (4.7.2)–(4.7.4), and (4.7.7) hold as well as*

$$\zeta<\min\left\{2\sqrt{1-\kappa},\,\frac{\theta\sqrt{1-\kappa}}{1+\frac{\mu}{2(m-\theta\mu)}}\right\}.\tag{4.7.13}$$

Then, for any initial data $\Phi_0\in\mathscr{H}$, the system (4.7.9) admits a unique solution

$$\Phi\in C(0,\infty;\mathscr{H}).$$

For $\Phi_0\in\mathscr{D}(\mathscr{A}(t))$, this problem admits a unique solution Φ with

$$\Phi\in C(0,\infty;\mathscr{D}(\mathscr{A}(t)))\cap C^1(0,\infty;\mathscr{H}).$$

Proof First, consider the following change of variable:

$$\rho := -\frac{1}{\mu}\xi + \frac{m-\theta\mu}{\mu}u(0) + \zeta u(1) = -\beta y(0) + \zeta u(1) - \theta u(0) + (ay_x)(0). \tag{4.7.14}$$

We aim to use the semigroups theory of nonautonomous systems [72, 158] so that we establish the existence and uniqueness of solutions of (4.7.6). To do so, the strategy is to use Kato's variable norm technique [122], which has been adopted for several cases such as the one-dimensional heat and wave equations [155], the multidimensional wave equation [156], and the disk-beam system [53, 55].

To begin, let K be a positive constant to be fixed later and define the time-dependent Kato's inner product

$$\begin{aligned}\langle (y, z, u, \xi, \eta), (\tilde{y}, \tilde{z}, \tilde{u}, \tilde{\xi}, \tilde{\eta})\rangle_t &= \int_0^1 (ay_x\tilde{y}_x + z\tilde{z})\,dx + K\tau(t)\int_0^1 u\tilde{u}\,dx \\ &\quad + \beta y(0)\tilde{y}(0) + \frac{1}{m-\theta\mu}\xi\tilde{\xi} + M\eta\tilde{\eta},\end{aligned} \tag{4.7.15}$$

with associated norm denoted by $\|.\|_t$.

Clearly, for $s, t \in [0, T]$, there exists $c \in (s, t)$ such that

$$\tau(t) = \tau(s) + \dot{\tau}(c)(t-s).$$

Recalling (4.7.2), it follows, for some positive constant ϖ, that

$$\frac{\tau(t)}{\tau(s)} \leq 1 + \frac{\varpi}{\tau_1}|t-s| \leq e^{\frac{\varpi}{\tau_1}|t-s|}. \tag{4.7.16}$$

On other hand, for all $\Phi = (y, z, u, \xi, \eta)$, we have

$$\begin{aligned}\|\Phi\|_t^2 - \|\Phi\|_s^2 e^{\frac{\varpi}{\tau_1}|t-s|} &= \left(1 - e^{\frac{\varpi}{\tau_1}|t-s|}\right) \\ &\left(\int_0^1 (a(x)y_x^2 + z^2)\,dx + \beta y^2(0) + \frac{1}{m-\theta\mu}\xi^2 + M\eta^2\right) \\ &+ K\left(\tau(t) - \tau(s)e^{\frac{\varpi}{\tau_1}|t-s|}\right)\int_0^1 u^2\,dx.\end{aligned}$$

Thanks to (4.7.16), we get

$$\tau(t) - \tau(s)e^{\frac{\varpi}{\tau_1}|t-s|} \leq 0,$$

which gives

$$\frac{\|\Phi\|_t}{\|\Phi\|_s} \le e^{\frac{\varpi}{2\tau_1}|t-s|}, \ \forall t, s \in [0, T]. \tag{4.7.17}$$

Now, let us fix $t > 0$ and take $\Phi = (y, z, u, \xi, \eta) \in \mathscr{D}(\mathscr{A}(t))$. Then

$$\begin{aligned}
\langle \mathscr{A}(t)\Phi, \Phi\rangle_t &= [\beta y(0) - (ay_x)(0)]\, z(0) + K \int_0^1 (x\dot{\tau}(t) - 1)uu_x \, dx + \frac{1}{m - \theta\mu}\rho\xi \\
&= \frac{\xi - mu(0)}{\mu} z(0) + \frac{K}{2}(\dot{\tau}(t) - 1)u^2(1) + \frac{K}{2}u^2(0) \\
&\quad - \frac{K}{2}\int_0^1 \dot{\tau}(t)u^2 \, dx + \frac{1}{m - \theta\mu}\rho\xi \\
&= -\theta u^2(0) + \zeta u(0)u(1) - \frac{\mu}{m - \theta\mu}\rho^2 + \frac{\zeta\mu}{m - \theta\mu}\rho u(1) \\
&\quad + \frac{K}{2}(\dot{\tau}(t) - 1)u^2(1) \\
&\quad + \frac{K}{2}u^2(0) + \frac{K|\dot{\tau}(t)|}{2\tau(t)}\tau(t)\int_0^1 u^2 \, dx.
\end{aligned}$$

Subsequently, using (4.7.4) and applying Young's inequality for the mixed products $u(0)u(1)$ and $\rho u(1)$, we deduce that

$$\begin{aligned}
\langle \mathscr{A}(t)\Phi, \Phi\rangle_t \le{}& \left(\frac{\zeta}{4\epsilon_1} + \frac{K}{2} - \theta\right) u^2(0) + \left(\frac{K}{2}(\kappa - 1) + \zeta\epsilon_1 + \frac{\zeta\mu}{m - \theta\mu}\epsilon_2\right) u^2(1) \\
&+ \left(\frac{\zeta}{4\epsilon_2} - 1\right) \frac{\mu}{m - \theta\mu}\rho^2 + r(t)\langle\Phi, \Phi\rangle_t,
\end{aligned}$$

where

$$r(t) = \frac{\sqrt{\dot{\tau}^2(t) + 1}}{2\tau(t)}. \tag{4.7.18}$$

Thereafter, choose

$$\epsilon_1 = \epsilon_2 = \sqrt{1 - \kappa}/2,$$

and then pick up a positive constant K such that

$$\frac{\zeta}{\sqrt{1 - \kappa}}\left(1 + \frac{\mu}{m - \theta\mu}\right) \le K \le 2\theta - \frac{\zeta}{\sqrt{1 - \kappa}}, \tag{4.7.19}$$

which is possible in view of (4.7.13). Owing to (4.7.7), (4.7.13), and (4.7.19), we have

$$\langle \mathscr{A}(t)\Phi, \Phi\rangle_t - r(t)\langle \Phi, \Phi\rangle_t \leq 0, \tag{4.7.20}$$

and hence, we have established the dissipativity of the operator

$$\widehat{\mathscr{A}}(t) := \mathscr{A}(t) - r(t)I,$$

with respect to the time-dependent Kato's inner product defined in (4.7.15).

Now, in light of (4.7.2) and (4.7.3), one can show that

$$\frac{d}{dt}\widehat{\mathscr{A}}(t) \in L^\infty(0, T; B(\mathscr{D}(\mathscr{A}(t)), \mathscr{H})) \tag{4.7.21}$$

the space of equivalence classes of essentially bounded, strongly measurable functions from $[0, T]$ into the set of bounded operators $B(\mathscr{D}(\mathscr{A}(t)), \mathscr{H})$. Indeed,

$$\dot{r}(t) = \frac{\tau(t)\dot{\tau}(t)\ddot{\tau}(t) - \dot{\tau}(t)(1 + \tau^2(t))}{2\tau^2(t)\sqrt{\dot{\tau}^2(t) + 1}},$$

which, after taking into account (4.7.2), is bounded on $[0, T]$, for all $T > 0$. Whereupon, we have

$$\frac{d}{dt}\mathscr{A}(t)\Phi = \left(0, 0, \frac{x\tau(t)\ddot{\tau}(t) - \dot{\tau}(t)(x\dot{\tau}(t) - 1)}{\tau^2(t)}, 0, 0\right),$$

where the term $\frac{x\tau(t)\ddot{\tau}(t) - \dot{\tau}(t)(x\dot{\tau}(t) - 1)}{\tau^2(t)}$ is bounded on $[0, T]$ by using (4.7.2) once more.

Next, the ultimate outcome is to prove the maximality of the operator $\mathscr{A}(t)$, i.e., $\lambda I - \mathscr{A}(t)$ is surjective for some $\lambda > 0$ and $t > 0$. Taking $(f, g, h, p, q) \in \mathscr{H}$, we look for an element $(y, z, u, \xi, \eta) \in \mathscr{D}(\mathscr{A}(0))$ such that $(\lambda I - \mathscr{A}(t))(y, z, u, \xi, \eta) = (f, g, h, p, q)$. Equivalently, we consider the following system:

$$\begin{cases} \lambda^2 y - (a y_x)_x = \lambda f + g, \\ \lambda \tau u + (1 - x\dot{\tau}(t))u_x = \tau h, \\ \mu\left[\beta(\lambda\mu + 1) + \lambda(\lambda m + \theta)\right] y(0) - \mu(\lambda\mu + 1)\left(a y_x\right)(0) - \zeta \mu u(1) \\ = \mu p + \mu(m\lambda + \theta) f(0), \\ \lambda^2 M y(1) + \left(a y_x\right)(1) = M q + \lambda M f(1), \\ z = \lambda y - f, \\ \xi = m u(0) + \mu[\beta y(0) - \left(a y_x\right)(0)], \; u(0) = z(0) = \lambda y(0) - f(0), \\ \eta = z(1) = \lambda y(1) - f(1). \end{cases} \tag{4.7.22}$$

Solving the equation of u in the above system, one obtains

$$u(x)=\begin{cases} e^{\lambda\frac{\tau(t)}{\dot\tau(t)}\ln(1-x\dot\tau(t))}\left[\lambda y(0)-f(0)+\tau(t)\displaystyle\int_0^x \frac{h(\rho)}{1-\rho\dot\tau(t)}e^{-\lambda\frac{\tau(t)}{\dot\tau(t)}\ln(1-\rho\dot\tau(t))}\,d\rho\right], & \text{if } \dot\tau(t)\neq 0,\\ e^{-\tau\lambda x}(\lambda y(0)-f(0))+\tau\displaystyle\int_0^x e^{-\tau\lambda(x-\rho)}h(\rho)\,d\rho, & \text{if } \tau(t)=\tau \text{ (constant)},\end{cases} \tag{4.7.23}$$

which gives

$$u(1)=\begin{cases} e^{\lambda\frac{\tau(t)}{\dot\tau(t)}\ln(1-\dot\tau(t))}\left[\lambda y(0)-f(0)+\tau(t)\displaystyle\int_0^1 \frac{h(\rho)}{1-\rho\dot\tau(t)}e^{-\lambda\frac{\tau(t)}{\dot\tau(t)}\ln(1-\rho\dot\tau(t))}\,d\rho\right], & \text{if } \dot\tau(t)\neq 0,\\ e^{-\tau\lambda}(\lambda y(0)-f(0))+\tau\displaystyle\int_0^1 e^{-\tau\lambda(1-\rho)}h(\rho)\,d\rho, & \text{if } \tau(t)=\tau \text{ (constant)}.\end{cases} \tag{4.7.24}$$

By virtue of (4.7.22) and (4.7.24), one has only to seek $y\in H^2(I)$ such that

$$\begin{cases} \lambda^2 y-(ay_x)_x=\lambda f+g,\\ \lambda^2 My(1)+(ay_x)(1)=Mq+\lambda Mf(1), \end{cases} \tag{4.7.25}$$

together with either

$$\begin{aligned} &\mu\left[\beta(\lambda\mu+1)+\lambda(\lambda m+\theta)-\zeta\lambda e^{\lambda\frac{\tau(t)}{\dot\tau(t)}\ln(1-\dot\tau(t))}\right]y(0)\\ &\quad-\mu(\lambda\mu+1)\,(ay_x)(0)=\mu p+ \end{aligned} \tag{4.7.26}$$

$$\left[\mu(\lambda\mu+\theta)-\zeta\mu e^{\lambda\frac{\tau(t)}{\dot\tau(t)}\ln(1-\dot\tau(t))}\right]f(0)+\zeta\tau(t)\mu e^{\lambda\frac{\tau(t)}{\dot\tau(t)}\ln(1-\dot\tau(t))}$$

$$\int_0^1 \frac{h(\rho)}{1-\rho\dot\tau(t)}e^{-\lambda\frac{\tau(t)}{\dot\tau(t)}\ln(1-\rho\dot\tau(t))}\,d\rho,$$

in the case where $\dot\tau(t)\neq 0$, or

$$\mu\left[\beta(\lambda\mu+1)+\lambda(\lambda m+\theta)-\zeta\lambda e^{-\tau\lambda}\right]y(0)-\mu(\lambda\mu+1)\,(ay_x)(0) \tag{4.7.27}$$

$$=\mu p+\left[\mu(\lambda\mu+\theta)-\zeta\mu e^{-\tau\lambda}\right]f(0)+\zeta\mu\tau\int_0^1 e^{-\tau\lambda(1-\rho)}h(\rho)\,d\rho,$$

if $\tau(t)=\tau$ (constant).

Multiplying the first equation in (4.7.25) by $\phi \in H^1(I)$ and integrating over $I = (0, 1)$, we obtain the following variational problem $\mathscr{R}(y, \phi) = \mathscr{T}(\phi)$, where $\mathscr{R}$ is the real bilinear form in $H^1(I) \times H^1(I)$ given by

$$\mathscr{R}(y,\phi) = \int_0^1 (\lambda^2 y\phi + a y_x \phi_x)\, dx + \begin{cases} \dfrac{y(0)\phi(0)}{\mu\lambda + 1}\left[\beta(\lambda\mu + 1) + \lambda(\lambda m + \theta) - \zeta\lambda e^{\lambda \frac{\tau(t)}{\dot\tau(t)}\ln(1-\dot\tau(t))}\right] \\ +\lambda^2 M y(1)\phi(1), \quad \text{if } \dot\tau \neq 0, \\ \dfrac{y(0)\phi(0)}{\mu\lambda + 1}\left[\beta(\lambda\mu + 1) + \lambda(\lambda m + \theta) - \zeta\lambda e^{-\tau\lambda}\right] \\ +\lambda^2 M y(1)\phi(1), \quad \text{if } \tau(t) = \tau \text{ (constant)}, \end{cases}$$

and

$$\mathscr{T}(\phi) = \int_0^1 (\lambda f + g)\phi\, dx + \frac{1}{\lambda\mu + 1} p\phi(0)$$

$$+ \begin{cases} \dfrac{1}{\lambda\mu + 1}\left(\lambda\mu + \theta - \zeta e^{\lambda \frac{\tau(t)}{\dot\tau(t)}\ln(1-\dot\tau(t))}\right) f(0)\phi(0) \\ + \dfrac{\zeta\tau(t)\phi(0)}{\lambda\mu + 1} e^{\lambda \frac{\tau(t)}{\dot\tau(t)}\ln(1-\dot\tau(t))} \displaystyle\int_0^1 \frac{h(\rho)}{1 - \rho\dot\tau(t)} e^{-\lambda \frac{\tau(t)}{\dot\tau(t)}\ln(1-\rho\dot\tau(t))}\, d\rho, \quad \text{if } \dot\tau \neq 0, \\ \dfrac{1}{\lambda\mu + 1}\left(\lambda\mu + \theta - \zeta e^{\tau\lambda}\right) f(0)\phi(0) + \dfrac{\zeta\tau}{\lambda\mu + 1}\displaystyle\int_0^1 e^{-\tau\lambda(1-\rho)} h(\rho)\, d\rho, \\ \text{if } \tau(t) = \tau \text{ (constant)}. \end{cases}$$

Thus, thanks to Lax-Milgram theorem (see Theorem 2.1.6), we can show, as long as $\lambda > 0$, the existence of a unique solution $y \in H^2(I)$ of (4.7.25) together with either (4.7.26) or (4.7.27). This implies that the operator $\lambda I - \mathscr{A}(t)$ is onto $\mathscr{H}$, for $\lambda > 0$ and for all $t > 0$. Since $r(t) > 0$ (see (4.7.18)), we immediately deduce that

$$\lambda I - \widehat{\mathscr{A}}(t) = (\lambda + r(t))I - \mathscr{A}(t) \quad \text{is surjective.} \tag{4.7.28}$$

Gathering all the facts (4.7.12), (4.7.17), (4.7.20), (4.7.21), and (4.7.28) and applying the evolution semigroups theory [72, 158], we can claim that for any initial data $\widehat{\Phi}_0 \in \mathscr{H}$ the Cauchy problem

$$\begin{cases} \widehat{\Phi}_t = \widehat{\mathscr{A}}(t)\widehat{\Phi}, \\ \widehat{\Phi}(0) = \widehat{\Phi}_0, \end{cases}$$

has a unique solution $\widehat{\Phi} \in C(0, \infty; \mathscr{H})$. Moreover, if $\widehat{\Phi}_0 \in \mathscr{D}(\mathscr{A}(0))$, then we have

$$\widehat{\Phi} \in C(0, \infty; \mathscr{D}(\mathscr{A}(t))) \cap C^1(0, \infty; \mathscr{H}).$$

Finally, thanks to the change of variable $\Phi(t) = e^{r(t)}\widehat{\Phi}(t)$, we can conclude the assertions of the theorem with respect to (4.7.9). □

4.7.3 Exponential Stability

The main objective of this subsection is to prove the exponential stability of the unique solution of the system (4.7.6), and a decay rate will be provided by proceeding as in [155, 156]. First, let us consider the following energy:

$$E(t) = \frac{1}{2}\left\{\int_0^1 \left[y_t^2(x,t) + a(x)y_x^2(x,t) + K\tau(t)y_t^2(0, t - x\tau(t))\right]dx\right\}$$
$$+ \frac{1}{2}\left\{\beta y^2(0,t) + \frac{1}{m - \theta\mu}\xi^2(t) + My_t^2(1,t)\right\}, \tag{4.7.29}$$

which corresponds to the inner product $\langle ., . \rangle_t$ defined in (4.7.15). Needless to recall that 4.1.5, (4.7.2)–(4.7.4), (4.7.7), and (4.7.13) are supposed to be satisfied, while K obeys (4.7.19). In order to reach our goal, we will use the following Lyapunov functional:

$$\mathscr{E}(t) = E(t) + \gamma\mathscr{P}(t), \tag{4.7.30}$$

where

$$\mathscr{P}(t) = \mathscr{P}_0(t) + \mathscr{P}_1(t) + \mathscr{P}_2(t)$$

with

$$\begin{cases} \mathscr{P}_0(t) = 2\displaystyle\int_0^1 F(x)y_t(x,t)y_x(x,t)\,dx, \\ \mathscr{P}_1(t) = K\tau(t)\displaystyle\int_0^1 e^{-2\varepsilon\tau(t)x}y_t^2(0, t - x\tau(t))\,dx, \ \varepsilon > 0, \\ \mathscr{P}_2(t) = C_0 y(0,t)\left(\displaystyle\int_0^1 y_t(x,t)\,dx + My_t(1,t)\right), \ C_0 > 0. \end{cases} \tag{4.7.31}$$

Note that γ and C_0 are positive constants that will be fixed later, while ε is an arbitrary positive constant. Moreover, F is a negative function defined by (see [167] and the previous sections for similar situations)

$$F(x) = -\exp\left(\int_x^1 \left|\frac{a_x}{a}\right|(s)\,ds\right)\left(M + \int_x^1 \exp\left(-\int_s^1 \left|\frac{a_x}{a}\right|(r)\,dr\right)ds\right),\quad \forall x \in [0,1]. \tag{4.7.32}$$

The latter satisfies

$$\begin{cases} F(1) = -M \text{ and } F(0) < 0, \\ F_x(x) \geq 1,\ (aF_x - |a_x|F)(x) \geq a(x),\ F(x) < 0,\ \forall x \in [0,1]. \end{cases} \tag{4.7.33}$$

First, observe that the energies $E(t)$ and $\mathscr{E}(t)$ are equivalent. Indeed, we have:

Lemma 4.7.1 *For γ small enough, there are two positive constants $A_1(\gamma)$ and $A_2(\gamma)$ such that*

$$A_1(\gamma)E(t) \leq \mathscr{E}(t) \leq A_2(\gamma)E(t). \tag{4.7.34}$$

Proof Applying Cauchy-Schwarz and Young's inequalities in $\mathscr{P}$, we have

$$|\mathscr{P}(t)| \leq \alpha E(t),$$

where

$$\alpha = \left(2\left\|\frac{F}{\sqrt{a}}\right\|_{L^\infty(I)} + 2 + \frac{\sqrt{M}+1}{\sqrt{\beta}}C_0\right).$$

Whereupon, we obtain (4.7.34) with

$$A_1(\gamma) = 1 - \gamma\alpha \quad \text{and} \quad A_2(\gamma) = 1 + \gamma\alpha, \tag{4.7.35}$$

as long as $\gamma < 1/\alpha$. □

The result on the decay of the energy $E(t)$ is stated below.

Theorem 4.7.2 *Suppose that 4.1.5, (4.7.2)–(4.7.4), (4.7.7), and (4.7.13) hold. Then, the energy E decays exponentially, and hence, there exist two positive constants C and υ such that*

$$E(t) \leq Ce^{-\upsilon t}E(0), \quad \forall t > 0.$$

Proof Our ultimate goal is to estimate the derivative of $\mathscr{E}(t)$ so that we can deduce the desired result. For sake of clarity, the proof is divided into three steps.

First Step Let us estimate $E'(t)$. To do so, multiplying the first equation in (4.7.6) by $y_t(x,t)$ and formally integrating with respect to $x \in I$, we get

$$\frac{1}{2}\frac{d}{dt}\int_0^1 \left[y_t^2(x,t)+a(x)y_x^2(x,t)\right]dx + (ay_x)(0,t)y_t(0,t) - (ay_x)(1,t)y_t(1,t) = 0. \tag{4.7.36}$$

Using (4.7.5), we have

$$(ay_x)(0,t) = \xi'(t) + \beta y(0,t) - \zeta y_t(0,t-\tau(t)) + \theta y_t(0,t). \tag{4.7.37}$$

Then, by (4.7.36)–(4.7.37) and the boundary condition at $x = 1$, we find

$$\frac{1}{2}\frac{d}{dt}\left(\int_0^1 \left[y_t^2(x,t)+a(x)y_x^2(x,t)\right]dx + \beta y^2(0,t) + My^2(1,t)\right) = -\xi'(t)y_t(0,t) + \zeta y_t(0,t-\tau(t))y_t(0,t) - \theta y_t^2(0,t). \tag{4.7.38}$$

On the other hand, using an integration by parts, it is easy to check that

$$\frac{K}{2}\frac{d}{dt}\tau(t)\int_0^1 y_t^2(0,t-x\tau(t))\,dx = \frac{K}{2}\left[y_t^2(0,t) - (1-\dot{\tau}(t))y_t^2(0,t-\tau(t))\right], \tag{4.7.39}$$

whereas (4.7.5) and the boundary conditions give

$$\xi'(t) = -\frac{1}{\mu}\xi(t) + \frac{m-\theta\mu}{\mu}y_t(0,t) + \zeta y_t(0,t-\tau(t)). \tag{4.7.40}$$

Then, (4.7.38)–(4.7.40) yield

$$E'(t) = \frac{K}{2}\left[y_t^2(0,t) - (1-\dot{\tau}(t))y_t^2(0,t-\tau(t))\right] - \theta y_t^2(0,t) - \frac{\mu}{m-\theta\mu}\xi'^2(t) + \zeta y_t(0,t-\tau(t))y_t(0,t) + \frac{\zeta\mu}{m-\theta\mu}y_t(0,t-\tau(t))\xi'(t).$$

Applying Young's inequality, the latter becomes

$$E'(t) \le \left(\frac{K}{2} + \frac{\zeta}{2\sqrt{1-\kappa}} - \theta\right)y_t^2(0,t) + \left(\frac{\zeta}{2\sqrt{1-\kappa}} - 1\right)\frac{\mu}{m-\theta\mu}\xi'^2(t) + \left(\zeta\frac{\sqrt{1-\kappa}}{2} + \frac{\zeta\mu\sqrt{1-\kappa}}{2(m-\theta\mu)} + \frac{K}{2}(\kappa-1)\right)y_t^2(0,t-\tau(t)). \tag{4.7.41}$$

Second Step The ultimate outcome is to estimate the second part $\mathscr{P}'(t)$ of $\mathscr{E}'(t)$. First, integrating by parts, using (4.7.32) and (4.7.33), and neglecting the negative term $F(1)(ay_x^2)(1,t)$, we obtain

$$\begin{aligned}\mathscr{P}_0'(t) &= F(1)(ay_x^2)(1,t) - F(0)(ay_x^2)(0,t) - \int_0^1 a\left(\frac{F}{a}\right)_x ay_x^2(x,t)\,dx \\ &\quad + F(1)y_t^2(1,t) - F(0)y_t^2(0,t) - \int_0^1 F_x y_t^2(x,t)\,dx \\ &\le -\int_0^1 y_t^2(x,t)\,dx - \int_0^1 ay_x^2(x,t)\,dx - My_t^2(1,t) \\ &\quad - F(0)y_t^2(0,t) - F(0)(ay_x^2)(0,t).\end{aligned} \tag{4.7.42}$$

In turn, by virtue of (4.7.37) and the boundary condition at $x = 0$ (see (4.7.6)), we have

$$\begin{aligned}&(ay_x)^2(0,t) - 2\beta(ay_x)(0,t)y(0,t) + \beta^2y^2(0,t) \\ &= \xi'^2(t) + \theta^2y_t^2(0,t) + \zeta^2y_t^2(0,t-\tau(t)) - 2\zeta\theta y_t(0,t)y_t(0,t-\tau(t)) \\ &\qquad + 2\theta\xi'(t)y_t(0,t) - 2\zeta\xi'(t)y_t(0,t-\tau(t)).\end{aligned}$$

Thanks to Young's inequality, the latter becomes

$$\begin{aligned}-F(0)(ay_x^2)(0,t) &\le -2\beta\frac{F(0)}{a(0)}(ay_x)(0,t)y(0,t) + \beta^2\frac{F(0)}{a(0)}y^2(0,t) \\ &\quad -3\frac{F(0)}{a(0)}\left(\xi'^2(t) + \theta^2y_t^2(0,t) + \zeta^2y_t^2(0,t-\tau(t))\right),\end{aligned}$$

which, together with (4.7.42), implies

$$\begin{aligned}\mathscr{P}_0'(t) &\le -\int_0^1 y_t^2(x,t)\,dx - \int_0^1 ay_x^2(x,t)\,dx - My_t^2(1,t) - F(0)y_t^2(0,t) \\ &\quad + \beta^2\frac{F(0)}{a(0)}y^2(0,t) \\ &\quad - - 2\beta\frac{F(0)}{a(0)}(ay_x)(0,t)y(0,t) \\ &\quad -3\tfrac{F(0)}{a(0)}\left(\xi'^2(t) + \theta^2y_t^2(0,t) + \zeta^2y_t^2(0,t-\tau(t))\right).\end{aligned} \tag{4.7.43}$$

Now, we turn to the estimation of the functional $\mathscr{P}_1'(t)$. A straightforward integration by parts leads to

$$\begin{aligned}\mathscr{P}_1'(t) &= K\Big[y^2(0,t) + (\dot\tau(t)-1)e^{-2\varepsilon\tau(t)}y_t^2(0,t-\tau(t)) \\ &\quad - 2\varepsilon\tau(t)\textstyle\int_0^1 e^{-2\varepsilon\tau(t)x}y_t^2(0,t-x\tau(t))\,dx\Big] \\ &\le Ky_t^2(0,t) - K(1-\kappa)e^{-2\varepsilon\tau_2}y_t^2(0,t-\tau(t)) \\ &\quad -2K\varepsilon\tau(t)e^{-2\varepsilon\tau_2}\textstyle\int_0^1 y_t^2(0,t-x\tau(t))\,dx,\end{aligned} \tag{4.7.44}$$

where (4.7.3)–(4.7.4) are invoked.

Next, we will estimate the last term of $\mathscr{P}'(t)$. Using (4.7.6), we get

$$\mathscr{P}_2'(t) = y_t(0,t)\left(\int_0^1 y_t(x,t)\,dx + My_t(1,t)\right) - C_2y(0,t)(ay_x)(0,t). \tag{4.7.45}$$

Applying Cauchy-Schwarz and Young's inequalities in (4.7.45), we obtain

$$\begin{aligned}\mathscr{P}_2'(t) &\le \frac{1}{2}\int_0^1 y_t^2(x,t)\,dx + \frac{M}{2}y_t^2(1,t) + C_0^2\frac{M+1}{2}y_t^2(0,t) \\ &\quad - C_0(ay_x)(0,t)y(0,t).\end{aligned} \tag{4.7.46}$$

Choosing $C_0 = -2\beta\frac{F(0)}{a(0)}$ and combining (4.7.43)–(4.7.44), (4.7.46) and the definition of $E(t)$ in (4.7.29), we find

$$\begin{aligned}\mathscr{P}'(t) &\le -\vartheta E(t) + \frac{1}{2(m-\theta\mu)}\xi^2(t) \\ &\quad + \left(-F(0) - 3\tfrac{F(0)}{a(0)}\theta^2 + C_2^2\tfrac{M+1}{2} + K\right)y_t^2(0,t) \\ &\quad + \left(-3\frac{F(0)}{a(0)}\zeta^2 - K(1-\kappa)e^{-2\varepsilon\tau_2}\right)y_t^2(0,t-\tau(t)) - 3\frac{F(0)}{a(0)}\xi'^2(t),\end{aligned} \tag{4.7.47}$$

where

$$\vartheta = \min\left\{1, 4\varepsilon e^{-2\varepsilon\tau_2}, -2\beta\frac{F(0)}{a(0)}\right\} > 0.$$

On the other hand, by virtue of the relationship between ξ and ξ' (see (4.7.40)) and Young's inequality once more, we have

$$\begin{aligned}
\frac{1}{2(m-\theta\mu)}\xi^2(t) &= \frac{m-\theta\mu}{2}y_t^2(0,t) + \frac{\mu^2\zeta^2}{2(m-\theta\mu)}y_t^2(0,t-\tau(t)) \\
&\quad + \tfrac{\mu^2}{2(m-\theta\mu)}\xi'^2(t) \\
&\quad - \frac{\mu^2\zeta}{m-\theta\mu}\xi'(t)y_t(0,t-\tau(t)) + \mu\zeta y_t(0,t)y_t(0,t-\tau(t)) \\
&\quad - \mu y_t(0,t)\xi'(t) \\
&\le \left(\frac{m-\theta\mu}{2}+1\right)y_t^2(0,t) \\
&\quad + \mu^2\zeta^2\left(\tfrac{1}{m-\theta\mu}+\tfrac{1}{2}\right)y_t^2(0,t-\tau(t)) \\
&\quad + \mu^2\left(\frac{1}{m-\theta\mu}+\frac{1}{2}\right)\xi'^2(t).
\end{aligned} \tag{4.7.48}$$

Inserting (4.7.48) into (4.7.47) yields

$$\begin{aligned}
\mathscr{P}'(t) &\le -\vartheta E(t) + \left(-F(0) - 3\frac{F(0)}{a(0)}\theta^2 + C_2^2\frac{M+1}{2} + K + \frac{m-\theta\mu}{2}+1\right)y_t^2(0,t) \\
&\quad + \left(-3\frac{F(0)}{a(0)}\zeta^2 - K(1-\kappa)e^{-2\varepsilon\tau_2} + \mu^2\zeta^2\left[\frac{1}{m-\theta\mu}+\frac{1}{2}\right]\right)y_t^2(0,t-\tau(t)) \\
&\quad + \left(\mu^2\left[\frac{1}{m-\theta\mu}+\frac{1}{2}\right] - 3\frac{F(0)}{a(0)}\right)\xi'^2(t).
\end{aligned} \tag{4.7.49}$$

Now, amalgamating (4.7.41) and (4.7.49) with (4.7.30), it follows that

$$\begin{aligned}
\mathscr{E}'(t) &= E'(t) + \gamma\mathscr{P}'(t) \\
&\le -\vartheta\gamma E(t) + \rho_{1,\gamma}y^2(0,t) + \rho_{2,\gamma}y^2(0,t-\tau(t)) + \rho_{3,\gamma}\xi'^2(t),
\end{aligned} \tag{4.7.50}$$

where

$$\begin{aligned}
\rho_{1,\gamma} &= \frac{K}{2} + \frac{\zeta}{2\sqrt{1-\kappa}} - \theta \\
&\quad + \gamma\left(-F(0) - 3\frac{F(0)}{a(0)}\theta^2 + C_2^2\frac{M+1}{2} + K + \frac{m-\theta\mu}{2}+1\right), \\
\rho_{2,\gamma} &= \zeta\frac{\sqrt{1-\kappa}}{2} + \frac{\zeta\mu\sqrt{1-\kappa}}{2(m-\theta\mu)} + \frac{K}{2}(\kappa-1) \\
&\quad + \gamma\left[-3\frac{F(0)}{a(0)}\zeta^2 - K(1-\kappa)e^{-2\varepsilon\tau_2} + \mu^2\zeta^2\left\{\frac{1}{m-\theta\mu}+\frac{1}{2}\right\}\right],
\end{aligned}$$

and

$$\rho_{3,\gamma} = \left(\frac{\zeta}{2\sqrt{1-\kappa}} - 1\right)\frac{\mu}{m-\theta\mu} + \gamma\left(\mu^2\left[\frac{1}{m-\theta\mu} + \frac{1}{2}\right] - 3\frac{F(0)}{a(0)}\right).$$

Third Step Subsequently, one can make the coefficient $\rho_{1,\gamma}$ nonpositive by proceeding as follows: Thanks to (4.7.19), there holds

$$\frac{K}{2} + \frac{\zeta}{2\sqrt{1-\kappa}} - \theta \le 0.$$

Next, we choose γ small enough so that

$$\frac{K}{2} + \frac{\zeta}{2\sqrt{1-\kappa}} - \theta + \gamma\left(-F(0) - 3\frac{F(0)}{a(0)}\theta^2 + C_2^2\frac{M+1}{2} + K + \frac{m-\theta\mu}{2} + 1\right) \le 0.$$

Similarly, using (4.7.19) and (4.7.13), we conclude that

$$\rho_{2,\gamma} \le 0 \quad \text{and} \quad \rho_{3,\gamma} \le 0,$$

for sufficiently small γ. This implies, thanks to (4.7.34), (4.7.50) and the fact that $\gamma < 1/\alpha$, that

$$\mathscr{E}'(t) \le -\vartheta\gamma E(t) \le -\frac{\vartheta\gamma}{A_2(\gamma)}\mathscr{E}(t),$$

where $A_2(\gamma)$ is defined in (4.7.35). Thereby

$$\mathscr{E}(t) \le \mathscr{E}(0)e^{-\upsilon t}, \quad t > 0,$$

with

$$\upsilon = \frac{\vartheta\gamma}{A_2(\gamma)}.$$

Finally, in light of (4.7.34), we can deduce the result of exponential decay rate of the system (4.7.6). □

Remark 4.7.1 There are numerous nonconstant functions $\tau(t)$ satisfying the requirements (4.7.2)–(4.7.4). For instance, one can take $\tau(t) = m_1 \sin t + m_2 \cos t + m_3$, where m_1, m_2, and m_3 are positive constants such that $m_1 + m_2 < \min\{1, m_3\}$. Specifically, the conditions (4.7.2)–(4.7.4) hold for $\tau(t) = \frac{1}{2}\sin t + \frac{1}{3}\cos t + 1$, with $\tau_1 = 1/6$, $\tau_2 = 11/6$, and $\kappa = 5/6$.

4.7.4 Concluding Remarks

The present section contains an exponential stabilization result of the dynamics of a flexible cable, which consists of a part of the overhead crane. Such a finding is obtained via Lyapunov method despite the presence of a time-dependent boundary delay but under reasonable conditions. Unlike the previous sections, the advantage of such an approach lies in its simplicity and, more importantly, the fact that an estimate of the decay rate is provided. Obviously, the fastest decay rate should be the subject of further study. One could also consider the same problem of this section but with an additional assumption in which the length of the cable varies with time. It is also worth mentioning that the time-dependent delay $\tau(t)$ is supposed not to degenerate ($\tau(t)$ is always positive). However, we believe that this condition could be relaxed by allowing the delay to vanish at some times $\hat{t}$. In fact, one could follow the same arguments as used in [153].

4.8 Nonlinear Boundary Delay: Nonlinear Boundary Stabilization

The main concern of this section is to investigate the stabilization problem of a variant of the crane system. Specifically, we assume that the mass m is small so that its dynamics can be neglected, that is, the dynamical term $my_{tt}(0,t)$ is disregarded. In turn, two nonlinearities occur. Specifically, the first nonlinearity is related to the velocity term, while the second one arises from the delayed term (see [167] for more details on the derivation of the model):

$$\begin{cases} y_{tt}(x,t) - (ay_x)_x(x,t) = 0, & x \in (0,1),\ t > 0, \\ (ay_x)(0,t) = U(t), & t > 0, \\ My_{tt}(1,t) + (ay_x)(1,t) = 0, & t > 0. \end{cases} \tag{4.8.1}$$

$U(t)$ is a *nonlinear* boundary *delayed* control to be designed so that the system is stable in a suitable functional space.

Despite such a situation, we aim to prove that the system has a unique global solution by means of semigroups theory of nonlinear operators (see Sect. 2.4). More importantly, our target is to make the energy of the system decay and provide an estimate of the decay rate of the energy depending on the type of the nonlinearities.

4.8.1 Introduction and Preliminary Results

First of all, it is noteworthy that, to the authors' best knowledge, no attempt has been made concerning the stabilization of the crane system (4.8.1) when the control $U(t)$

is nonlinear and at the same time subject to a nonlinear delay term. This constitutes the main motivation of the present section. To be more precise, the system (4.8.1) is shown to be exponentially stabilized in [167] under the action of the nondelayed boundary control

$$U(t) = ky(0,t) - g(y_t(0,t)),\ k > 0, \tag{4.8.2}$$

where g is a nonlinear function satisfying the following conditions:
g is continuous and increases on $\mathbb{R}$. Moreover, $g(0) = 0$, $sg(s) > 0$, for all $s \neq 0$, and it has a linear or power growth close to the origin, that is, there exist positive constants η_1 and η_2 such that

$$\eta_1 \mid x \mid \leq \mid g(x) \mid \leq \eta_2 \mid x \mid, \quad \text{for all } x \text{ in } \mathbb{R}.$$

However, if there exist positive constants η_1, η_2 such that for all $x \in \mathbb{R}$,

$$\eta_1 \min\left\{\mid x \mid, \mid x \mid^p\right\} \leq \mid g(x) \mid \leq \eta_2 \mid x \mid, \quad \text{for some } p > 1,$$

or

$$\eta_1 \mid x \mid \leq \mid g(x) \mid \leq \eta_2 \max\left\{\mid x \mid, \mid x \mid^q\right\}, \quad \text{for some } q \in (0, 1),$$

then the closed-loop system (4.8.1) and (4.8.2) is polynomially stabilized. The same result is established in [70] with an interior damping control $y(x,t)$ and a boundary control $U(t) = g(y_t(0,t))$. Note that both papers [167] and [70] do not treat the case of arbitrary growing g close to the origin. It turns out that when the dynamic of the platform is not neglected, then the system is strongly stable by means of the nonlinear control (4.8.2). This result is improved in [76], where it is proven that the solutions of the system asymptotically converge to a steady state when the nonlinear boundary control in (4.8.2) depends only on the velocity term $y_t(0,t)$, that is, $k = 0$ in (4.8.2).

The reason for introducing nonlinear controls in the feedback law is twofold (at least): First, it is highly recommended to have in practice robust controls, for which the stability property of the system is conserved regardless of the presence of small perturbations. These perturbations could result from modeling errors in the controls or moderate disturbances and uncertainties in the controls. Considering a class of nonlinear controls permits some degrees of freedom and so moderate errors, disturbances and uncertainties could be tolerated in contrast to linear controls that are too restrictive. Second, recall that most real-world controls are inherently nonlinear. Furthermore, the input amplitudes of controls are usually constrained by the power of the actuators which go into a nonlinear saturation.

Based on the above discussion, we propose the following boundary control:

$$U(t) = \sigma y(0,t) - \alpha f(y_t(0,t-\tau)) + \beta g(y_t(0,t)), \quad t > 0, \tag{4.8.3}$$

where $\sigma > 0$, $\alpha \geq 0$, $\beta > 0$ are feedback gains, $\tau > 0$ is the constant time delay, whereas f and g are nonlinear functions that obey the following requirements:

(H1) $\begin{cases} f(0) = 0, \ f \text{ is increasing and there exists a real positive constant} \\ \quad \kappa_1 \text{ such that} \\ |f(p) - f(q)| \leq \kappa_1 |p - q|, \quad \text{for all } p, \ q \in \mathbb{R}. \end{cases}$

(H2) $\begin{cases} \text{there exists a positive real constant } r_1 > 0 \text{ such that} \\ (g(p) - g(q))\,(p - q) \geq r_1 (p - q)^2, \quad \text{for all } p, \ q \in \mathbb{R}. \end{cases}$

Subsequently, given a solution y of the system (4.8.1) and (4.8.3), we introduce the associated energy

$$\begin{aligned} E_0(t) = {} & \frac{1}{2} \int_0^1 \left(y_t^2(x,t) + a(x) y_x^2(x,t) \right) dx + \frac{K}{2} \tau \int_0^1 y_t^2(0, t - \tau\rho)\, d\rho \\ & + \frac{1}{2} \left[\sigma y^2(0,t) + M y_t^2(1,t) \right], \end{aligned} \tag{4.8.4}$$

where K is a positive constant to be determined. A formal integration by parts yields

$$\dot{E}_0(t) = \frac{K}{2} y_t^2(0,t) - \frac{K}{2} y_t^2(0, t-\tau) - \alpha f(y_t(0, t-\tau)) y_t(0,t) - \beta g(y_t(0,t)) y_t(0,t),$$

where the dot denotes the time derivative. By virtue of *(H1)–(H2)* and Young's inequality, we reach that

$$\dot{E}_0(t) \leq \frac{\alpha\kappa_1 - K}{2} y_t^2(0, t - \tau) + \left(\frac{K + \alpha\kappa_1}{2} - \beta r_1 \right) y_t^2(0,t). \tag{4.8.5}$$

Now, we suppose that

$$\alpha\kappa_1 < \beta r_1, \tag{4.8.6}$$

and

$$\alpha\kappa_1 < K < 2\beta r_1 - \alpha\kappa_1. \tag{4.8.7}$$

Owing to (4.8.5), (4.8.6), and (4.8.7), we deduce that the energy $E_0(t)$ is decreasing.

As in [149, 184], we introduce the auxiliary variable $z(\rho, t)$ by

$$z(\rho, t) = y_t(0, t - \rho\tau), \quad \rho \in (0,1), \ t > 0,$$

which yields $z(0,t) = y_t(0,t)$, for $t > 0$ and

$$\tau z_t(\rho, t) + z_\rho(\rho, t) = 0. \ \rho \in (0,1), \ t > 0.$$

Whereupon, the closed-loop system (4.8.1) and (4.8.3) has the following form:

$$\begin{cases} y_{tt}(x,t) - (ay_x)_x(x,t) = 0, & x \in (0,1),\ t > 0, \\ (ay_x)(0,t) = \sigma y(0,t) - \alpha f(z(1,t)) + \beta g(y_t(0,t)), & t > 0, \\ My_{tt}(1,t) + (ay_x)(1,t) = 0, & t > 0, \\ \tau z_t(\rho,t) + z_\rho(\rho,t) = 0, & \rho \in (0,1),\ t > 0, \\ z(0,t) = y_t(0,t), & t > 0, \\ y(x,0) = y_0(x),\ y_t(x,0) = y_1(x), & x \in (0,1), \\ z(\rho,0) = y_t(0,-\rho\tau) = \chi(-\rho\tau), & \rho \in (0,1), \end{cases} \tag{4.8.8}$$

where y_0, y_1 and χ are initial data to be chosen in appropriate functional spaces.

The major contribution of this section is twofold: First, we prove that the system (4.8.8) is well-posed by using semigroups theory of nonlinear operators (see Sect. 2.4). It is noteworthy that we have tried to use Faedo-Galerkin method but encountered a difficulty to get an a priori estimate due to the dynamical term $y_{tt}(0,t)$. Also, it seems that one can set the problem as a linear infinite-dimensional port-Hamiltonian system with nonlinear boundary conditions (see, for instance, [29, 116]). Since we are not considering our problem in boundary control and observation forms, such an approach could be considered in another work. Second, the system energy is shown to decay. Note that the decay rate will be explicitly provided, depending on the properties satisfied by the nonlinear functions f and g. In particular, convexity arguments will be used. The idea behind such an approach goes back to [130], where a semilinear wave equation with nonlinear boundary conditions and nonlinear velocity dissipative boundary feedback is considered. It is important to mention that the authors in [130] did not use Young's inequality [186] in the context of nonlinearly damped systems, with arbitrary growing feedbacks close to the origin. Additionally, no constructive examples of concave functions are provided. It seems that the method of combining convexity arguments, Young's inequality, and perturbed energy technics has been introduced in [142] for the wave equation with a nonlinear internal damping. It turned out that the convexity properties can be explored for a general class of dissipative systems [7, 8]. We also point out that the importance of the works [7, 8] lies in the fact that simple sharp optimal or quasi-optimal upper energy decay rates have been established in [7, 8]. This concise methodology established in [7, 8] (see also [9, 10]) gave rise to many works where such an approach has been used for numerous systems.

Our contribution considerably improves earlier results of [24–26, 70, 74, 76, 92–95, 146, 148, 167, 172, 185], where the *time delay is disregarded* and also of [12, 17, 63–65, 184], where both delayed and nondelayed terms in the proposed controls are *linear*. It is also worth mentioning that unlike the previous works [167] and [70], we are able, thanks to the general case developed in [7], to treat a wider class of nonlinearities and get some decay estimates of the energy $E_0(t)$.

Before ending this discussion, we point out that the modulus of tension of the cable, namely, $a(x)$, is supposed to be a positive function (see (4.1.5)). Notwithstanding, one could consider a degenerate case, where $a(x)$ vanishes at

$x_0 = 0$ or $x_0 = 1$. We believe that such an interesting problem could be solved by using semigroups theory together with the energy method and convexity arguments but in *weighted Sobolev spaces*. This approach is frequently used when dealing with degenerate operators (see, for instance, [11, 109] for a one-dimensional degenerate wave equation). Of course, the proof of such a claim remains a promising research avenue.

Now, let us provide a brief outline of the content of the current section. In Sect. 4.8.1, we show that the system (4.8.8) has a global unique solution. The main ingredient of the proof is the utilization of semigroups theory. Section 4.8.3 is dedicated to the proof of the energy decay of the system (4.8.8). Several estimates of the decay rate are put forward thanks to the energy method. Lastly, the section ends with some concluding remarks.

4.8.2 Global Existence of Solutions to the Problem (4.8.8)

This subsection deals with the global existence of the problem (4.8.8). To do so, semigroups theory of nonlinear operators (see Sect. 2.4 and [39]) will be utilized.

First, consider the state variable $\varphi = (y, v, z, \xi)$ and the following state space

$$\mathscr{H} = H^1(0,1) \times \left(L^2(0,1)\right)^2 \times \mathbb{R},$$

equipped with the following real inner product:

$$\begin{aligned} \langle (y, v, z, \xi), (\tilde{y}, \tilde{v}, \tilde{z}, \tilde{\xi}) \rangle_{\mathscr{H}} &= \int_0^1 (a y_x \tilde{y}_x + v\tilde{v})\, dx \\ +K\tau \int_0^1 z\tilde{z}\, dx + \sigma y(0)\tilde{y}(0) + M\xi\tilde{\xi}, \end{aligned} \tag{4.8.9}$$

where $K > 0$ satisfies (4.8.7). It is noteworthy that $\mathscr{H}$ endowed with the inner product (4.8.9) is a Hilbert space thanks to the conditions (4.1.5), (4.8.6), and (4.8.7). This claim can be easily ascertained.

Subsequently, we define a nonlinear operator $\mathscr{A}$ as follows:

$$\mathscr{D}(\mathscr{A}) = \left\{ \begin{array}{l} (y, v, z, \xi) \in \mathscr{H};\ y \in H^2(I),\ v, z \in H^1(I),\ (a y_x)(0) = \sigma y(0) + \alpha f(z(1)) + \beta g(v(0)), \\ z(0) = v(0),\ \xi = v(1) \end{array} \right\}, \tag{4.8.10}$$

and for all $(y, v, z, \xi) \in \mathscr{D}(\mathscr{A})$,

$$\mathscr{A}(y, v, z, \xi) = \left(-v, -(a y_x)_x, \frac{z_\rho}{\tau}, \frac{(a y_x)(1)}{M} \right).$$

Whereupon, the system (4.8.8) is written in $\mathscr{H}$ as follows:

$$\begin{cases} \dot{\varphi}(t) + \mathscr{A}\varphi(t) = 0, \\ \varphi(0) = \varphi_0, \end{cases} \tag{4.8.11}$$

whereas $\varphi = (y(.,t), y_t(.,t), z(.,t), \xi(t))$ and $\varphi_0 = (y_0, y_1, \chi(-\tau\cdot), \xi_0)$.

Our first main result is:

Theorem 4.8.1 *Assume that the conditions (4.1.5), (H1)–(H2), and (4.8.6)–(4.8.7) hold. Then, we have:*

(a) *The operator $\mathscr{A}$ generates a strongly continuous semigroup of contractions $S(t)$ on $\mathscr{H}$.*
(b) *For any initial data $(y_0, y_1, \chi(-\tau\cdot), \xi_0) \in \mathscr{D}(\mathscr{A})$, the problem (4.8.11) admits a unique strong solution*
$(y(.,t), y_t(.,t), z(.,t), \xi(t)) = S(t)(y_0, y_1, \chi(-\tau\cdot), \xi_0) \in \mathscr{D}(\mathscr{A}), \ \forall t \geq 0$ such that

$$y \in W^{1,\infty}(0,\infty; H^1(0,1)) \cap L^\infty(0,\infty; H^2(0,1)),$$
$$z \in W^{1,\infty}(0,\infty; L^2(0,1)) \cap L^\infty(0,\infty; H^1(0,1)), \quad \xi \in W^{1,\infty}(0,\infty; \mathbb{R}).$$

Moreover, the function $t \to \|\mathscr{A}(y(.,t), y_t(.,t), z(.,x), \xi(t))\|_{\mathscr{H}}$ is decreasing.
(c) *For any initial data $(y_0, y_1, \chi(-\tau\cdot), \xi_0) \in \mathscr{H}$, the problem admits a unique weak solution*
$(y(.,t), y_t(.,t), z(.,t), \xi(t))$ such that

$$y \in C^1(0,\infty; L^2(0,1)) \cap C(0,\infty; H^1(0,1)), \quad z \in C(0,\infty; L^2(0,1)),$$
$$\xi \in L^\infty(0,\infty; \mathbb{R}).$$

Proof We first prove that $\mathscr{A}$ is monotone (see Definition 2.4.1). To this end, let (y_1, v_1, z_1, ξ_1), $(y_2, v_2, z_2, \xi_2) \in \mathscr{D}(\mathscr{A})$. One can readily check that

$$\begin{aligned}
&\langle \mathscr{A}(y_1, v_1, z_1, \xi_1) - \mathscr{A}(y_2, v_2, z_2, \xi_2), (y_1, v_1, z_1, \xi_1) - (y_2, v_2, z_2, \xi_2)\rangle_{\mathscr{H}} = \\
&\frac{K}{2}\left[(z_1(1) - z_2(1))^2 - (v_1(0) - v_2(0))^2\right] \\
&-\alpha(v_1(0) - v_2(0))\Big(f(z_2(1)) - f(z_1(1))\Big) \\
&+\beta(v_1(0) - v_2(0))\Big(g(v_1(0)) - g(v_2(0))\Big).
\end{aligned}$$

Invoking *(H1)* and Young's inequality, we obtain

$$\alpha\Big(f(z_2(1)) - f(z_1(1))\Big)(v_1(0) - v_2(0)) \le \frac{\alpha}{2}\kappa_1(z_1(1) - z_2(1))^2 + \frac{\alpha}{2}\kappa_1(v_1(0) - v_2(0))^2. \tag{4.8.12}$$

On the other hand, *(H2)* yields

$$\beta\,(g(v_1(0)) - g(v_2(0)))\,(v_1(0) - v_2(0)) \ge \beta r_1(v_1(0) - v_2(0))^2. \tag{4.8.13}$$

Inserting the estimates (4.8.12) and (4.8.13) in the last identity, it follows that

$$\langle \mathscr{A}(y_1, v_1, z_1, \xi_1) - \mathscr{A}(y_2, v_2, z_2, \xi_2), (y_1, v_1, z_1, \xi_1) - (y_2, v_2, z_2, \xi_2)\rangle_{\mathscr{H}}$$

$$\ge \frac{K - \alpha\kappa_1}{2}(z_1(1) - z_2(1))^2 + \left(\beta r_1 - \frac{K + \alpha\kappa_1}{2}\right)(v_1(0) - v_2(0))^2.$$

In view of (4.8.6)–(4.8.7), we conclude that the operator $\mathscr{A}$ is monotone.

Next, the immediate task is to show that $-\mathscr{A}$ is maximal, that is, $(I + \mathscr{A})$ is onto $\mathscr{H}$. To proceed, let $(p_1, p_2, p_3, p_4) \in \mathscr{H}$, and we seek $(y, v, z, \xi) \in \mathscr{D}(\mathscr{A})$ such that

$$\begin{cases} v = y - p_1, \\ v - (ay_x)_x = p_2, \\ z_\rho + \tau z = \tau p_3, \\ (ay_x)(0) = \sigma y(0) + \alpha f(z(1)) + \beta g(v(0)), \\ M\xi + (ay_x)(1) = Mp_4, \\ z(0) = v(0),\ \xi = v(1). \end{cases} \tag{4.8.14}$$

Whereupon, the z equation gives

$$z(x) = e^{-\tau x}(y(0) - p_1(0)) + \tau \int_0^x e^{-\tau(x-s)} p_3(s)\, ds, \tag{4.8.15}$$

and hence,

$$z(1) = e^{-\tau}(y(0) - p_1(0)) + \tau \int_0^1 e^{-\tau(1-s)} p_3(s)\, ds. \tag{4.8.16}$$

Hence, it amounts to finding y in $H^2(0, 1)$ as a solution of the elliptic problem

$$\begin{cases} y - (ay_x)_x = p_1 + p_2, \\ (ay_x)(0) = \sigma y(0) + \alpha f\left(e^{-\tau}(y(0) - p_1(0)) + \tau \int_0^1 e^{-\tau(1-s)} p_3(s)\, ds\right) \\ -\beta g(y(0) - p_1(0)), \\ My(1) + (ay_x)(1) = M(p_4 + p_1(1)). \end{cases} \tag{4.8.17}$$

To fulfill this task, we shall use the convex minimum method [187]. First, consider the functional $J(\cdot)$ on $H^1(0, 1)$

$$\begin{aligned} J(\phi) = \frac{1}{2}\left\{\int_0^1 \left[\phi^2 + a\phi_x^2\right] dx + \sigma\phi^2(0) + M\phi^2(1)\right\} - \int_0^1 (p_1 + p_2)\,\phi\, dx \\ +\alpha F\left(e^{-\tau}(\phi(0) - p_1(0)) + \tau \int_0^1 e^{-\tau(1-s)} p_3(s)\, ds\right) \\ +\beta G(\phi(0) - p_1(0)) - M[p_1(1) + p_4]\phi(1), \end{aligned}$$

where

$$F(s) = \int_0^s f(r)\, dr \quad \text{and} \quad G(s) = \int_0^s g(r)\, dr. \tag{4.8.18}$$

By virtue of *(H1)* and *(H2)*, the functions F and G are convex, nonnegative, and continuously differentiable on $\mathbb{R}$. This implies that $J(\cdot)$ is convex and coercive. In turn, it is easy to verify that $J(\cdot) : H^1(0, 1) \to \mathbb{R}$ is continuous. Consequently, there exists a unique function $y \in H^1(0, 1)$ such that for any $\phi \in H^1(0, 1)$, we have (see [187, Theorem 25. E., p. 516])

$$\begin{aligned} &\int_0^1 (y\phi + ay_x\phi_x - (p_1 + p_2)\phi)\, dx \\ &+ \left[\sigma y(0) + \alpha f\left(e^{-\tau}(y(0) - p_1(0)) + \tau \int_0^1 e^{-\tau(1-s)} p_3(s)\, ds\right)\right. \\ &\left. +\beta g(y(0) - p_1(0))\right]\phi(0) \\ &+ [My(1) - M(p_4 + p_1(1))]\,\phi(1) = 0. \end{aligned} \tag{4.8.19}$$

Taking ϕ in $\mathscr{D}(0, 1)$, (4.8.19) yields

$$\int_0^1 (y\phi + ay_x\phi_x - (p_1 + p_2)\phi)\, dx = 0, \tag{4.8.20}$$

and hence, $y \in H^2(0,1)$ satisfies the equation $y - (ay_x)_x - p_1 + p_2 = 0$. This, together with an integration by parts in (4.8.19), gives

$$\begin{cases} \sigma y(0) + \alpha f\left(e^{-\tau}(y(0) - p_1(0)) + \tau \displaystyle\int_0^1 e^{-\tau(1-s)} p_3(s)\, ds\right) \\ +\beta g(y(0) - p_1(0)) = (ay_x)(1), \\ M(p_4 + p_1(1)) - My(1) = (ay_x)(1). \end{cases} \tag{4.8.21}$$

Consequently, y is a solution of (4.8.17), and the operator $-\mathscr{A}$ is maximal monotone in the Hilbert space $\mathscr{H}$. Lastly, since $f(0) = g(0) = 0$ and $\mathscr{A}$ is maximal monotone, we conclude that $D(\mathscr{A})$ in $\mathscr{H}$. This achieves the proof of the first assertion (a).

The proof of the other assertions directly follows from the semigroups theory of nonlinear operators [39] (see also Sect. 2.4). □

4.8.3 *Stability of the Solutions to the System (4.8.8)*

This subsection is intended to study the long-time behavior of the solutions to (4.8.8) despite the presence of the nonlinear delayed boundary. First of all, we assume in this subsection that the conditions (4.1.5), *(H1)–(H2)*, and (4.8.6)–(4.8.7) are fulfilled so that the outcomes of Theorem 4.8.1 are valid.

4.8.3.1 Asymptotic Stability

We start by using Lasalle's invariance principle (see Theorem 2.4.3) to show that the solutions to (4.8.8) are strongly asymptotically stable. The result is stated below:

Theorem 4.8.2 *Given an initial data* $(y_0, y_1, \chi(-\tau\cdot), \xi_0) \in \mathscr{H}$, *the unique solution* $S(t)(y_0, y_1, \chi(-\tau\cdot), \xi_0)$ *of (4.8.11) is strongly asymptotically stable in* $\mathscr{H}$. *Thereby, the energy* $E_0(t)$ *defined by (4.8.4) asymptotically tends to zero in* $\mathscr{H}$ *as* $t \longrightarrow +\infty$.

Proof Thanks to a standard density argument and the contraction of the semigroup $S(t)$, we just need to prove the result for an initial data $\Psi_0 = (y_0, y_1, \chi(-\tau\cdot), \xi_0) \in D(\mathscr{A})$ (see [167] for more details). Then, by virtue of Theorem 4.8.1, that the trajectory of solution $\{(y(\cdot,t), y_t(\cdot,t), y_t(0, t-\tau\cdot), \xi(t))\}_{t\geq 0}$ is bounded for the graph norm. On the other hand, Sobolev embedding permits to claim that the canonical embedding $i : D(\mathscr{A}) \to \mathscr{H}$ is compact, where $D(\mathscr{A})$ is endowed with the graph norm. Thenceforth, the set $\{(y(\cdot,t), y_t(\cdot,t), y_t(0, t-\tau\cdot), \xi(t))\}_{t\geq 0}$

is precompact in $\mathscr{H}$. Thanks to Lasalle's invariance principle for nonlinear systems (see Theorem 2.4.3), we deduce that the ω-limit set defined by

$$\omega(\psi_0) = \left\{ \psi = (y, v, z, \xi) \in \mathscr{H};\ \psi = \lim_{t \to +\infty} S(t_n)\psi_0, \text{ for some } t_n \to +\infty \right\}$$

is non-empty. Furthermore, $\omega(\psi_0) \subset D(\mathscr{A})$ is compact, invariant under the semigroup $(S(t))_{t \geq 0}$, and

$$S(t)\psi_0 \longrightarrow \omega(\psi_0), \qquad \text{as } t \to +\infty.$$

Clearly, it remains to show that $\omega(\psi_0) = \{0_{\mathscr{H}}\}$. To do so, we pick up $\widehat{\psi}_0 = (\widehat{y}_0, \widehat{y}_1, \widehat{\chi}(-\tau \cdot), \widehat{\xi}_0) \in \omega(\psi_0)$. Whereupon, the element $\widehat{\psi}(t) := (\widehat{y}(\cdot, t), \widehat{y}_t(\cdot, t), \widehat{y}_t(0, t - \tau \cdot), \widehat{\xi}(t)) = S(t)\widehat{\psi}_0 \subset \omega(\psi_0)$ is a solution in the strong sense of the following problem:

$$\begin{cases} \widehat{y}_{tt}(x,t) - (a\widehat{y}_x)_x(x,t) = 0, & 0 < x < 1,\ t > 0, \\ (a\widehat{y}_x)(0,t) = \sigma \widehat{y}(0,t) + \alpha f(\widehat{y}_t(0, t - \tau)) + \beta g(\widehat{y}_t(0,t)), & t > 0, \\ M\widehat{y}_{tt}(1,t) + (a\widehat{y}_x)(1,t) = 0, & t > 0, \\ (\widehat{y}(.,0), \widehat{y}_t(.,0), \widehat{y}_t(0, -\tau), \widehat{\xi}(0)) = \widehat{\psi}_0 \in \omega(\psi_0), & \text{on } \in (0,1). \end{cases} \tag{4.8.22}$$

In turn, thanks to Theorem 2.1.3 in [114], we have

$$\frac{d}{dt} \left\| \widehat{\psi}(t) \right\|_{\mathscr{H}} = 0,$$

and hence, the energy $E_0(t)$ corresponding to the solution $\widehat{\psi}(t)$ is constant, for all $t \geq 0$. This, together with (4.8.5)–(4.8.7), implies that $\widehat{y}_t(0, t - \tau) = 0$ and $\widehat{y}_t(0, t) = 0$, for all $t \geq 0$. Inserting these findings in (4.8.22), we conclude that $\widehat{y}$ is a solution of

$$\begin{cases} \widehat{y}_{tt}(x,t) - (a\widehat{y}_x)_x(x,t) = 0, & 0 < x < 1,\ t > 0, \\ (a\widehat{y}_x)(0,t) = \sigma \widehat{y}(0,t), & t > 0, \\ M\widehat{y}_{tt}(1,t) + (a\widehat{y}_x)(1,t) = 0, & t > 0, \\ \widehat{y}_t(0,t) = 0, & t > 0. \end{cases} \tag{4.8.23}$$

In light of [167, Lemma 4.2], the problem (4.8.23) admits only the trivial solution $\widehat{y} = 0$, for all $t \geq 0$, and hence, one can deduce that $\widehat{\psi}_0 = 0$. This achieves the proof of the theorem. □

4.8.3.2 Energy Decay Estimates

Owing to (4.8.5) and (4.8.6)–(4.8.7), we have:

Lemma 4.8.1 *The energy functional $E_0(t)$ is nonincreasing and satisfies*

$$\dot{E}_0(t) \leq \frac{\alpha\kappa_1 - K}{2} y_t^2(0, t-\tau) + \left(\frac{K + \alpha\kappa_1}{2} - \beta r_1\right) y_t^2(0, t). \quad (4.8.24)$$

The following result will be useful:

Lemma 4.8.2 *Define the functional*

$$G(t) = 2\int_0^1 \Theta(x) y_t(x, t) y_x(x, t) dx, \quad (4.8.25)$$

where the function $\Theta(x)$ is defined by [167]

$$\begin{cases} \Theta_x + \Theta \left|\dfrac{a_x}{a}\right| = 1, \; 0 < x < 1, \\ \Theta(1) = -M. \end{cases}$$

Then the functional $G(t)$ satisfies for any $t > 0$,

$$\begin{aligned} \dot{G}(t) \leq & -\int_0^1 y_t^2 dx - \int_0^1 a(x) y_x^2 dx - M y_t^2(1, t) - \Theta(0) y_t^2(0, t) \\ & + \mu\sigma^2 \frac{\Theta(0)}{a(0)} y^2(0, t) - 2\mu\sigma \frac{\Theta(0)}{a(0)} y(0, t)(a y_x)(0, t) \\ & - 2\beta^2 \mu \frac{\Theta(0)}{a(0)} g^2(y_t(0, t)) - 2\alpha^2 \mu \frac{\Theta(0)}{a(0)} f^2(y_t(0, t-\tau)), \quad (4.8.26) \end{aligned}$$

where $\mu \geq 1$ is a constant to be determined later.

Proof Although the arguments used here run on the same lines as in [167], with some changes born out of necessity due to the delay term, we shall provide some details for the sake of completeness.

First, we know from [167] that Θ satisfies the following properties:

$$\begin{cases} \Theta_x = 1 - \Theta \left|\dfrac{a_x}{a}\right| \geq 1, \\ a\left(\dfrac{\Theta}{a}\right)_x \geq 1, \\ \Theta(0) \leq -M, \end{cases} \quad (4.8.27)$$

and whence

$$\Theta(1)(ay_x^2)(1,t) = -M(ay_x^2)(1,t) \le 0, \;\; \Theta(1)y_t^2(1,t) = -My_t^2(1,t), \tag{4.8.28}$$

$$-\int_0^1 a\left(\frac{\Theta}{a}\right)_x ay_x^2\,dx \le -\int_0^1 ay_x^2\,dx, \;\; -\int_0^1 \Theta_x y_t^2\,dx \le -\int_0^1 y_t^2\,dx. \tag{4.8.29}$$

By virtue of (4.8.8) and (4.8.28)–(4.8.29), we have

$$\begin{aligned}\dot{G}(t) &= 2\int_0^1 \Theta(x)(a(x)y_x)_x y_x\,dx + 2\int_0^1 \Theta y_t y_{xt}\,dx\\ &= \Theta(1)(ay_x^2)(1,t) - \Theta(0)(ay_x^2)(0,t) - \int_0^1 a\left(\frac{\Theta}{a}\right) ay_x^2\,dx\\ &\quad +\Theta(1)y_t^2(1,t) - \Theta(0)y_t^2(0,t) - \int_0^1 \Theta_x y_t^2\,dx\\ &\le -\int_0^1 y_t^2\,dx - \int_0^1 a(x)y_x^2\,dx - My_t^2(1,t)\\ &\quad -\Theta(0)y_t^2(0,t) - \mu\Theta(0)(ay_x^2)(0,t).\end{aligned} \tag{4.8.30}$$

From the boundary condition at $x = 0$, we infer

$$\begin{aligned}&(ay_x^2)(0,t) + \frac{\sigma^2}{a(0)}y^2(0,t) - 2\frac{\sigma}{a(0)}(ay_x)(0,t)y(0,t)\\ &= \frac{1}{a(0)}\left(\beta g(y_t(0,t)) + \alpha f(y_t(0,t-\tau))\right)^2\\ &\le 2\frac{\beta^2}{a(0)}g^2(y_t(0,t)) + 2\frac{\alpha^2}{a(0)}f^2(y_t(0,t-\tau)).\end{aligned} \tag{4.8.31}$$

Inserting (4.8.31) into (4.8.30), we can get the desired estimate (4.8.26). □

Now, inspired by [167], we define the functional $J(t)$ as follows:

$$J(t) = \varpi y(0,t)\int_0^1 y_t dx + \varpi M y(0,t)y_t(1,t), \quad t > 0, \tag{4.8.32}$$

where ϖ is a positive constant to be determined. Arguing as in [167], we can get the following lemma.

Lemma 4.8.3 *The functional $J(t)$ satisfies the following:*

$$\dot{J}(t) \le \frac{1}{2}\int_0^1 y_t^2 dx + \frac{M}{2}y_t^2(1,t) + \frac{\varpi^2}{2}(1+M)y_t^2(0,t)$$

$$- \varpi y(0,t)(ay_x)(0,t), \text{ for any } t > 0. \tag{4.8.33}$$

Combining (4.8.24), (4.8.34), and (4.8.36), we have:

Lemma 4.8.4 *There exists a constant* $C_i > 0$ $(i = 1, 2, 3)$ *such that for any* $t > 0$, *we have*

$$\begin{aligned}\dot{G}(t) + \dot{J}(t) \le & -E_0(t) + C_1 y_t^2(0,t) + C_2 g^2(y_t(0,t)) + C_3 f^2(y_t(0,t-\tau)) \\ & + \frac{K}{2}\tau \int_0^1 y_t^2(0, t-\tau\rho)\, d\rho.\end{aligned} \tag{4.8.34}$$

Proof Noting (4.8.4), it follows from (4.8.26) and (4.8.33) that

$$\begin{aligned}\dot{G}(t) + \dot{J}(t) \le & -E_0(t) + \left[-\Theta(0) + \frac{\varpi^2}{2}(1+M)\right] y_t^2(0,t) \\ & + \left[\frac{\sigma}{2} + \mu\sigma^2 \frac{\Theta(0)}{a(0)}\right] y^2(0,t) \\ & - \left[\varpi + 2\mu\sigma \frac{\Theta(0)}{a(0)}\right] y(0,t)(ay_x)(0,t) + \frac{K}{2}\tau \int_0^1 y_t^2(0, t-\tau\rho) \\ & - 2\beta^2\mu \frac{\Theta(0)}{a(0)} g^2(y_t(0,t)) - 2\alpha^2\mu \frac{\Theta(0)}{a(0)} f^2(y_t(0,t-\tau)).\end{aligned}$$

At this stage, we take

$$\mu = \max\left\{1, -\frac{a(0)}{2\sigma\Theta(0)}\right\}, \quad \varpi = -2\mu\sigma\frac{\Theta(0)}{a(0)}.$$

This gives

$$\mu \ge 1, \quad \frac{\sigma}{2} + \mu\sigma^2\frac{\Theta(0)}{a(0)} \le 0,$$

and

$$\varpi + 2\mu\sigma\frac{\Theta(0)}{a(0)} = 0.$$

Therefore, we get

$$\begin{aligned}\dot{G}(t) + \dot{J}(t) \le & -E_0(t) + C_1 y_t^2(0,t) + C_2 g^2(y_t(0,t)) + C_3 f^2(y_t(0,t-\tau)) \\ & + \frac{K}{2}\tau \int_0^1 y_t^2(0, t-\tau\rho),\end{aligned}$$

with

$$C_1 = -\Theta(0) + 2\left(\mu\sigma\frac{\Theta(0)}{a(0)}\right)^2 (1+M) > 0,$$

and

$$C_2 = -2\beta^2\mu\frac{\Theta(0)}{a(0)} > 0, \quad C_3 = -2\alpha^2\mu\frac{\Theta(0)}{a(0)} > 0.$$

□

Subsequently, we have, as in [156], the following result:

Lemma 4.8.5 *Define the functional $I_0(t)$ by*

$$I_0(t) = \tau \int_0^1 e^{-\delta\tau\rho} y_t^2(0, t-\tau\rho)\, d\rho, \tag{4.8.35}$$

where δ is a positive constant. Then the functional $I_0(t)$ satisfies for any $t > 0$,

$$\dot{I}_0(t) = -\delta I(t) - e^{-\delta\tau} y_t^2(0, t-\tau) + y_t^2(0, t). \tag{4.8.36}$$

Proof It suffices to use $(4.8.8)_4$ and integrate by parts. □

Now, we are able to state our second main result.

Theorem 4.8.3 *Given an initial data $(y_0, y_1, \chi(-\tau\cdot), \xi_0) \in \mathscr{H}$, consider the energy $E_0(t)$ defined by (4.8.4) corresponding to the unique solution $S(t)(y_0, y_1, \chi(-\tau\cdot), \xi_0)$ of (4.8.11). Then, we have:*
If there exists a positive real constant r_2 such that

$$|g(s)| \le r_2|s|, \ \forall s \in \mathbb{R}, \tag{4.8.37}$$

then given $M > 1$, it follows that

$$E_0(t) \le M e^{-\omega t} E_0(0), \quad \forall t \ge 0,$$

for some positive constant ω independent of $E_0(0)$.

Proof Once again, we prove the result for regular solutions, and then in view of a density argument and the contraction of the semigroup, one can extend the result to weak solutions. Thereafter, recalling (4.8.25), (4.8.32), and (4.8.35), we introduce the perturbed energy similarly to [126]

$$\mathscr{E}_\epsilon(t) = E_0(t) + \epsilon\left(G(t) + J(t) + m I_0(t)\right), \tag{4.8.38}$$

where m is a positive constant. It is simple to check the existence of a positive constant B independent of $E_0(0)$ such that

$$|G(t) + J(t) + mI_0(t)| \leq BE_0(t), \quad \text{for any } t \geq 0.$$

This, together with (4.8.38) and the fact that $E_0(t)$ is decreasing, implies that for any $M > 1$

$$\mathscr{E}_\epsilon(t) \leq \sqrt{M} E_0(t) \leq M \mathscr{E}_\epsilon(t), \tag{4.8.39}$$

provided that $\epsilon \leq \frac{\sqrt{M}-1}{B\sqrt{M}}$.

Differentiating (4.8.38) and using (4.8.24), (4.8.34), and (4.8.36), we obtain

$$\begin{aligned}
\dot{\mathscr{E}}_\epsilon(t) \leq & -\epsilon E_0(t) + \epsilon\Big((m + C_1)y_t^2(0,t) + C_2 g^2(y_t(0,t))\Big) \\
& + \tau\epsilon \left(\int_0^1 \left[\frac{K}{2} - m\delta e^{-\delta\tau\rho}\right] y_t^2(0, t - \tau\rho)\, d\rho\right) \\
& + \epsilon \left(C_3 f^2(y_t(0, t-\tau)) - m e^{-\delta\tau} y_t^2(0, t-\tau)\right) \\
& + \frac{\alpha\kappa_1 - K}{2} y_t^2(0, t-\tau) + \left(\frac{K + \alpha\kappa_1}{2} - \beta r_1\right) y_t^2(0,t).
\end{aligned}$$

Invoking *(H1)*, the last estimate becomes

$$\begin{aligned}
\dot{\mathscr{E}}_\epsilon(t) \leq & -\epsilon E_0(t) + \epsilon\Big((m + C_1)y_t^2(0,t) + C_2 g^2(y_t(0,t))\Big) \\
& + \epsilon\tau\epsilon \left(\int_0^1 \left[\frac{K}{2} - m\delta e^{-\delta\tau\rho}\right] y_t^2(0, t - \tau\rho)\, d\rho\right) \\
& + \epsilon \left(C_3 \kappa_1^2 - m e^{-\delta\tau}\right) y_t^2(0, t-\tau) \\
& + \frac{\alpha\kappa_1 - K}{2} y_t^2(0, t-\tau) + \left(\frac{K + \alpha\kappa_1}{2} - \beta r_1\right) y_t^2(0,t).
\end{aligned} \tag{4.8.40}$$

Now, in view of (4.8.37), we deduce from (4.8.40) that

$$\begin{aligned}
\dot{\mathscr{E}}_\epsilon(t) \leq & -\epsilon \left(E_0(t)\right)^{\frac{p+1}{2}} + \left(\frac{K + \alpha\kappa_1}{2} - \beta r_1 + \epsilon(m + C_1 + C_2 r_2^2)\right) y_t^2(0,t) \\
& + \left(\frac{K + \alpha\kappa_1}{2} \epsilon(C_3\kappa_1^2 - m e^{-\delta\tau})\right) y_t^2(0, t-\tau) \\
& + \epsilon\tau \left((K\tau/2) - m\delta\tau e^{-\delta\tau}\right) \int_0^1 y_t^2(0, t-\tau\rho)\, d\rho.
\end{aligned} \tag{4.8.41}$$

Choosing m sufficiently large so that

$$m \geq e^{\delta\tau} \max\left\{C_3\kappa_1^3, \frac{K}{2}\right\}$$

and then ϵ small enough such as

$$\epsilon \leq \frac{m + C_1 + C_2 r_2^2}{\beta r_1 - \frac{K+\alpha\kappa_1}{2}},$$

one can obtain the desired result. □

In the second part of this subsection, we shall obtain more general energy decay estimates by using convexity arguments. First, we assume that:

(A1) The function $f : \mathbb{R} \to \mathbb{R}$ is an odd function of class $C^1(\mathbb{R})$. Additionally, there exist constants $0 < \nu_1 < 1$ and $\nu_2 > 0$ such that

$$\nu_1 s f(s) \leq F(s) \leq \nu_2 s g(s), \tag{4.8.42}$$

where

$$F(s) = \int_0^s f(r)dr.$$

(A2) The function $g \in C(\mathbb{R})$ is a monotone increasing function. There exist constants $c_1, c_2 > 0$ and a strictly increasing odd function $h \in C(\mathbb{R})$ such that

$$\begin{cases} h(|s|) \leq |g(s)| \leq h^{-1}(|s|), & \text{for all } |s| \leq \epsilon, \\ c_1|s| \leq |g(s)| \leq c_2|s|, & \text{for all } |s| \geq \epsilon. \end{cases}$$

In addition, we assume that there exists a constant $r_0 > 0$ such that $H(x)$ defined by

$$H(x) = \sqrt{x}h(\sqrt{x}) \tag{4.8.43}$$

is a strictly convex function on $[0, r_0^2]$, for $r_0 \in (0, 1]$.

Remark 4.8.1

(1) The function H defined above is introduced in [7].
(2) Define the function $F(y)$ by

$$L(y) = \frac{H^*(y)}{y}, \quad y \in (0, H'(r_0^2)],$$

where H^* is the convex conjugate function of H defined by

$$H^*(y) = \sup_{x \in \mathbb{R}}(xy - H(x)).$$

By Alabau-Boussouira [7, Proposition 2.2], we know that L is a strictly increasing and continuous function. In addition,

$$L(y) = \frac{H^*(y)}{y} \leq (H')^{-1}(y), \quad y \in (0, H'(r_0^2)]. \tag{4.8.44}$$

Finally, the weight of weak dissipation and the delay terms satisfy

$$\alpha < \frac{\beta \nu_1}{\nu_2}. \tag{4.8.45}$$

In turn, the energy $E(t)$ of the problem (4.8.8) can be defined by

$$\begin{aligned} E(t) = &\frac{1}{2}\int_0^1 \left(y_t^2(x,t) + a(x)y_x^2(x,t)\right) dx + \mathscr{K}\tau \int_0^1 F\left(y_t(0, t - \tau\rho)\right) d\rho \\ &+\frac{1}{2}\left[\sigma y^2(0,t) + M y_t^2(1,t)\right], \end{aligned} \tag{4.8.46}$$

where $F(s) = \int_0^s f(r)dr$ and the constant $\mathscr{K}$ satisfies (4.8.48). The reader can check that this new energy is equivalent to $E_0(t)$ (see (4.8.4)). Furthermore, we assume that the solution to (4.8.8) is stemmed from a regular initial condition so that the computations are valid.

Multiplying the y equation in (4.8.8) by y_t and integrating the result over $(0, 1)$ with respect to x, we get

$$\begin{aligned} &\frac{1}{2}\frac{d}{dt}\int_0^1 (y_t^2 + a(x)y_x^2)dx + \frac{1}{2}\frac{d}{dt}[M y_t^2(1,t) + \sigma y^2(0,t)] \\ &\quad + \beta g(y_t(0,t))y_t(0,t) + \alpha f(z(1,t))y_t(0,t) = 0. \end{aligned} \tag{4.8.47}$$

Similarly, we multiply the z equation of (4.8.8) by $\mathscr{K} f(z(\rho, t))$, where $\mathscr{K}$ is a positive constant satisfying

$$\frac{\alpha(1-\nu_1)}{\nu_1} \leq \mathscr{K} \leq \frac{\beta - \alpha\nu_2}{\nu_2}. \tag{4.8.48}$$

Then, integrating the result on $(0, 1)$ with respect to ρ, and noting that $z(0,t) = y_t(0,t)$, we reach

$$\mathscr{K}\tau \frac{d}{dt}\int_0^1 F(z(\rho,t))d\rho = -\mathscr{K}(F(z(1,t)) - F(y_t(0,t))). \tag{4.8.49}$$

Combining (4.8.47) and (4.8.49) and using (4.8.42), we obtain

$$\dot{E}(t) \leq -(\beta - \mathscr{K}\nu_2)g(y_t(0,t))y_t(0,t) - \alpha f(z(1,t))y_t(0,t) - \mathscr{K}F(z(1,t)). \tag{4.8.50}$$

On the other hand, as in [28] (see also [7–9, 37, 45]), we introduce the conjugate function F^* of the convex function F, i.e., $F^*(s) = \sup_{t\in\mathbb{R}^+}(st - F(t))$. Then, for any $s \geq 0$,

$$F^*(s) = s(F')^{-1}(s) - F((F')^{-1}(s))$$

is the Legendre transform of F, and satisfies the following Young's inequality [186]: For any $s, t \geq 0$,

$$st \leq F^*(s) + F(t).$$

According to the definition of F, we have

$$F^*(s) = sf^{-1}(s) - F(f^{-1}(s)),$$

which yields

$$F^*(f(z(1,t))) = f(z(1,t))z(1,t) - F(z(1,t)).$$

Therefore,

$$\begin{aligned} f(z(1,t))y_t(0,t) &\leq F^*(f(z(1,t))) + F(y_t(0,t)) \\ &= f(z(1,t))z(1,t) - F(z(1,t)) + F(y_t(0,t)). \end{aligned} \tag{4.8.51}$$

Substituting (4.8.51) into (4.8.50), we get

$$\begin{aligned} \dot{E}(t) \leq &-(\beta - \mathscr{K}\nu_2)g(y_t(0,t))y_t(0,t) + (\alpha - \mathscr{K}\nu_1)f(z(1,t))z(1,t) \\ &-\alpha F(z(1,t)) + \alpha F(y_t(0,t)). \end{aligned}$$

Combining the latter with (4.8.42), it follows that

$$\begin{aligned} \dot{E}(t) \leq &-(\beta - \mathscr{K}\nu_2 - \alpha\nu_2)g(y_t(0,t))y_t(0,t) \\ &-(\mathscr{K}\nu_1 - \alpha + \alpha\nu_1)f(z(1,t))z(1,t). \end{aligned} \tag{4.8.52}$$

Due to the range of $\mathscr{K}$ and the relation of α and β (see (4.8.48)), we know

$$\beta - \mathscr{K}\nu_2 - \alpha\nu_2 \geq 0 \text{ and } \mathscr{K}\nu_1 - \alpha + \alpha\nu_1 \geq 0,$$

from which (4.8.52) leads to

$$\dot{E}(t) \le 0. \tag{4.8.53}$$

Our third main outcome is:

Theorem 4.8.4 *Suppose (A1)–(A2) hold. Then, there exist constants $\varepsilon_0 > 0$, $\lambda > 0$, $\gamma_1 > 0$, and $\gamma_2 > 0$ such that the energy $E(t)$ satisfies for any $t > 0$,*

$$E(t) \le \lambda H_2^{-1}(\gamma_1 t + \gamma_2), \tag{4.8.54}$$

where

$$H_2(t) = \int_t^1 \frac{1}{H_1(s)} ds, \quad \text{and} \quad H_1(t) = tH'(\varepsilon_0 t).$$

Here, H_1 is a strictly decreasing convex function on $(0, r_0^2]$, with $\lim_{t\to 0} H_1(t) = +\infty$.

For sake of clarity, the proof of Theorem 4.8.4 will be achieved through several steps by using the general approach due to [7]. Indeed, numerous technical results will be needed.

First, by virtue of (4.8.52) and (4.8.53), the energy functional $E(t)$ is nonincreasing and satisfies

$$\dot{E}(t) \le -(\beta - \mathscr{K}\nu_2 - \alpha\nu_2) g(y_t(0,t)) y_t(0,t) - (\mathscr{K}\nu_1 - \alpha + \alpha\nu_1) f(z(1,t)) z(1,t). \tag{4.8.55}$$

Thereafter, arguing as in Lemma 4.8.1 and Lemma 4.8.2, we have

$$\begin{aligned} \dot{G}(t) + \dot{J}(t)) \le & -E(t) + C_1 y_t^2(0,t) + C_2 g^2(y_t(0,t)) + C_3 f^2(z(1,t)) \\ & + \mathscr{K}\tau \int_0^1 F(z(\rho,t)) d\rho, \end{aligned} \tag{4.8.56}$$

where $C_i > 0$ $(i = 1, 2, 3)$ is defined in Lemma 4.8.1. In fact, it suffices to use (4.8.46) along with (4.8.26) and (4.8.33).

The first technical lemma is:

Lemma 4.8.6 *Define the functional $I(t)$ by*

$$I(t) = \int_0^1 e^{-2\tau\rho} F(z(\rho,t)) d\rho.$$

Then, we have

$$\dot{I}(t) \leq -2I(t) - \frac{e^{-2\tau}}{\tau}\nu_1 f(z(1,t))z(1,t) + \frac{\nu_2}{\tau} g(y_t(0,t))y_t(0,t), \textit{ for all } t > 0. \tag{4.8.57}$$

Proof Using $(4.8.8)_4$, we see that

$$\begin{aligned}
\dot{I}(t) &= \int_0^1 e^{-2\tau\rho} f(z(\rho,t))z_t(\rho,t)\,d\rho \\
&= -\frac{1}{\tau}\int_0^1 e^{-2\tau\rho} f(z(\rho,t))z_\rho(\rho,t)\,d\rho \\
&= -\frac{1}{\tau}\int_0^1 \left[\frac{\partial}{\partial\rho}\left(e^{-2\tau\rho}F(z(\rho,t))\right) + 2\tau e^{-2\tau\rho}F(z(\rho,t))\right] d\rho \\
&= -\frac{1}{\tau}\left[e^{-2\tau}F(z(1,t)) - F(y_t(0,t))\right] - 2\int_0^1 e^{-2\tau\rho}F(z(\rho,t))d\rho \\
&= -2I(t) - \frac{e^{-2\tau}}{\tau}F(z(1,t)) + \frac{1}{\tau}F(y_t(0,t)),
\end{aligned}$$

which, along with (4.8.42), implies (4.8.57). □

Now, we define a functional by

$$\mathscr{L}(t) = NE(t) + (G(t) + J(t)) + N_1 I(t),$$

where N and N_1 are positive constants to be determined later. The second technical lemma is:

Lemma 4.8.7 *There exist positive constants $\beta_1 > 0$ and $\beta_2 > 0$ such that*

$$\beta_1 E(t) \leq \mathscr{L}(t) \leq \beta_2 E(t). \tag{4.8.58}$$

In addition, for any $t > 0$, we have

$$\begin{aligned}
\dot{\mathscr{L}}(t) \leq{} & -E(t) - \left[N(\beta - K\nu_2 - \alpha\nu_2) - \frac{\nu_2}{\tau}N_1\right] g(y_t(0,t))y_t(0,t) \\
& + C_1 y_t^2(0,t) + C_2 g^2(y_t(0,t)).
\end{aligned} \tag{4.8.59}$$

Proof In light of Young's and Hölder's inequalities, there exists a positive constant C_4 such that

$$|\mathscr{L}(t) - NE(t)| = |G(t) + J(t) + N_1 I(t)| \leq C_4 E(t),$$

which, taking $\beta_1 = N - C_4 > 0$ and $\beta_2 = N + C_4 > 0$ for large $N > 0$, yields (4.8.58).

Moreover, owing to (4.8.55) and (4.8.56)–(4.8.57), we have

$$\begin{aligned}\dot{\mathscr{L}}(t) \leq & -E(t) - \left[N(\beta - K\nu_2 - \alpha\nu_2) - \frac{\nu_2}{\tau}N_1\right] g(y_t(0,t))y_t(0,t)\\ & - \left[N(K\nu_1 - \alpha + \alpha\nu_1) + \frac{e^{-2\tau}}{\tau}\nu_1 N_1\right] f(z(1,t))z(1,t)\\ & - \int_0^1 (2N_1 e^{-2\tau\rho} - \mathscr{K}\tau)F(z(\rho,t))d\rho + C_1 y_t^2(0,t)\\ & + C_2 g^2(y_t(0,t)) + C_3 f^2(z(1,t)).\end{aligned}$$

In turn, using the fact that $|f(s)| \leq \kappa_1|s|$ (see *(H1)*), we have

$$f^2(s) \leq \kappa_1 s f(s).$$

Consequently, the above estimate of $\dot{\mathscr{L}}$ writes

$$\begin{aligned}\dot{\mathscr{L}}(t) \leq & -E(t) - \left[N(\beta - \mathscr{K}\nu_2 - \alpha\nu_2) - \frac{\nu_2}{\tau}N_1\right] g(y_t(0,t))y_t(0,t)\\ & - \left[N(\mathscr{K}\nu_1 - \alpha + \alpha\nu_1) + \frac{e^{-2\tau}}{\tau}\nu_1 N_1 - \kappa_1 C_3\right] f(z(1,t))z(1,t)\\ & - \int_0^1 (2N_1 e^{-2\tau\rho} - \mathscr{K}\tau)F(z(\rho,t))d\rho + C_1 y_t^2(0,t) + C_2 g^2(y_t(0,t)).\end{aligned} \tag{4.8.60}$$

Now, we choose $N_1 > 0$ sufficiently large so that

$$2N_1 e^{-2\tau\rho} - \mathscr{K}\tau > 0 \ \text{ and } \ \frac{e^{-2\tau}}{\tau}\nu_1 N_1 - \kappa_1 C_3 > 0,$$

and (4.8.59) follows. □

Proof of Theorem 4.8.4 We consider the following two cases.
Case I H is linear.
We infer from (4.8.59) that there exists $k_5 > 0$ such that

$$\dot{\mathscr{L}}(t) \leq -E(t) - \left[N(\beta - \mathscr{K}\nu_2 - \alpha\nu_2) - \frac{\nu_2}{\tau}N_1 - k_5 C_1 - k_5 C_2\right] g(y_t(0,t))y_t(0,t).$$

Taking $N > 0$ large enough so that

$$N(\beta - \mathcal{K}\nu_2 - \alpha\nu_2) - \frac{\nu_2}{\tau}N_1 - k_5C_1 - k_5C_2 > 0,$$

then we obtain

$$\dot{\mathcal{L}}(t) \leq -E(t).$$

In view of (4.8.58), we have

$$E(t) \leq \beta_2 E(0)e^{-\frac{1}{\beta_1}t}.$$

This result is in accordance with Theorem 4.8.3.
Case II H is nonlinear over $[0, \epsilon]$.
It follows from (4.8.59) that there exists a constant $c > 0$ such that

$$\dot{\mathcal{L}}(t) \leq -E(t) + c(y_t^2(0,t) + g^2(y_t(0,t))). \tag{4.8.61}$$

Now, we choose $\varepsilon_1 = \min\{r_0, h(r_0)\}$. Assume that $\max\{r_0, h(r_0)\} < \epsilon$ and

$$\frac{h(\varepsilon_1)}{\epsilon}|s| \leq \frac{h(|s|)}{|s|}|s| \leq |g(s)| \leq \frac{h^{-1}(|s|)}{|s|}|s| \leq \frac{h^{-1}(\epsilon)}{\varepsilon_1}|s|,$$

for all s satisfying $\varepsilon_1 \leq |s| \leq \epsilon$. Then (A2) changes

$$\begin{cases} h(|s|) \leq |g(s)| \leq h^{-1}(|s|), & \text{for all } |s| \leq \varepsilon_1, \\ c_1|s| \leq |g(s)| \leq c_2|s|, & \text{for all } |s| \geq \varepsilon_1. \end{cases} \tag{4.8.62}$$

Thereby

$$H(s^2) \leq sg(s) \text{ and } H(g^2(s)) \leq sg(s) \text{ for all } |s| \leq \varepsilon_1,$$

which implies

$$s^2 + g^2(s) \leq 2H^{-1}(sg(s)) \text{ for all } |s| \leq \varepsilon_1. \tag{4.8.63}$$

To estimate the term $y_t^2(0,t) + g^2(y_t(0,t))$, we invoke (4.8.62)–(4.8.63), to get

$$\begin{aligned} & y_t^2(0,t) + g^2(y_t(0,t)) \\ & = (y_t^2(0,t) + g^2(y_t(0,t)))1_{\{|y_t| \leq \varepsilon_1\}} + (y_t^2(0,t) + g^2(y_t(0,t)))1_{\{|y_t|| \geq \varepsilon_1\}} \\ & \leq 2H^{-1}(y_t(0,t)g(y_t(0,t)))1_{\{|y_t| \leq \varepsilon_1\}} + (y_t^2(0,t) + g^2(y_t(0,t)))1_{\{|y_t|| \geq \varepsilon_1\}}, \end{aligned}$$

which, along with (4.8.61), yields

$$\begin{aligned}\dot{\mathscr{L}}(t) &\leq -E(t) + 2cH^{-1}(y_t(0,t)g(y_t(0,t)))1_{\{|y_t|\leq\varepsilon_1\}} \\ &\quad + c(y_t^2(0,t) + g^2(y_t(0,t)))1_{\{|y_t||\geq\varepsilon_1\}} \\ &\leq -E(t) + 2cH^{-1}(y_t(0,t)g(y_t(0,t)))1_{\{|y_t|\leq\varepsilon_1\}} - cE'(t). \quad (4.8.64)\end{aligned}$$

Letting $\mathscr{L}_1(t) = \mathscr{L}(t) + cE(t) \sim E(t)$, we get from (4.8.64) that

$$\dot{\mathscr{L}}_1(t) \leq -E(t) + 2cH^{-1}(y_t(0,t)g(y_t(0,t)))1_{\{|y_t|\leq\varepsilon_1\}}. \quad (4.8.65)$$

In turn, we define the functional $\mathscr{L}_2(t)$ by

$$\mathscr{L}_2(t) = H'\left(\varepsilon_0\frac{E(t)}{E(0)}\right)\mathscr{L}_1(t) + rE(t),$$

where $0 < \varepsilon_0 < \epsilon$ and $r > 0$. Clearly, the functional $\mathscr{L}_2(t)$ is equivalent to $E(t)$.

On the other hand, noting that H', $H'' \geq 0$ over $(0, r_0^2]$ and $E'(t) \leq 0$, we obtain

$$\begin{aligned}\dot{\mathscr{L}}_2(t) &\leq -E(t)H'\left(\varepsilon_0\frac{E(t)}{E(0)}\right) \\ &\quad +2cH'\left(\varepsilon_0\frac{E(t)}{E(0)}\right)H^{-1}(y_t(0,t)g(y_t(0,t)))1_{\{|y_t|\leq\varepsilon_1\}} + rE'(t). \quad (4.8.66)\end{aligned}$$

Recall that the conjugate function H^*, of the convex function H, satisfies Young's inequality

$$AB \leq H^*(A) + H(B), \quad \text{for } A \in (0, H'(r_0^2)],\ B \in (0, r_0^2].$$

Now, we take

$$A = H'\left(\varepsilon_0\frac{E(t)}{E(0)}\right), \quad B = H^{-1}(y_t(0,t)g(y_t(0,t)))1_{\{|y_t|\leq\varepsilon_1\}}.$$

Since $\varepsilon_1 = \min\{r_0, h(r_0)\}$, we deduce that if $r_0 > h(r_0)$, then

$$y_t(0,t)g(y_t(0,t))1_{\{|y_t|\leq\varepsilon_1\}} \leq \varepsilon_1 h^{-1}(\varepsilon_1) = r_0 h(r_0) = H(r_0^2),$$

while, if $r_0 \leq h(r_0)$, then we have

$$y_t(0,t)g(y_t(0,t))1_{\{|y_t|\leq\varepsilon_1\}} \leq r_0 h^{-1}(r_0) \leq r_0 h(r_0) = H(r_0^2).$$

This shows that $B \in (0, r_0^2]$. Taking $\varepsilon_0 > 0$ sufficiently small and noting that E is decreasing, we can get $A \in (0, H'(r_0^2)]$.

Then, we infer from (4.8.66) and (4.8.44) that there exists a constant $c_1 > 0$ such that

$$\dot{\mathscr{L}}_2(t) \le -(E(0) - 2c\varepsilon_0)H'\left(\varepsilon_0 \frac{E(t)}{E(0)}\right)\frac{E(t)}{E(0)} - c_1 E'(t) + rE'(t). \quad (4.8.67)$$

Taking ε_0 small and r large to get from (4.8.67), there exists a constant $k > 0$ such that

$$\dot{\mathscr{L}}_2(t) \le -k\frac{E(t)}{E(0)}H'\left(\varepsilon_0 \frac{E(t)}{E(0)}\right). \quad (4.8.68)$$

Then, we denote

$$H_1(t) = tH'(\varepsilon_0 t).$$

Whereupon, it follows from (4.8.68) that

$$\dot{\mathscr{L}}_2(t) \le -kH_1\left(\frac{E(t)}{E(0)}\right).$$

Furthermore, given a constant $\theta_1 > 0$, let

$$\mathscr{L}_3(t) = \theta_1 \frac{\mathscr{L}_2(t)}{E(0)}.$$

It is easy to verify that

$$\mathscr{L}_3(t) \sim \mathscr{L}_2(t) \sim E(t). \quad (4.8.69)$$

Then, there exists a constant $\theta_2 > 0$ such that

$$\dot{\mathscr{L}}_3(t) \le -\theta_2 H_1(\mathscr{L}_3(t)),$$

which gives

$$\frac{d}{dt}H_2(\mathscr{L}_3(t)) \ge \theta_2, \quad (4.8.70)$$

where

$$H_2(s) = \int_s^1 \frac{1}{H_1(t)}dt.$$

Thus, we obtain from (4.8.70) that for any $t > 0$,

$$\mathscr{L}_3(t) \le H_2^{-1}(\theta_2 t + \theta_3), \tag{4.8.71}$$

where θ_3 is a constant. Lastly, (4.8.54) is an immediate consequence of (4.8.69) and (4.8.71). □

We end this section by giving some examples of g borrowed from [10, 37] to illustrate the energy decay rates.

Example 4.8.1 Let g be a function that satisfies

$$c_1 \min\{s, s^p\} \le |g(s)| \le c_2 \max\{s, s^{\frac{1}{p}}\}, \text{ on } (0, \epsilon],$$

where $c_1, c_2 > 0$ and $p \ge 1$. For $h(s) = cs^p$, then $h(s)$ is linear if $p = 1$, and hence, $f(s)$ and $g(s)$ are also linear. We can take $f(s) = c_1 s$, $g(s) = c_2 s$ for suitable constants $c_1, c_2 > 0$ that satisfy (A1) and (A2). Next, we can see that $H_2^{-1}(t) = e^{-ct}$. Then it follows from (4.8.54) that

$$E(t) \le \lambda e^{-c(\gamma_1 t + \gamma_2)}.$$

Now, if $p > 1$, then $h(s)$ is nonlinear on $(0, \epsilon]$. We can take $g(s) = c_1 s^{p-1}$ and $f(s) = c_2 s^p$ for suitable constants $c_1, c_2 > 0$ satisfying (A1)–(A2). Then, we can obtain $H_1(t) = ct^{\frac{p+1}{2}}$, and $H_2^{-1}(t) = (ct + 1)^{-\frac{2}{p-1}}$. We infer from (4.8.54) that

$$E(t) \le [c(\gamma_1 t + \gamma_2) + 1]^{-\frac{2}{p-1}}.$$

Example 4.8.2 Let $h(s) = cs^p(-\ln s)^q$, where $c > 0$, $p \ge 1$, $q \in \mathbb{R}$ on $(0, \epsilon]$. A direct calculation implies

$$H(s) = cs^{\frac{p+1}{2}}(-\ln\sqrt{s})^q.$$

Then, we have

$$H'(s) = cs^{\frac{p-1}{2}}(-\ln\sqrt{s})^{q-1}\left[\frac{p+1}{2}(-\ln\sqrt{s}) - \frac{q}{2}\right],$$

and consequently,

$$H_1(t) = cs^{\frac{p+1}{2}}(-\ln\sqrt{s})^{q-1}\left[\frac{p+1}{2}(-\ln\sqrt{s}) - \frac{q}{2}\right].$$

Thereby, if t is near 0, then we have

$$H_2(t) = c\int_t^1 \frac{1}{s^{\frac{p+1}{2}}(-\ln\sqrt{s})^{q-1}\left[\frac{p+1}{2}(-\ln\sqrt{s}) - \frac{q}{2}\right]}ds$$
$$= c\int_1^{\frac{1}{\sqrt{t}}} \frac{z^{p-2}}{(\ln z)^{q-1}\left(\frac{p+1}{2}\ln z - \frac{q}{2}\right)}dz.$$

In turn, in a neighborhood of 0, we have

$$H_2(t) = \begin{cases} \frac{c}{t^{\frac{p-1}{2}}(-\ln t)^q}, & \text{if } p > 1,\\ c(-\ln t)^{1-q}, & \text{if } p = 1,\ q < 1,\\ c[\ln(-\ln t)], & \text{if } p = 1,\ q = 1. \end{cases}$$

Hence, in the neighborhood of $+\infty$, we have

$$H_2^{-1}(t) = \begin{cases} ct^{-\frac{2}{p-1}}(\ln t)^{-\frac{2q}{p-1}}, & \text{if } p > 1,\\ ce^{-t^{\frac{1}{1-q}}}, & \text{if } p = 1,\ q < 1,\\ ce^{-e^t}, & \text{if } p = 1,\ q = 1. \end{cases}$$

We get from (4.8.54) that

$$E(t) \le \begin{cases} c(\gamma_1 t + \gamma_2)^{-\frac{2}{p-1}}(\ln(\gamma_1 t + \gamma_2))^{-\frac{2q}{p-1}}, & \text{if } p > 1,\\ ce^{-(\gamma_1 t+\gamma_2)^{\frac{1}{1-q}}}, & \text{if } p = 1,\ q < 1,\\ ce^{-e^{\gamma_1 t+\gamma_2}}, & \text{if } p = 1,\ q = 1. \end{cases}$$

4.8.4 Concluding Discussion

The main concern of this section is to deal with the stabilization of a flexible structure (overhead crane system) with a nonlinear boundary delayed control. Using general conditions on the nonlinearities involved in the control, the energy of the system is shown to decay thanks to convexity arguments introduced in [142] (see also [7–9]). This outcome generalizes the results obtained in [24–26, 70, 74, 76, 92–95, 146, 148, 167, 172, 185], where no delay occurs and also in [12, 17, 63–65, 184], where the suggested boundary feedback law consists of delayed and nondelayed *linear* controls.

It is noteworthy that the decay rate of the energy is derived for the crane system by assuming that the dynamic of the platform is not neglected. This situation occurs if the mass m is small, and hence, the dynamical term $my_{tt}(0, t)$ could be disregarded. Notwithstanding, one should consider the more realistic situation when m is not small and so its dynamics must be considered.

We also believe that an interesting problem to be treated is when $a(x)$, the modulus of cable tension degenerates.

Lastly, motivated by practical reasons, where the material used for some flexible structure is viscoelastic, it would be desirable to investigate the stabilization of the same mechanical system but when a nonlinear (finite or infinite) memory term occurs in the boundary control.

4.9 Boundary Memory-Type Control: Exponential Stabilization

This section is concerned with the boundary stabilization problem of an overhead crane with infinite memory. We show, under some assumptions of the memory kernel, that the problem is well-posed in the sense of semigroups theory (see Sect. 2.3). Furthermore, the uniform exponential stability of the solutions to the system is established by means of the frequency-domain method.

In order to highlight the motivation of our current section, recall that the system (4.1.4) can be asymptotically stabilized by means of the boundary control

$$\mathscr{F}(t) = -\alpha y(0, t) - \beta y_t(0, t),$$

where $\alpha > 0$ and β are positive feedback gains. However, this control is not strong enough to ensure the *exponential* stability of the closed-loops system [74]. This means that in order to guarantee the exponential stability of the system, then one has to design a control $\mathscr{F}(t)$ of higher order such as $y_{xt}(0, t)$. Indeed, if

$$\mathscr{F}(t) = -\zeta y(0, t) - \alpha\beta y_t(0, t) + \beta y_{xt}(0, t), \quad \zeta, \alpha, \beta \text{ are positive constants}$$

or

$$\mathscr{F}(t) = -\zeta y(0, t) - (b + \zeta e) y_t(0, t) + e\,(a y_{xt})\,(0, t),$$

then the closed-loop system is exponentially stable provided that ζ, δ and e are positive constants that satisfy $m > be$ [145]. In turn, the natural question that arises is: what is the effect of the presence of a memory term involving the boundary velocity $y_t(0, t)$? The answer to such a question represents the principal motivation of the present section. On the other hand, needless to mention that the memory phenomenon arises in many control systems, where at least one of the equations depends on the past values of one or more variables involved in the problem. Indeed, viscoelastic cables, population dynamics, heat flow in real conductors are modeled by differential equations with memory [13]. Furthermore, it has been noticed that the presence of a memory term in numerous physical systems may have a deleterious effect on either the stability or the performance of controlled systems [51, 69, 85]. Therefore, it is crucial to investigate the impact of a memory phenomenon in the control $\mathscr{F}(t)$. Such a memory phenomenon could be justified by the fact that the stress at a time t depends on the history of the small deformation of the cable. Motivated by this discussion, we propose the following feedback control:

$$\mathscr{F}(t) = -\sigma y(0,t) + \omega\,(ay_{xt})\,(0,t) - \alpha y_t(0,t) - \nu \int_0^\infty \xi(s) y_t(0,t-s)\,ds,\ t>0, \tag{4.9.1}$$

in which α, σ, ν, ω are the feedback gains of the memory-type boundary control. Moreover, setting $\alpha = \delta + \sigma\omega$, where $\delta > 0$ and assuming that $m > \delta\omega$, the memory kernel ξ should obey the following conditions:

- **A1.** $\xi \in L^1(\mathbb{R}_+) \cap H^2(\mathbb{R}_+)$ such that $\xi(s) \geq 0$ for all $s \in \mathbb{R}_+$ and $\int_0^\infty \xi(s)\,ds > 0$.
- **A2.** $\xi'(s) < 0$ and $\xi''(s) \geq 0$ for any $s \in \mathbb{R}_+$. Additionally, we suppose that there exists a positive constant $C \geq \kappa + \frac{\omega\nu\xi(0)}{m-\delta\omega}$, for some $\kappa > 0$ such that

$$\xi''(s) \geq -C\xi'(s), \qquad \forall s \in \mathbb{R}_+. \tag{4.9.2}$$

The simplest and typical example of the kernel ξ that satisfies the assumptions *A1.–A2* is $\xi = Ke^{-Cs}$, where K is an arbitrary positive constant and C is a positive constant involved in *A2*.

Whereupon, the closed-loop system (4.1.4) and (4.9.1) is

$$\begin{cases} y_{tt} - (ay_x)_x = 0, & x \in \Omega,\ t > 0, \\ my_{tt}(0,t) - (ay_x)\,(0,t) = -\sigma y(0,t) + \omega\,(ay_{xt})\,(0,t) & \\ -\alpha y_t(0,t) - \nu \displaystyle\int_0^\infty \xi(s) y_t(0,t-s)\,ds, & t > 0, \\ My_{tt}(1,t) + (ay_x)\,(1,t) = 0, & t > 0, \\ y(x,0) = y_0(x),\ y_t(x,0) = y_1(x), & x \in \Omega, \\ y(0,t-\tau) = f(t-\tau), & t \in (0,\tau), \end{cases} \tag{4.9.3}$$

where f is an initial data corresponding to the presence of the past history of y. Then, we can introduce the so-called minimal state variable (see [98])

$$\eta(t,s) = \int_0^\infty \xi'(s+\tau)\,[y(0,t-\tau) - y(0,t)]\,d\tau. \tag{4.9.4}$$

As a result of the assumptions *A1* and *A2*, there holds

$$\xi(\infty) := \lim_{s\to\infty} \xi(s) = 0.$$

Therewith, it follows (formally) from the definition of the minimal state variable $\eta(t, s)$ given by (4.9.4) that (see [98] and also [52])

$$\eta(\cdot, \infty) := \lim_{s\to\infty} \eta(\cdot, s) = 0 \quad \text{and} \quad \eta_t(t, s) = \eta_s(t, s) + \xi(s) y_t(0, t). \qquad (4.9.5)$$

It is noteworthy that in view of these conditions (and in particular $\eta(\infty) = 0$), one can introduce the following space

$$\mathscr{S} = \left\{ \eta(\cdot) \in H^1(\mathbb{R}_+);\ \ \eta(\infty) = 0, \ \ -\int_0^\infty \frac{\eta_s^2(s)}{\xi'(s)}\, ds < \infty \right\},$$

equipped with the inner product

$$\langle \eta, \widetilde{\eta} \rangle = -\int_0^\infty \frac{1}{\xi'(s)} \eta_s(s) \widetilde{\eta}_s(s)\, ds.$$

A simple integration by parts leads to rewriting the boundary condition at $x = 0$ in (4.9.3) as follows:

$$m y_{tt}(0, t) - (a y_x)(0, t) = -\sigma y(0, t) - \nu \eta(t, 0) - (\delta + \sigma\omega) y_t(0, t) + \omega\, (a y_{xt})\, (0, t),$$

which, together with (4.9.3), gives the following system:

$$\begin{cases} y_{tt} - (a y_x)_x = 0, & x \in \Omega,\ t > 0, \\ \eta_t(t, s) = \eta_s(t, s) + \xi(s) y_t(0, t), & t,\ s > 0, \\ m y_{tt}(0, t) - (a y_x)(0, t) = -\sigma y(0, t) - \nu \eta(t, 0) & \\ -(\delta + \sigma\omega) y_t(0, t) + \omega\, (a y_{xt})\, (0, t), & t > 0, \\ M y_{tt}(1, t) + (a y_x)\, (1, t) = 0, & t > 0, \\ y(x, 0) = y_0(x),\ y_t(x, 0) = y_1(x), & x \in \Omega, \\ \eta(0, s) = \eta_0(s), & s > 0, \end{cases} \qquad (4.9.6)$$

where

$$\eta_0(s) = y_1(0) \xi(s) + \int_0^\infty \xi'(s + \tau) f(\tau)\, d\tau, \quad s > 0.$$

The novelty of this section is fourfold: First, unlike the previous works [145, 146], an infinite memory term is taken into account in the proposed higher-order boundary control (4.9.1). To the authors' best knowledge, this is the first time that such a case is considered for the crane system. Second, the minimal state approach

[98] is adopted so as to treat the infinite memory term and analyze the problem. Third, in order to establish the exponential stability of the closed-loop system and contrary to [52], where the Lyapunov method is exploited, the frequency-domain method is used in this section. Note here that the introduction of the minimal state variable $\eta(t, s)$ given by (4.9.4) does not permit the compactness of the trajectories set since the compactness argument used in this chapter fails. This means that the resolvent of the system's operator is not compact, and thus its spectrum does not consist only of eigenvalues. Last but not least, in contrast to the previous sections of this chapter, we do not require the usual smallness condition of the feedback gain ν of the memory term versus the feedback gain α of the velocity term.

Next, the state space $\mathscr{H}$ is defined as follows:

$$\mathscr{H} = H^1(\Omega) \times L^2(\Omega) \times \mathscr{S} \times \mathbb{R}^2,$$

endowed with the inner product

$$\begin{aligned}\langle (y, z, \eta, \mu, \theta), (\tilde{y}, \tilde{z}, \tilde{\eta}, \tilde{\mu}, \tilde{\theta}) \rangle_{\mathscr{H}} = & \int_\Omega (a y_x \tilde{y}_x + z\tilde{z})\, dx \\ & -\nu \textstyle\int_0^\infty \frac{1}{\xi'(s)} \eta_s(s) \tilde{\eta}_s(s)\, ds \\ & +\sigma y(0)\tilde{y}(0) + \dfrac{1}{m - \delta\omega} \mu\tilde{\mu} + M\theta\tilde{\theta}.\end{aligned} \tag{4.9.7}$$

Under the conditions *A1* and *A2*, one can readily verify that $\mathscr{H}$ is a Hilbert space.

Subsequently, let us consider the state variable $\Phi = (y, z, \eta, \mu, \theta)$, where $z(\cdot, t) = y_t(\cdot, t)$, η is defined by (4.9.4), $\theta = y_t(1, t)$, and

$$\mu = m y_t(0, t) + \omega[\sigma y(0, t) - (a y_x)(0, t)]. \tag{4.9.8}$$

Then, one can write the system (4.9.6) in the abstract form as follows:

$$\begin{cases} \Phi_t(t) = \mathscr{A}\Phi(t), \\ \Phi(0) = \Phi_0 = (y_0, y_1, \eta_0(\cdot), \mu_0, \theta_0), \end{cases} \tag{4.9.9}$$

in which $\mathscr{A}$ is a linear operator defined by

$$\mathscr{D}(\mathscr{A}) = \left\{ \begin{array}{l} (y, z, \eta, \mu, \theta) \in \mathscr{H};\ y \in H^2(\Omega), \quad z \in H^1(\Omega), \eta_s \in \mathscr{S}, \\ \mu = m z(0) + \omega[\sigma y(0) - (a y_x)(0)],\ \theta = z(1) \end{array} \right\}, \tag{4.9.10}$$

$$\begin{aligned}\mathscr{A}(y, z, \eta, \mu, \theta) = \Bigg(& z, (a y_x)_x, \eta_s + \xi(s) z(0), -\frac{\mu}{\omega} + \frac{m - \delta\omega}{\omega} z(0) \\ & -\nu\eta(0), -\frac{(a y_x)(1)}{M} \Bigg).\end{aligned} \tag{4.9.11}$$

Let us point out that in order to handle the memory term in our problem, we used the minimal state approach [13, 98], in contrast to the history approach [79]. The latter consists in introducing the auxiliary variable defined by $y(t) - y(t-s)$, where $t, s \geq 0$. This expression is called the past history. In turn, it has been noticed that approach has a major drawback: It does not discern between solutions stemmed from two different initial past histories [13]. Consequently, an alternative method must be used. This is how the minimal state variable was introduced (see [13] for more details about this concept).

Before ending this discussion, let us briefly highlight some additional points. The reader can certainly see from the proposed boundary control (4.9.1) that the memory term acts on the *time variable t* in an *infinite interval* $(0, \infty)$. This consists of a major difference when we compare such a control with the one designed by several authors via the backstepping method [30, 31, 86, 127, 181, 182]. Furthermore, the main concern in the backstepping method is to show the existence of the memory kernel of the integral transformation, while the memory kernel ξ in our case is assumed to be given and satisfy the assumptions *A1–A2*. Finally, a quick look at the closed-loop system (4.9.6) or its abstract form (4.9.9) shows that the nonlocal character (the presence of the infinite integral) of the problem is concealed thanks to the minimal state approach adopted in the present section.

Now, let us describe an outline of this section. In Sect. 4.9.1, we show that the Cauchy problem (4.9.9) is well-posed in $\mathscr{H}$. Section 4.9.2 is devoted to the proof of the exponential stability of (4.9.9). Lastly, the section closes with concluding discussions.

4.9.1 Well-Posedness of the Problem

This subsection is consecrated to the study of existence and uniqueness of the solutions of the system (4.9.9) under the conditions *A1* and *A2*.

The main result of this subsection is stated below:

Theorem 4.9.1 *Suppose that A1 and A2 hold. Then, for any initial data $\Phi_0 \in \mathscr{D}(\mathscr{A})$, the Cauchy problem (4.9.9) has a unique strong solution Φ that belongs to $C((0,\infty); \mathscr{D}(\mathscr{A})) \cap C^1((0,\infty); \mathscr{H})$. In particular, we have*

$$y \in C^2((0,\infty); L^2(\Omega)) \cap C^1((0,\infty); H^1(\Omega)) \cap C^0((0,\infty); H^2(\Omega)),$$

$$\eta \in C^1((0,\infty); \mathscr{S}) \cap C^0((0,\infty); S^1), \quad \mu, \theta \in C^1((0,\infty); \mathbb{R}),$$

where $\mathscr{S}^1 = \{\eta \in \mathscr{S} \quad \text{such that} \quad \eta_s \in S\}$.

However, if the initial data $\Phi_0 \in \mathscr{H}$, then the system (4.9.9) has a unique mild solution $\Phi \in C^1((0,\infty);\mathscr{H})$. In this case, we have

$$y \in C^1((0,\infty); L^2(\Omega)) \cap C((0,\infty); H^1(\Omega)),$$

$$\eta \in C((0,\infty); \mathscr{S}), \quad \mu,\ \theta \in C((0,\infty); \mathbb{R}).$$

Proof The proof is divided into three steps.

Step 1 Let us show that the operator $\mathscr{A}$ is dissipative in $\mathscr{H}$. To do so, let $\Phi = (y, z, \eta, \mu, \theta) \in \mathscr{D}(\mathscr{A})$. Subsequently, we introduce an auxiliary variable

$$\rho := -\frac{1}{\omega}\mu + \frac{m-\delta\omega}{\omega} z(0) - \nu\eta(0) = -\sigma y(0) - \nu\eta(0) - \delta z(0) + (ay_x)(0). \tag{4.9.12}$$

Using the latter together with (4.9.7) and (4.9.10)–(4.9.11), we obtain after a straightforward computation that is based on integrations by parts and Young's inequality

$$\begin{aligned}
\langle \mathscr{A}\Phi, \Phi\rangle_{\mathscr{H}} &= \int_\Omega \{a z_x y_x + (ay_x)_x z\}\, dx - \nu \int_0^\infty \frac{1}{\xi'(s)} (\eta_s + \xi(s) z(0))_s \eta_s\, dx \\
&\quad + \sigma z(0) y(0) + \frac{1}{m-\delta\omega}\rho\mu - (ay_x)(1)\theta \\
&= (\sigma y(0) - (ay_x)(0)) z(0) - \nu \int_0^\infty \frac{1}{\xi'(s)} \eta_{ss}(s)\eta_s(s)\, ds \\
&\quad - \nu z(0) \textstyle\int_0^\infty \eta_s(s)\, ds + \dfrac{\rho\mu}{m-\delta\omega} \\
&= \left(-z(0) + \frac{1}{m-\delta\omega}\mu\right)\rho - \delta z^2(0) + \frac{\nu}{2}\frac{\eta_s^2(0)}{\xi'(0)} \\
&\quad - \tfrac{\nu}{2} \textstyle\int_0^\infty \frac{\xi''(s)}{(\xi'(s))^2} \eta_s^2(s)\, ds \\
&= -\frac{\omega}{m-\delta\omega}\rho^2 - \delta z^2(0) - \frac{\omega\nu}{m-\delta\omega}\rho\eta(0) + \frac{\nu}{2}\frac{\eta_s^2(0)}{\xi'(0)} \\
&\quad - \tfrac{\nu}{2} \textstyle\int_0^\infty \frac{\xi''(s)}{(\xi'(s))^2} \eta_s^2(s)\, ds \\
&\le \frac{\omega}{m-\delta\omega}\left(\frac{\nu}{2}\varepsilon - 1\right)\rho^2 - \delta z^2(0) + \frac{\omega\nu}{m-\delta\omega}\frac{\eta^2(0)}{2\varepsilon} + \frac{\nu}{2}\frac{\eta_s^2(0)}{\xi'(0)} \\
&\quad - \tfrac{\nu}{2} \textstyle\int_0^\infty \frac{\xi''(s)}{(\xi'(s))^2} \eta_s^2(s)\, ds,
\end{aligned} \tag{4.9.13}$$

for any $\varepsilon > 0$. On the other hand, owing to Hölder's inequality, we have

$$\eta^2(0) = \left(\int_0^\infty \sqrt{-\xi'(s)} \frac{\eta_s(s)}{\sqrt{-\xi'(s)}}\, ds\right)^2 \le \xi(0) \int_0^\infty \frac{1}{-\xi'(s)} \eta_s^2\, ds. \tag{4.9.14}$$

Inserting (4.9.14) into (4.9.13), we get

$$\langle \mathscr{A}\Phi, \Phi\rangle_{\mathscr{H}} \leq \frac{\omega}{m-\delta\omega}\left(\frac{\nu}{2}\varepsilon - 1\right)\rho^2 - \delta z^2(0) + \frac{\nu}{2}\frac{\eta_s^2(0)}{\xi'(0)}$$
$$-\frac{\nu}{2}\int_0^\infty \frac{1}{\xi'(s)}\left(\frac{\omega\xi(0)}{\varepsilon(m-\delta\omega)} + \frac{\xi''(s)}{\xi'(s)}\right)\eta_s^2\, ds. \quad (4.9.15)$$

Lastly, choosing $\varepsilon = \frac{1}{\nu}$ and invoking the assumptions *A1–A2*, we conclude that the operator $\mathscr{A}$ is dissipative in $\mathscr{H}$.

Step 2 The task ahead is to verify that the operator $\mathscr{A}$ is injective. To proceed, let us consider $\Phi = (y, z, \eta, \mu, \theta) \in \mathscr{D}(\mathscr{A})$ so that $\mathscr{A}\Phi = 0$. Equivalently, we consider the following system:

$$z = 0, \quad (ay_x)_x = 0, \quad (4.9.16)$$
$$\eta_s + \xi(s)z(0) = 0, \quad (4.9.17)$$
$$-\frac{1}{\omega}\mu + \frac{m-\delta\omega}{\omega}z(0) - \nu\eta(0) = 0, \quad (4.9.18)$$
$$(ay_x)(1) = 0. \quad (4.9.19)$$

In light of (4.9.16), we have $\theta = z(1) = 0$ as well as $z(0) = 0$. This, together with (4.9.17) and (4.9.5), implies that $\eta = 0$. This gives $\mu = 0$ due to (4.9.18). Hence, using (4.9.16),(4.9.19) and the fact that $\Phi \in \mathscr{D}(\mathscr{A})$, we conclude that $(ay_x)_x = 0$ and $y(0) = (ay_x)(0) = (ay_x)(1) = 0$, which, obviously, implies that $y = 0$. Consequently, we obtain $\Phi = 0_{\mathscr{H}}$.

Step 3 We claim that $\mathscr{A}$ is onto $\mathscr{H}$. To ascertain the correctness of this claim, let $(f, g, v, p, q) \in \mathscr{H}$, and we look for an element $(y, z, \eta, \mu, \theta) \in \mathscr{D}(\mathscr{A})$ such that

$$\mathscr{A}(y, z, \eta, \mu, \theta) = (f, g, v, p, q).$$

It is easy to verify that the latter can be written as follows:

$$z = f, \ (ay_x)_x = g, \quad (4.9.20)$$
$$\eta_s + \xi(s)z(0) = v, \quad (4.9.21)$$
$$\mu - (m-\delta\omega)z(0) + \nu\omega\eta(0) = -\omega p, \quad (4.9.22)$$
$$(ay_x)(1) = -Mq, \quad (4.9.23)$$
$$\mu = mz(0) + \omega[\sigma y(0) - (ay_x)(0)], \ \theta = z(1). \quad (4.9.24)$$

In light of (4.9.20) and (4.9.21), a direct computation yields

$$\eta(s) = -\int_s^\infty (v(\tau) - \xi(\tau)z(0))\, d\tau.$$

Combining the latter with (4.9.20), we obtain

$$\eta(0) = \int_0^\infty (f(0)\xi(s) - v(s))\, ds. \tag{4.9.25}$$

In view of (4.9.20), (4.9.22)–(4.9.25), we need to seek $y \in H^2(\Omega)$ such that

$$\begin{cases} (ay_x)_x = g, \\ \sigma y(0) - (ay_x)(0) = -p - \delta f(0) + \nu \displaystyle\int_0^\infty (v(s) - f(0)\xi(s))\, ds, \\ (ay_x)(1) = -Mq. \end{cases} \tag{4.9.26}$$

Multiplying the first equation in (4.9.26) by $\psi \in H^1(\Omega)$, we get the weak formulation of (4.9.26):

$$\begin{aligned} \int_\Omega ay_x\psi_x\, dx + \sigma y(0)\psi(0) = &\left[-p - \delta f(0) + \nu \int_0^\infty (v(s) - f(0)\xi(s))\, ds\right]\psi(0) \\ &-Mq\psi(1) - \int_\Omega g\psi\, dx. \end{aligned} \tag{4.9.27}$$

It is not difficult to verify that the left-hand side of (4.9.27) is a continuous and coercive bilinear form on $H^1(\Omega) \times H^1(\Omega)$, while the right-hand side is a continuous linear form on $H^1(\Omega)$. Applying Lax-Milgram theorem (see Theorem 2.1.6) and using classical elliptic theory, one can deduce the existence of a unique solution $y \in H^2(\Omega)$ of (4.9.26). This establishes that the range of the operator $\mathscr{A}$ is $\mathscr{H}$, that is, $\mathscr{A}$ is onto $\mathscr{H}$.

Finally, one can apply Theorem 2.3.4 to claim that $\mathscr{A}$ generates a C_0-semigroup of contractions on $\mathscr{H}$. Thereby, semigroups theory of linear operators (see Sect. 2.3 or [158]) allows us to claim the existence of a unique solution of (4.9.9) as stated in Theorem 4.9.1. □

4.9.2 Exponential Stability

In this subsection, we show, under the assumptions on the memory kernel ξ (see *A1* and *A2*), that the semigroup $e^{t\mathscr{A}}$ is uniformly exponentially stable. The proof of this result is based on the resolvent method (see Theorem 2.3.8) combined with the multiplier technique. Special attention is given to the memory term in order to establish some estimations for the resolvent on the imaginary axis.

Now, we can state our main result:

Theorem 4.9.2 *Suppose that the assumptions on the memory kernel ξ given by A1 and A2 are fulfilled. Then, the associated semigroup $e^{t\mathscr{A}}$ of (4.9.9) is exponentially stable in $\mathscr{H}$.*

Proof The task ahead is to prove that the two properties of Theorem 2.3.8 are fulfilled, namely, (2.3.16) and (2.3.17).
(i) Proof of (2.3.16): Recall that in view of the space $\mathscr{S}$, the embedding $\mathscr{D}(\mathscr{A})$ into $\mathscr{H}$ is not necessarily compact, which implies that the spectrum $\sigma(\mathscr{A})$ of $\mathscr{A}$ does not consist only of eigenvalues of $\mathscr{A}$. Keeping in mind that $0 \in \rho(\mathscr{A})$, to overcome this problem, it suffices to prove that

$$i\gamma I - \mathscr{A} \text{ is injective, } \gamma \in \mathbb{R} \text{ and } i\varpi I - \mathscr{A} \text{ is surjective, } \varpi \in \mathbb{R}. \tag{4.9.28}$$

The proof of (4.9.28) will be done by a contradiction argument.

Assume that the first assertion of (4.9.28) is not true, that is, $\ker\{i\gamma I - \mathscr{A}\} \neq 0$. Thus, there is a nonzero real number γ such that $i\gamma$ is an element of $\sigma(\mathscr{A})$, that is, there exists $\Phi = (y, z, \eta, \mu, \theta) \in \mathscr{D}(\mathscr{A})$ such that $\|\Phi\|_{\mathscr{H}} = 1$ and

$$\mathscr{A}\Phi = i\gamma\Phi. \tag{4.9.29}$$

Equivalently, we have the following identities:

$$z = i\gamma y, \tag{4.9.30}$$

$$(ay_x)_x = i\gamma z, \tag{4.9.31}$$

$$\eta_s + \xi(s)z(0) = i\gamma\eta, \tag{4.9.32}$$

$$-\frac{\mu}{\omega} + \frac{m - \delta\omega}{\omega} z(0) - \nu\eta(0) = i\gamma\mu, \tag{4.9.33}$$

$$(ay_x)(1) = -i\gamma M\theta. \tag{4.9.34}$$

Using (4.9.15) and taking the inner product of both sides of (4.9.29) with Φ, and using *A1*–*A2*, we reach

$$\rho = z(0) = \eta_s(0) = \int_0^\infty \frac{1}{\xi'(s)} \left(\frac{\omega\nu\xi(0)}{(m - \delta\omega)} + \frac{\xi''(s)}{\xi'(s)} \right) \eta_s^2 \, ds = 0. \tag{4.9.35}$$

On the other hand, taking into account (4.9.2), we immediately find

$$\int_0^\infty \frac{-\kappa}{\xi'(s)} \eta_s^2 \, ds \leq \int_0^\infty \frac{1}{\xi'(s)} \left(\frac{\omega\nu\xi(0)}{(m - \delta\omega)} + \frac{\xi''(s)}{\xi'(s)} \right) \eta_s^2 \, ds. \tag{4.9.36}$$

This obviously gives $\eta = 0$ in $\mathscr{S}$, and hence, $\eta(0) = 0$. Thenceforth, invoking (4.9.30)–(4.9.31), (4.9.33), and (4.9.36), we have

$$\begin{cases} (ay_x)_x + \gamma^2 y = 0, \\ y(0) = (ay_x)(0) = 0, \quad (ay_x)(1) = \gamma^2 M y(1). \end{cases} \tag{4.9.37}$$

Thereafter, consider the function [167]

$$\varphi(x) = \exp\left(\int_x^1 \left|\frac{a'}{a}\right|(s)\,ds\right)\left(-M - \int_s^1 \exp\left(-\int_x^1 \left|\frac{a'}{a}\right|(r)\,dr\right)\,ds\right),$$
$$\forall s \in [0,1], \tag{4.9.38}$$

which satisfies

$$\varphi(1) = -M,\ \ \varphi'(x) \geq 1,\ \ (a\varphi' - |a'|\varphi)(x) \geq a(x),\ \ \text{and}\ \varphi(x) < 0,\ \ \forall x \in [0,1]. \tag{4.9.39}$$

Multiplying the equation in (4.9.37) by φy_x and integrating the resulting equation over Ω, one can conclude that the only solution of the problem (4.9.37) is the trivial one, that is, $y = 0$. This clearly implies, because of (4.9.30)–(4.9.34), that all the components of Φ vanish, which contradicts the fact that $\|\Phi\| = 1$. Thus, the first assertion of (4.9.28) is verified, and hence, the operator $\mathscr{A}$ has no purely imaginary eigenvalues.

Suppose that the second assertion of (4.9.28) does not hold, which means that there is a nonzero real γ and a vector $V \in \mathscr{H}$ satisfying $V \notin \Im m(i\gamma I - \mathscr{A})$. Thereby, there exists a sequence $(V_n)_n$ in $\mathscr{H}$ such that

$$V_n \to V \text{ and } (i\gamma I - \mathscr{A})^{-1}V_n = U_n \to +\infty \text{ in } \mathscr{H}.$$

Whereupon, we have

$$(i\gamma I - \mathscr{A})\left(\frac{U_n}{\|U_n\|}\right) = \frac{V_n}{\|U_n\|} \text{ in } \mathscr{H}.$$

In other words, one can find a unit sequence $\Phi_n = (y_n, z_n, \eta_n, \mu_n, \theta_n)$ in the domain of $\mathscr{A}$ such that

$$(i\gamma I - \mathscr{A})\Phi_n = \mathscr{F}_n = (f_n, g_n, v_n, p_n, q_n) \to 0,\ \text{in } \mathscr{H}. \tag{4.9.40}$$

This later is equivalent to dealing with the following limits:

$$i\gamma y_n - z_n = f_n \to 0,\ \text{in } H^1(\Omega), \tag{4.9.41}$$
$$i\gamma z_n - (ay_{n,x})_x = g_n \to 0,\ \text{in } L^2(\Omega), \tag{4.9.42}$$
$$i\gamma \eta_n - \eta_{n,x} - \xi(s)z_n(0) = v_n \to 0,\ \text{in } \mathscr{S}, \tag{4.9.43}$$
$$i\gamma \mu_n + \frac{1}{\omega}\mu_n - \frac{m - \delta\omega}{\omega}z_n(0) + \nu\eta_n(0) = p_n \to 0,\ \text{in } \mathbb{C}, \tag{4.9.44}$$
$$i\gamma M\theta_n + (ay_{n,x})(1) = Mq_n \to 0,\ \text{in } \mathbb{C}. \tag{4.9.45}$$

In light of (4.9.40) and (4.9.15), we can conclude that

$$\rho_n \to 0, \quad z_n(0) \to 0, \quad \eta_{n,s}(0) \to 0, \tag{4.9.46}$$

and

$$\int_0^\infty \frac{1}{\xi'(s)} \left(\frac{\omega\nu\xi(0)}{m-\delta\omega} + \frac{\xi''(s)}{\xi'(s)} \right) \eta_{n,s}^2 \, ds \to 0. \tag{4.9.47}$$

According to (4.9.2), we have

$$\int_0^\infty \frac{-\kappa}{\xi'(s)} \eta_{n,s}^2 \, ds \le \int_0^\infty \frac{1}{\xi'(s)} \left(\frac{\omega\nu\xi(0)}{(m-\delta\omega)} + \frac{\xi''(s)}{\xi'(s)} \right) \eta_{n,s}^2 \, ds. \tag{4.9.48}$$

This estimate, together with (4.9.47), yields

$$\eta_n \to 0 \text{ in } \mathscr{S}. \tag{4.9.49}$$

Thereafter, using Holder's inequality, we obtain

$$\begin{aligned} \eta_n^2(0) &= \left(\int_0^\infty \eta_{n,s}(s) \, ds \right)^2 = \left(\int_0^\infty \sqrt{-\xi'(s)} \frac{\eta_{n,s}(s)}{\sqrt{-\xi'(s)}} \, ds \right)^2 \\ &\le \xi(0) \, \|\eta_n\|_{\mathscr{S}}^2 , \end{aligned} \tag{4.9.50}$$

which gives, thanks to (4.9.49), that

$$\eta_n(0) \to 0 \text{ in } \mathbb{C}. \tag{4.9.51}$$

Now, using the definition in (4.9.7) of the inner product of $\mathscr{H}$ and the fact that $\|\mathscr{F}_n\|_{\mathscr{H}} \to 0$, it follows that

$$f_n(0) \to 0, \tag{4.9.52}$$

and, thus, by (4.9.41) and (4.9.46), we find

$$y_n(0) \to 0 \text{ in } \mathbb{C}. \tag{4.9.53}$$

In view of (4.9.44), (4.9.46), and (4.9.51), we get

$$\mu_n \to 0. \tag{4.9.54}$$

Then, combining (4.9.46), (4.9.51), (4.9.53), and (4.9.54), we deduce from the definition of μ_n that

$$y_{n,x}(0) \to 0. \tag{4.9.55}$$

Now, writing z_n in terms of y_n from (4.9.41) and inserting it into (4.9.42), we get

$$(ay_{n,x})_x + \gamma^2 y_n = -i\gamma f_n - g_n.$$

Next, multiplying the latter by $\varphi y_{n,x}$, integrating the resulting equation over Ω, using (4.9.53), (4.9.55) and the fact that $\mathscr{F}_n \to 0$ in $\mathscr{H}$ and the boundedness of the components y_n, z_n in their appropriate spaces, a straightforward calculation yields

$$\gamma^2\varphi(1)y_n^2(1) + a(1)\varphi(1)y_{n,x}^2(1) - \int_\Omega (a\varphi_x - a_x\varphi)(x)y_{n,x}^2\, dx \\ - \gamma^2 \int_\Omega \varphi_x(x)y_n^2\, dx = o(1). \tag{4.9.56}$$

Invoking (4.9.39), the identity (4.9.56) implies

$$y_n(1) \to 0, \quad y_{n,x}(1) \to 0, \tag{4.9.57}$$

and

$$y_n \to 0 \text{ in } H^1(\Omega). \tag{4.9.58}$$

Thenceforth, combining (4.9.57) and (4.9.58) respectively with (4.9.45) and (4.9.41), it follows that

$$z_n \to 0 \text{ in } L^2(\Omega) \quad \text{and} \quad \theta_n \to 0 \text{ in } \mathbb{C}. \tag{4.9.59}$$

Lastly, the findings in (4.9.49), (4.9.54), and (4.9.58)–(4.9.59) contradict the property $\|\Phi_n\|_{\mathscr{H}} = 1,\ \forall n \in \mathbb{N}$. Whereupon, we managed to prove (4.9.28), and hence, (2.3.16) holds.

(ii) Proof of (2.3.17): If this were not true, then, thanks to Banach-Steinhaus theorem (see Theorem 2.1.7), there exist a sequence of real numbers $\gamma_n \to +\infty$ and a sequence of vectors $\Phi_n = (y_n, z_n, \eta_n, \mu_n, \theta_n) \in \mathscr{D}(\mathscr{A})$ with $\|\Phi_n\|_{\mathscr{H}} = 1$ such that

$$(i\gamma_n I - \mathscr{A})\Phi_n = \mathscr{F}_n = (f_n, g_n, v_n, p_n, q_n) \to 0, \text{ in } \mathscr{H}. \tag{4.9.60}$$

Equivalently, (4.9.60) can be written as follows:

$$i\gamma_n y_n - z_n = f_n \to 0, \text{ in } H^1(\Omega), \tag{4.9.61}$$

$$i\gamma_n z_n - (ay_{n,x})_x = g_n \to 0, \text{ in } L^2(\Omega), \tag{4.9.62}$$

$$i\gamma_n \eta_n - \eta_{n,x} - \xi(s)z_n(0) = v_n \to 0, \text{ in } \mathscr{S}, \tag{4.9.63}$$

$$i\gamma_n \mu_n + \frac{1}{\omega}\mu_n - \frac{m - \delta\omega}{\omega} z_n(0) + \nu\eta_n(0) = p_n \to 0, \text{ in } \mathbb{C}, \tag{4.9.64}$$

$$i\gamma_n M\theta_n + (ay_{n,x})(1) = Mq_n \to 0, \text{ in } \mathbb{C}. \tag{4.9.65}$$

The ultimate objective is to prove

$$\Phi_n \to 0 \text{ in } \mathscr{H}, \tag{4.9.66}$$

which gives a contradiction. The proof is divided into three steps.

First Step The aim is to show the convergence of $\eta_n, \mu_n, \theta_n, \gamma_n y_n(0)$ and $\gamma_n y_n(1)$ in their respective spaces. Taking into consideration the inequality

$$\|(i\gamma_n I - \mathscr{A})\Phi_n\|_{\mathscr{H}} \ge |\langle(i\gamma_n I - \mathscr{A})\Phi_n, \Phi_n\rangle|, \tag{4.9.67}$$

and recalling (4.9.15) and (4.9.60), we obtain

$$\rho_n \to 0, \quad z_n(0) \to 0, \quad \eta_{n,s}(0) \to 0, \tag{4.9.68}$$

and

$$\int_0^\infty \frac{1}{\xi'(s)}\left(\frac{\omega\nu\xi(0)}{m-\delta\omega} + \frac{\xi''(s)}{\xi'(s)}\right)\eta_{n,s}^2\, ds \to 0. \tag{4.9.69}$$

Thereby, arguing as in part (i), one can verify that (4.9.48) and (4.9.50) also hold, and hence,

$$\eta_n \to 0 \text{ in } \mathscr{S} \quad \text{and} \quad \eta_n(0) \to 0 \text{ in } \mathbb{C}. \tag{4.9.70}$$

In view of (4.9.61) and the boundedness of Φ_n, it follows that

$$y_n = \frac{1}{i\gamma_n}(z_n + f_n) \to 0 \text{ in } L^2(\Omega). \tag{4.9.71}$$

On the other hand, by virtue of (4.9.52), (4.9.61) and the second limit in (4.9.68), we obtain

$$\gamma_n y_n(0) \to 0 \text{ and } y_n(0) \to 0. \tag{4.9.72}$$

Furthermore, amalgamating (4.9.12) with (4.9.68), (4.9.70), and (4.9.72), we find

$$(ay_{n,x})(0) \to 0. \tag{4.9.73}$$

Therewith, multiplying (4.9.62) by $\frac{1}{\gamma_n^2}a(x)y_{n,x}$ and integrating, we have

$$\frac{i}{\gamma_n}\int_\Omega z_n a(x)y_{n,x}\, dx - \frac{1}{\gamma_n^2}\int_\Omega (a(x)y_{n,x})_x a(x)y_{n,x}\, dx = \frac{1}{\gamma_n^2}\int_\Omega g_n a(x)y_{n,x}\, dx. \tag{4.9.74}$$

Taking into account that each component of Φ_n is bounded in its corresponding space, the first and third integrals in (4.9.74) are also bounded. A simple integration by parts for the second integral shows that

$$\frac{(ay_{n,x})(1)}{\gamma_n} \to 0.$$

Hence, by virtue of (4.9.65), we have

$$\theta_n = z_n(1) \to 0. \tag{4.9.75}$$

Multiplying (4.9.64) by $\frac{1}{i\gamma_n}$ and using the fact that μ_n is bounded as well as $z_n(0) \to 0$ and $\eta_n(0) \to 0$, we immediately obtain

$$\mu_n \to 0. \tag{4.9.76}$$

Recalling the two limits for f_n given by (4.9.52) and (4.9.61), one can readily verify that

$$f_n(1) \to 0. \tag{4.9.77}$$

Consequently, from (4.9.61), (4.9.75), and (4.9.77), we get

$$\gamma_n y_n(1) \to 0 \text{ and } y_n(1) \to 0. \tag{4.9.78}$$

Second Step The goal is to improve the results given by (4.9.71) and (4.9.75). Particularly, it consists in proving

$$y_n \to 0 \text{ in } H_0^1(\Omega), \tag{4.9.79}$$

and

$$(ay_{n,x})(1) \to 0. \tag{4.9.80}$$

To do this end, we first substitute (4.9.61) in (4.9.62), multiply the resulting identity by $\varphi y_{n,x}$, where φ is the function defined by (4.9.38), and then integrate over Ω. We obtain

$$\int_\Omega (-\gamma_n^2 y_n - (ay_{n,x})_x)\varphi(x) y_{n,x}\, dx = \int_\Omega (i\gamma_n f_n + g_n)\varphi(x) y_{n,x}\, dx. \tag{4.9.81}$$

In light of the convergence of f_n, g_n, $f_n(0)$, $f_n(1)$, $\gamma_n y_n(0)$, and $\gamma_n y_n(1)$ to zero respectively in $H^1(\Omega)$, $L^2(\Omega)$, and $\mathbb{C}$, and recalling the boundedness of y_n in $H^1(\Omega)$, a straightforward computation shows that the right-hand side of (4.9.81) converges to zero.

Next, let us return to the left-hand side of (4.9.81). Simple integrations by parts lead to

$$\begin{aligned}
&\int_\Omega (-\gamma_n^2 y_n - (a y_{n,x})_x)\varphi(x) y_{n,x}\, dx \\
&= \frac{1}{2}\int_\Omega \varphi'(x)|\gamma_n y_n(x)|^2\, dx + \frac{1}{2}\int_\Omega (a\varphi' - a'\varphi)(x)|(y_n)_x|^2\, dx \\
&-\frac{\varphi(1)}{2}|\gamma_n y_n(1)|^2 + \frac{\varphi(0)}{2}|\gamma_n y_n(0)|^2 - \frac{a(1)\varphi(1)}{2}|y_{n,x}(1)|^2 \\
&+\frac{a(0)\varphi(0)}{2}|y_{n,x}(0)|^2.
\end{aligned} \tag{4.9.82}$$

Thereafter, inserting in (4.9.82) the estimates

$$\frac{\varphi(1)}{2}|\gamma_n y_n(1)|^2 = \frac{\varphi(0)}{2}|\gamma_n y_n(0)|^2 = \frac{a(0)\varphi(0)}{2}|y_{n,x}(0)|^2 = o(1)$$

that we established in the first step, we have

$$\begin{aligned}
\int_\Omega (-\gamma_n^2 y_n - (a y_{n,x})_x)\varphi(x) y_{n,x}\, dx = {}& \frac{1}{2}\int_\Omega \varphi'(x)|\gamma_n y_n(x)|^2\, dx \\
&+\tfrac{1}{2}\textstyle\int_\Omega (a\varphi' - a'\varphi)(x)|(y_n)_x|^2\, dx \\
&-\frac{a(1)\varphi(1)}{2}|y_{n,x}(1)|^2 + o(1).
\end{aligned}$$

This, combined with (4.9.81), implies that

$$\frac{1}{2}\int_\Omega \varphi'(x)|\gamma_n y_n(x)|^2\, dx + \frac{1}{2}\int_\Omega (a\varphi' - a'\varphi)(x)|(y_n)_x|^2\, dx - \frac{a(1)\varphi(1)}{2}|y_{n,x}(1)|^2 = o(1),$$

which gives, thanks to (4.9.39), that

$$\int_\Omega |\gamma_n y_n(x)|^2\, dx + \int_\Omega a(x)|(y_n)_x|^2\, dx + Ma(1)|y_{n,x}(1)|^2 = o(1).$$

Lastly, owing to (4.9.4), we deduce from the last estimate our desired results (4.9.79) and (4.9.80).

Third Step The immediate target is to prove that the component z_n of Φ_n also converges to 0 in $L^2(\Omega)$, which completes the proof of the contradiction (4.9.66). To do so, we use (4.9.61) to express z_n in terms of y_n, substitute it in (4.9.62), and take the multiplier y_n in the resulting equation. We find

$$\int_\Omega z_n(z_n + f_n)\, dx - \int_\Omega (a y_{n,x})_x y_n\, dx = \int_\Omega g_n y_n\, dx.$$

Next, by virtue of the boundedness of y_n and z_n and the convergence of f_n and g_n to 0 in $L^2(\Omega)$, it follows that

$$\int_\Omega |z_n|^2\, dx + \int_\Omega a|y_{n,x}|^2\, dx = o(1).$$

Therewith, (4.9.79) gives $z_n \to 0$ in $L^2(\Omega)$, as desired.

□

Remark 4.9.1 The problem (4.9.9) is a weak form of (4.9.6) in the sense that if Φ_0 belongs to $\mathscr{H}$ or even $\mathscr{D}(\mathscr{A})$, then μ_t and $my_{tt}(0,t) + \omega[\sigma y_t(0,t) - (ay_{xt})(0,t)]$ are not necessarily equal. In turn, the equality holds when $\Phi_0 \in \mathscr{D}(\mathscr{A}^2)$, and hence, in this case, both problems (4.9.6) and (4.9.9) are equivalent.

On the other hand, the stability result stated in Theorem 4.9.2 implies in particular that the state variable

$$\mu = my_t(0,t) + \omega[\sigma y(0,t) - (ay_x)(0,t)]$$

exponentially decays, but not necessarily for the cable tension $y_x(0,t)$ nor for the mass velocity $y_t(0,t)$.

4.9.3 Concluding Remarks

Using the so-called minimal state approach, we managed to show that the overhead crane system can be exponentially stabilized by a boundary control with infinite memory. It is noteworthy that such a desirable result is obtained under some conditions on the memory kernel. In turn, unlike the previous sections of this chapter, no smallness condition of the feedback gain ν of the memory term is required. This represents a major advantage of adopting the minimal state approach.

It would be interesting to weaken the exponential decay requirement on the memory kernel stated in the assumption *A2* (see (4.9.2)).

Another promising research avenue is to consider the case where the memory kernel is also time dependent.

4.10 Interior and Boundary Memory-Type Controls: Exponential Stabilization

In the current section, we are interested in the stabilization of the crane system studied in Sect. 4.4, but when the boundary control takes into consideration the presence of a memory term. In this chapter, we suggest a moment control subject to the presence of an infinite memory term. This means that the feedback law consists of one localized interior control and a memory-type boundary control, namely,

$$\begin{cases} \text{interior control} : b(x)y_t(x,t), \\ \text{boundary control} : -\sigma y(0,t) - \alpha y_t(0,t) - \beta \displaystyle\int_0^\infty \kappa(s) y_t(0,t-s)\, ds, \end{cases} \tag{4.10.1}$$

where $\sigma > 0$, while $\alpha > 0$, $\beta > 0$ are the feedback gains corresponding to the memory-type boundary force control. Furthermore, the space variable coefficient $b(x)$ is the distributed actuator function, which has local support on $(0, 1)$ and is strictly positive on *only a subinterval* $(c, d) \subset (0, 1)$. On the other hand, κ is the memory kernel that satisfies the following assumptions:

- A1. $\kappa \in L^1(\mathbb{R}_+) \cap H^2(\mathbb{R}_+)$ such that $\kappa(s) \geq 0$ for all $s \in \mathbb{R}_+$ and $\displaystyle\int_0^\infty \kappa(s)\, ds > 0$.
- A2. $\kappa'(s) < 0$ and $\kappa''(s) \geq 0$ for any $s \in \mathbb{R}_+$.

Whereupon, we deal with the following problem:

$$\begin{cases} y_{tt}(x,t) - (ay_x)_x(x,t) + b(x)y_t(x,t) = 0, & 0 < x < 1,\ t > 0, \\ my_{tt}(0,t) - (ay_x)(0,t) = -\sigma y(0,t) & \\ -\beta \int_0^\infty \kappa(s) y_t(0,t-s)\, ds - \alpha y_t(0,t), & t > 0, \\ My_{tt}(1,t) + (ay_x)(1,t) = 0, & t > 0, \\ y(x,0) = y_0(x),\ y_t(x,0) = y_1(x), & x \in (0,1), \\ y_t(0,t-s) = f(t-s), & t \in (0,s). \end{cases} \tag{4.10.2}$$

The main objective is to compensate the effect of the memory term in the boundary feedback law through the action of the interior control $b(x)y_t(x,t)$.

It is worth noting that this case is different from that studied in Sect. 4.9. Indeed, the boundary control in (4.10.1) is of low order unlike (4.9.1), where $y_{xt}(0,t)$ is needed. This means that the control (4.10.1) is simpler to implement in practice. Last but not least, the requirement on the memory kernel κ is weakened, since we do not impose an extra condition on the constant involved in (4.9.2).

The rest of this section is organized as follows. In Sect. 4.10.1, we shall establish the well-posedness to our problem by means of semigroups theory of linear operators. Section 4.10.2 is consecrated to the uniform exponential stabilization result for our hybrid system. Lastly, some concluding remarks are given in the last subsection.

4.10.1 Well-Posedness of the Problem

Let us introduce the so-called minimal state variable (see [13, 98])

$$\eta(t,s) = \int_0^\infty \kappa'(s+\tau)\,[y(0,t-\tau) - y(0,t)]\; d\tau. \tag{4.10.3}$$

As a result of the assumptions A1 and A2, we have

$$\kappa(\infty) := \lim_{s\to\infty} \kappa(s) = 0. \tag{4.10.4}$$

Additionally, in view of A1 and A2, one can introduce the following space:

$$\mathscr{S} = \left\{ \eta(\cdot) \in H^1(\mathbb{R}_+);\ \ \eta(\infty) = 0,\ \ -\int_0^\infty \frac{\eta_s^2(s)}{\kappa'(s)}\, ds < \infty \right\},$$

equipped with the inner product

$$\langle \eta, \widetilde{\eta} \rangle = -\int_0^\infty \frac{1}{\kappa'(s)} \eta_s(s) \widetilde{\eta}_s(s)\, ds.$$

Therewith, a formal calculation in (4.10.3) yields

$$\eta(\cdot,\infty) := \lim_{s\to\infty} \eta(\cdot,s) = 0 \quad \text{and} \quad \eta_t(t,s) = \eta_s(t,s) + \kappa(s) y_t(0,t), \tag{4.10.5}$$

and

$$m y_{tt}(0,t) - (a y_x)(0,t) = -\sigma y(0,t) - \beta \eta(t,0) - \alpha y_t(0,t).$$

Accordingly, one can rewrite the system (4.10.2) as follows:

$$\begin{cases} y_{tt}(x,t) - (a y_x)_x(x,t) + b(x) y_t(x,t) = 0, & 0 < x < 1,\ t > 0, \\ \eta_t(t,s) = \eta_s(t,s) - \kappa(s) y_t(0,t), & t,\ s > 0, \\ m y_{tt}(0,t) - (a y_x)(0,t) = -\sigma y(0,t) - \beta \eta(t,0) - \alpha y_t(0,t), & t > 0, \\ M y_{tt}(1,t) + (a y_x)(1,t) = 0, & t > 0, \\ y(x,0) = y_0(x),\ \ y_t(x,0) = y_1(x), & x \in (0,1), \\ \eta(0,s) = \eta_0(s), & s > 0, \end{cases} \tag{4.10.6}$$

where η_0 depends on the initial conditions y_1 and f given (4.10.2). In fact, using (4.10.3) and (4.10.4), we get

$$\eta_0(s) = y_1(0)\kappa(s) - \int_0^\infty \kappa'(s+\tau) f(\tau)\, d\tau, \quad s > 0.$$

Next, let $\Phi = (y, z, \eta, \mu, \theta)$, where $z(\cdot, t) = y_t(\cdot, t)$, $\mu = y_t(0, t)$, and $\theta = y_t(1, t)$. Now, consider the state space $\mathscr{H}$ defined by

$$\mathscr{H} = H^1(0, 1) \times L^2(0, 1) \times \mathscr{S} \times \mathbb{R}^2,$$

endowed with the inner product

$$\begin{aligned} \langle (y, z, \eta, \mu, \theta), (\tilde{y}, \tilde{z}, \tilde{\eta}, \tilde{\mu}, \tilde{\theta}) \rangle_{\mathscr{H}} = & \int_0^1 (a y_x \tilde{y}_x + z\tilde{z})\, dx \\ & -\beta \int_0^\infty \tfrac{1}{\kappa'(s)} \eta_s(s) \widetilde{\eta}_s(s)\, ds \\ & +\sigma y(0)\tilde{y}(0) + m\mu\tilde{\mu} + M\theta\tilde{\theta}. \end{aligned} \tag{4.10.7}$$

Under the condition (4.1.5) as well as the assumptions A1 and A2, we can readily verify that $\mathscr{H}$ is a Hilbert space.

Thereafter, we can write the closed-loop system (4.10.6) in the abstract form

$$\begin{cases} \Phi_t(t) = \mathscr{A}\Phi(t), \\ \Phi(0) = \Phi_0 = (y_0, y_1, \eta_0(s), \mu_0, \theta_0), \end{cases} \tag{4.10.8}$$

in which $\mathscr{A}$ is a linear operator defined by

$$\begin{aligned} \mathscr{D}(\mathscr{A}) = \Big\{ & (y, z, \eta, \mu, \theta) \in \mathscr{H};\ y \in H^2(0, 1),\quad z \in H^1(0, 1), \\ & \eta_s \in \mathscr{S}\ \ \mu = z(0),\ \theta = z(1) \Big\}, \end{aligned} \tag{4.10.9}$$

and

$$\begin{aligned} \mathscr{A}(y, z, u, \kappa, \eta) = \Big(& z,\ (a y_x)_x - b(x)z,\ \eta_s + \kappa(s)\mu,\ \frac{1}{m}\left[-\sigma y(0) + (a y_x)(0)\right. \\ & \left. -\beta\eta(0) - \alpha\mu\right],\ -\frac{(a y_x)(1)}{M} \Big). \end{aligned} \tag{4.10.10}$$

The ultimate outcome is to prove that the Cauchy problem (4.10.8) is well-posed in the sense of semigroups theory of linear operators.

Firstly, we have the following lemma:

Lemma 4.10.1 *Assume that* (4.1.5), A1 *and* A2 *hold. Then, both operators* $\mathscr{A}$ *and its adjoint* $\mathscr{A}^*$ *are dissipative in* $\mathscr{H}$.

Proof Consider $\Phi = (y, z, \eta, \mu, \theta) \in \mathscr{D}(\mathscr{A})$. In light of (4.10.4)–(4.10.5), (4.10.7), and (4.10.9), we obtain after several integrations by parts

$$\begin{aligned}
\langle \mathscr{A}\Phi, \Phi\rangle_{\mathscr{H}} &= \int_0^1 \left\{ a y_x z_x + (a y_x)_x z - b(x) z^2 \right\} dx \\
&\quad -\beta \textstyle\int_0^\infty \frac{1}{\kappa'(s)} (\eta_s + \kappa(s)\mu)_s \eta_s \, dx \\
&\quad +(a y_x)(0)\mu - \alpha\mu^2 - \beta\eta(0)\mu - (a y_x)(1)\theta \\
&= -\alpha\mu^2 - \beta\eta(0)\mu - \int_0^1 b(x) z^2 \, dx - \beta \int_0^\infty \frac{1}{\kappa'(s)} \eta_{ss}(s)\eta_s(s)\, ds \\
&\quad -\beta\mu \textstyle\int_0^\infty \eta_s(s)\, ds \\
&= -\alpha\mu^2 - \int_0^1 b(x) z^2 \, dx + \frac{\beta}{2}\frac{\eta_s^2(0)}{\kappa'(0)} - \beta \int_0^\infty \frac{\kappa''(s)}{(\kappa'(s))^2} \eta_s^2(s)\, ds.
\end{aligned} \tag{4.10.11}$$

By virtue of A1 and A2, we conclude that $\langle \mathscr{A}\Phi, \Phi\rangle_{\mathscr{H}} \le 0$, for all $\Phi \in \mathscr{D}(\mathscr{A})$, and hence, $\mathscr{A}$ is dissipative in $\mathscr{H}$.

Next, we shall determine the adjoint operator $\mathscr{A}^*$. Let $\Phi = (y, z, \eta, \mu, \theta) \in \mathscr{D}(\mathscr{A})$ and $\widetilde{\Phi} = (\widetilde{y}, \widetilde{z}, \widetilde{\eta}, \widetilde{\mu}, \widetilde{\theta}) \in \mathscr{H}$. Using (4.10.7) and (4.10.9)–(4.10.10) and putting $\widetilde{z}(0) = \widetilde{\mu}$ and $\widetilde{z}(1) = \widetilde{\theta}$, a straightforward calculation yields

$$\begin{aligned}
\left\langle \mathscr{A}\Phi, \widetilde{\Phi}\right\rangle_{\mathscr{H}} &= \int_0^1 \{ a z_x \widetilde{y}_x + (a y_x)_x \widetilde{z} - b(x) z\widetilde{z} \} \, dx \\
&\quad -\beta \textstyle\int_0^\infty \frac{1}{\kappa'(s)} (\eta_s + \kappa(s)\mu)_s \widetilde{\eta}_s \, dx \\
&\quad +\sigma z(0)\widetilde{y}(0) - \sigma y(0)\widetilde{\mu} + (a y_x)(0)\widetilde{\mu} - \alpha\mu\widetilde{\mu} \\
&\quad -\beta\eta(0)\widetilde{\mu} - (a y_x)(1)\widetilde{\theta} \\
&= -\int_0^1 \{ a y_x \widetilde{z}_x + z(a\widetilde{y}_x)_x + b(x) z\widetilde{z} \} \, dx \\
&\quad +z(1)(a\widetilde{y}_x)(1) - z(0)(a\widetilde{y}_x)(0) \\
&\quad +\beta \frac{\eta_s(0)\widetilde{\eta}_s(0)}{\kappa'(0)} - \beta \int_0^\infty \frac{1}{\kappa'(s)} \eta_s \widetilde{\eta}_{ss}(s)\, ds \\
&\quad +\beta \textstyle\int_0^\infty \frac{\kappa''(s)}{(\kappa'(s))^2} \eta_s \widetilde{\eta}_s \, ds + \beta\mu\widetilde{\eta}(0) \\
&\quad +\sigma z(0)\widetilde{y}(0) - \sigma y(0)\widetilde{\mu} - \alpha\mu\widetilde{\mu} - \beta\eta(0)\widetilde{\mu}.
\end{aligned}$$

Writing the sum of the two integrals as follows:

$$\begin{aligned}
&\int_0^\infty \frac{\kappa''(s)}{(\kappa'(s))^2} \eta_s \widetilde{\eta}_s \, ds - \int_0^\infty \frac{1}{\kappa'(s)} \eta_s \widetilde{\eta}_{ss}(s)\, ds \\
&\quad = \int_0^\infty \frac{1}{\kappa'(s)} \eta_s \left\{ \frac{\kappa''(s)}{\kappa'(s)} \widetilde{\eta}_s - (\widetilde{\eta}_s + \kappa(s)\widetilde{\mu})_s \right\} ds \\
&\qquad +\widetilde{\mu}\eta(0),
\end{aligned}$$

we find

$$\begin{aligned}\left\langle \mathscr{A}\Phi, \widetilde{\Phi}\right\rangle_{\mathscr{H}} &= \int_0^1 \{az_x\widetilde{y}_x + (ay_x)_x\widetilde{z} - b(x)z\widetilde{z}\}\, dx \\ &-\beta \int_0^\infty \frac{1}{\kappa'(s)}\eta_s \{\mathscr{L}(\widetilde{\eta}) - (\widetilde{\eta_s} + \kappa(s)\widetilde{\mu})_s\}\, ds \\ &+\sigma y(0)(-\widetilde{\mu}) + \mu\left[\sigma\widetilde{y}(0) - (a\widetilde{y}_x)(0) - \alpha\widetilde{\mu} + \beta\widetilde{\eta}(0)\right] + \theta(a\widetilde{y}_x)(1),\end{aligned}$$

where $\mathscr{L}(\eta(s)) = -\kappa'(s)\left(\frac{H(s)}{\kappa'(s)}\right)' \eta_s(s)$ in which H is the Heaviside function.

Whereupon, in view of the identity $\left\langle \mathscr{A}\Phi, \widetilde{\Phi}\right\rangle_{\mathscr{H}} = \left\langle \Phi, \mathscr{A}^*\widetilde{\Phi}\right\rangle_{\mathscr{H}}$, we obtain

$$\mathscr{A}^*(y, z, \eta, \mu, \theta) =$$

$$\left(-z, -(ay_x)_x - b(x)z, \mathscr{L}(\eta) - \eta_s - \kappa(s)\mu, \frac{1}{m}\left[\sigma y(0) - (ay_x)(0) - \alpha\mu + \beta\eta(0)\right], \frac{(ay_x)(1)}{M}\right),$$

and

$$\mathscr{D}(\mathscr{A}) = \mathscr{D}(\mathscr{A}^*).$$

Subsequently, let us prove that the operator $\mathscr{A}^*$ is also dissipative. Arguing as for (4.10.11), we find

$$\begin{aligned}\left\langle \mathscr{A}^*\Phi, \Phi\right\rangle_{\mathscr{H}} &= -\alpha\mu^2 - \int_0^1 b(x)z^2\, dx - \frac{\beta}{2}\frac{\eta_s^2(0)}{\kappa'(0)} + \beta\int_0^\infty \frac{\kappa''(s)}{(\kappa'(s))^2}\eta_s^2(s)\, ds \\ &- \beta\int_0^\infty \frac{1}{\kappa'(s)}\mathscr{L}_s(\eta)\eta_s\, ds. \qquad (4.10.12)\end{aligned}$$

Using the definition of $\mathscr{L}(\eta)$ together with a simple integration by parts, we have

$$\int_0^\infty \frac{1}{\kappa'(s)}\mathscr{L}_s(\eta)\eta_s\, ds = \frac{\eta_s^2(0)}{\kappa'(0)} - \int_0^\infty \frac{\kappa''(s)}{(\kappa'(s))^2}\eta_s^2(s)\, ds. \qquad (4.10.13)$$

Thenceforth, owing to (4.10.12)–(4.10.13), it follows that

$$\left\langle \mathscr{A}^*\Phi, \widetilde{\Phi}\right\rangle_{\mathscr{H}} = -\alpha\mu^2 - \int_0^1 b(x)z^2\, dx + \frac{\beta}{2}\frac{\eta_s^2(0)}{\kappa'(0)} - \frac{\beta}{2}\int_0^\infty \frac{\kappa''(s)}{(\kappa'(s))^2}\eta_s^2(s)\, ds,$$

for all $\Phi \in \mathscr{D}(\mathscr{A}^*)$. Hence, by the assumptions `A1` and `A2`, the operator $\mathscr{A}^*$ is dissipative. □

Consequently, as a direct result of Corollary 4.4. in [158], the operator $\mathscr{A}$ is the infinitesimal generator of a C_0-semigroup of contractions $e^{t\mathscr{A}}$ on $\mathscr{H}$. Additionally, the semigroups theory of linear operators yields [158]:

Proposition 4.10.1 *Assume that the assumptions* A1 *and* A2 *are fulfilled. Then, for any initial data* $\Phi_0 \in \mathscr{H}$*, the Cauchy problem (4.10.8) has a unique mild global solution* $\Phi(t) \in \mathscr{H}$*. Moreover, if* $\Phi_0 \in \mathscr{D}(\mathscr{A})$*, there exists a unique classical global solution* $\Phi(t) \in \mathscr{A}$*.*

4.10.2 Exponential Stability of the System (4.10.2)

The main objective of this subsection is to show the exponential stability result of the system (4.10.2). As previously done, the strategy consists in applying the following frequency-domain theorem (see Theorem 2.3.8) and combining a contradiction argument with the multiplier technique to carry out a special analysis for the resolvent.

Now, we are ready to state our goal that gives the exponential stability result of the semigroup $e^{t\mathscr{A}}$ in the space $\mathscr{H}$.

Theorem 4.10.1 *Assume that the assumptions* A1 *and* A2 *hold. Moreover, suppose that the memory kernel satisfies the following additional condition: There exists a positive constant* ϖ *such that*

$$\kappa''(s) \geq -\varpi\kappa'(s), \quad \text{for all} \quad s \in \mathbb{R}_+. \tag{4.10.14}$$

Then, the system (4.10.2) is exponentially stable in $\mathscr{H}$*, that is, the associated semigroup* $e^{t\mathscr{A}}$ *exponentially tends toward zero in* $\mathscr{H}$*.*

The proof of this theorem relies on some results given in the following lemma and two propositions.

Lemma 4.10.2 *Under the condition (4.1.5) and the assumptions* A1 *and* A2*, we have*

$$0 \in \rho(\mathscr{A}).$$

Proof To prove this lemma, consider $\Phi = (y, z, \eta, \mu, \theta) \in \mathscr{D}(\mathscr{A})$ such that $\mathscr{A}\Phi = 0$. Equivalently, we have

$$z = 0, (ay_x)_x = 0, \tag{4.10.15}$$

$$\eta_s + \kappa(s)\mu = 0, \tag{4.10.16}$$

$$(ay_x)(0) - \sigma y(0) - \alpha\mu - \beta\eta(0) = 0, \tag{4.10.17}$$

$$(ay_x)(1) = 0. \tag{4.10.18}$$

In view of (4.10.15) and (4.10.18), we have $ay_x = 0$, which implies that $(ay_x)(0) = 0$ and $y = c \in \mathbb{R}$. Thanks to (4.10.15), it results that $\mu = \theta = 0$. Then, (4.10.16) and (4.10.5) give $\eta = 0$, and thus $y(0) = 0$ by (4.10.17). Therefore, we get $y = 0$, and hence, $\Phi = 0$.

Next, we claim that the operator $\mathscr{A}$ is onto $\mathscr{H}$. For that, one has to show that given $F = (f, g, v, p, q) \in \mathscr{H}$, there exists $U = (u, v, \eta, \mu, \theta) \in \mathscr{D}(\mathscr{A})$ such that $\mathscr{A}U = F$. One can readily verify that this is equivalent to solving the following problem:

$$\begin{cases} z = f, \\ (ay_x)_x - b(x)z = g, \\ \eta_s + \kappa(s)\mu = v, \\ (ay_x)(0) - \sigma y(0) - \alpha\mu - \beta\eta(0) = mp, \\ (ay_x)(1) = -Mq. \end{cases} \tag{4.10.19}$$

Solving the equation of η in the above system, we find

$$\eta(s) = \int_s^\infty (v(\tau) - \kappa(\tau) f(1))\, d\tau, \tag{4.10.20}$$

and thus

$$\eta(0) = \int_0^\infty (v(s) - \kappa(s) f(1))\, ds. \tag{4.10.21}$$

Inserting (4.10.20) and (4.10.21) in (4.10.19), it amounts to considering

$$\begin{cases} (ay_x)_x = b(x) f + g, \\ \eta_s + \kappa(s)\mu = v, \\ (ay_x)(0) - \sigma y(0) = \alpha f(0) + \beta \displaystyle\int_0^\infty (v(s) - \kappa(s) f(1))\, ds + mp, \\ (ay_x)(1) = -Mq. \end{cases} \tag{4.10.22}$$

Multiplying the first equation in (4.10.22) by $\psi \in H_0^1(0, 1)$ and then integrating from 0 to 1, we find the variational problem

$$\mathscr{F}(y, \psi) = \mathscr{K}(\psi),$$

where $\mathscr{F}$ is a real bilinear form in $H_0^1(0, 1) \times H_0^1(0, 1)$ given as

$$\mathscr{F}(y, \psi) = \int_0^1 ay_x\psi_x\, dx + \sigma y(0)\psi(0),$$

and $\mathscr{K}$ is a linear form defined as follows:

$$\mathscr{K}(\psi) = Mq\psi(1) + \left[\alpha f(0) + \beta \int_0^\infty (v(s) - \kappa(s) f(1))\, ds + mp\right] \psi(0)$$
$$- \int_0^1 (b(x) f + g)\psi\, dx.$$

In light of (4.1.5), A1 and A2, and applying Lax-Milgram theorem (see Theorem 2.1.6), one can deduce the existence of a unique solution $y \in H^2(0, 1)$ of (4.10.22), and hence, $\mathscr{A}$ is onto $\mathscr{H}$. Therefore, $0 \in \rho(\mathscr{A})$. □

Next, we show that $\mathscr{A}$ has no continuous spectrum on the imaginary axis.

Proposition 4.10.2 *Suppose that (4.1.5),* A1*, and* A2 *hold. If λ is a real number, then $i\lambda$ belongs to the resolvent set $\rho(\mathscr{A})$.*

Proof First of all, it is important to mention that due to the space $\mathscr{S}$, the injection $\mathscr{D}(\mathscr{A}) \hookrightarrow \mathscr{H}$ is not compact, which implies that the operator $\mathscr{A}$ does not have a compact resolvent and its spectrum $\sigma(\mathscr{A})$ does not consist only of eigenvalues of $\mathscr{A}$. In turn, recalling that $0 \in \rho(\mathscr{A})$, we shall show that

$$\ker(i\lambda I - \mathscr{A}) = \{0\} \quad \text{and} \quad \Im(i\lambda I - \mathscr{A}) = \mathscr{H}, \quad \forall \lambda \in \mathbb{R}^*. \tag{4.10.23}$$

To prove the first requirement in (4.10.23), we will use a contradiction argument. Indeed, assume that this property is false. Then, there exists $\lambda \in \mathbb{R}^*$ such that $i\lambda$ is an element of $\sigma(\mathscr{A})$, that is, there exists $\Phi = (y, z, \eta, \mu, \theta) \in \mathscr{D}(\mathscr{A})$ verifying $\|\Phi\|_{\mathscr{H}} = 1$ and

$$\mathscr{A}\Phi = i\lambda\Phi. \tag{4.10.24}$$

Hence, we have

$$z = i\lambda y, \tag{4.10.25}$$

$$(ay_x)_x - b(x)z = i\lambda z, \tag{4.10.26}$$

$$\eta_s + \kappa(s)\mu = i\lambda\eta, \tag{4.10.27}$$

$$(ay_x)(0) - \sigma y(0) - \alpha\mu - \beta\eta(0) = i\lambda m\mu, \tag{4.10.28}$$

$$(ay_x)(1) = -i\lambda M\theta. \tag{4.10.29}$$

Taking the inner product with Φ on both sides of (4.10.24) together with the dissipativity of $\mathscr{A}$ given in (4.10.11), it follows that

$$\mu = \eta_s(0) = \int_0^1 b(x)z^2\,dx = \int_0^\infty \frac{\kappa''(s)}{(\kappa'(s))^2}\eta_s^2(s)\,ds = 0. \tag{4.10.30}$$

Next, in view of (4.10.14), we have the inequality

$$0 \le -\varpi \int_0^\infty \frac{1}{\kappa'(s)}\eta_s^2(s)\,ds \le \int_0^\infty \frac{\kappa''(s)}{(\kappa'(s))^2}\eta_s^2(s)\,ds = 0, \tag{4.10.31}$$

which implies that $\|\eta\|_{\mathscr{S}} = 0$ and thus $\eta = 0$. From this, together with (4.10.25)–(4.10.26), (4.10.28), and (4.10.30)–(4.10.31), we can deduce that

$$\begin{cases} (ay_x)_x + \lambda^2 y = 0, \\ y(0) = (ay_x)(0) = 0; \quad (ay_x)(1) = \lambda^2 My(1). \end{cases} \tag{4.10.32}$$

Thereafter, consider the function (see [17, 64, 70, 74] for similar cases)

$$q(x) = \exp\left(\int_x^1 \left|\frac{a'}{a}\right|(s)\,ds\right)\left(-M - \int_x^1 \exp\left(\int_x^1 \left|\frac{a'}{a}\right|(r)\,dr\right)\,ds\right),$$
$$\forall s \in [0, 1], \tag{4.10.33}$$

which satisfies the following properties:

$$q(1) = -M,\ \ q'(x) \ge 1,\ \ (aq' - |a'|q)(x) \ge a(x) \ge \hat{a} > 0,\ \ \text{and}\ q(x) < 0,$$
$$\forall x \in [0, 1]. \tag{4.10.34}$$

Multiplying the equation in (4.10.32) by qy_x and integrating by parts, one can readily verify that the boundary value problem (4.10.32) has only the trivial solution $y = 0$, and hence, $\theta = 0$ since (4.10.29). Thereby, these findings give $\Phi = (y, z, \eta, \mu, \theta) = 0$, which contradicts the fact that $\|\Phi\| = 1$. Consequently, the first assumption in (4.10.23) is verified.

Now, suppose that the second condition of (3.12) is not true. Then, there exists at least one nonzero $\lambda \in \mathbb{R}$ and $F \in \mathscr{H}$ such that $F \notin \Im(i\lambda I - \mathscr{A})$. This implies the existence of a sequence $(F_n) \subset \mathscr{H}$ where $F_n \to F$ and $(i\lambda I - \mathscr{A})^{-1}F_n = G_n \to \infty$ in $\mathscr{H}$ from which we have

$$(i\lambda I - \mathscr{A})\frac{G_n}{\|G_n\|} = \frac{F_n}{\|G_n\|} \to 0, \quad \text{in} \ \ \mathscr{H}.$$

Equivalently, there exists a sequence $\Phi_n = (y_n, z_n, \eta_n, \mu_n, \theta_n) \in \mathscr{D}(\mathscr{A})$ with $\|\Phi_n\|_{\mathscr{H}} = 1$ such that

$$(i\lambda I - \mathscr{A})\Phi_n = (f_n, g_n, v_n, p_n, q_n) \to 0, \quad \text{in} \quad \mathscr{H}, \tag{4.10.35}$$

that is,

$$i\lambda y_n - z_n = f_n \to 0, \quad \text{in} \quad H_0^1(\Omega), \tag{4.10.36}$$

$$i\lambda z_n - (a y_{n,x})_x + b(x) z_n = g_n \to 0, \quad \text{in} \quad L^2(\Omega), \tag{4.10.37}$$

$$i\lambda \eta_n - \eta_{n,s} - \kappa(s)\mu_n = v_n \to 0, \quad \text{in} \quad \mathscr{S}, \tag{4.10.38}$$

$$i\lambda m\mu_n - (a y_{n,x})(0) + \sigma y_n(0) + \alpha\mu_n + \beta\eta_n(0) = m p_n \to 0, \quad \text{in} \quad \mathbb{C}, \tag{4.10.39}$$

$$i\lambda M\theta_n + (a y_{n,x})(1) = M q_n \to 0, \quad \text{in} \quad \mathbb{C}. \tag{4.10.40}$$

Thanks to (4.10.11) and (4.10.35), we can deduce

$$\mu_n \to 0, \quad \frac{\eta_{n,s}(0)}{\kappa'(0)} \to 0, \tag{4.10.41}$$

$$\int_0^1 b(x) z_n^2 \, dx \to 0, \tag{4.10.42}$$

$$\text{and} \qquad \int_0^\infty \frac{\kappa''(s)}{(\kappa'(s))^2} \eta_{n,s}^2(s) \, ds \to 0. \tag{4.10.43}$$

Using once more (4.10.14), we obtain

$$0 \leq -\varpi \int_0^\infty \frac{1}{\kappa'(s)} \eta_{n,s}^2(s) \, ds \leq \int_0^\infty \frac{\kappa''(s)}{(\kappa'(s))^2} \eta_{n,s}^2(s) \, ds \to 0,$$

and hence,

$$\eta_n \to 0 \quad \text{in} \quad \mathscr{S}. \tag{4.10.44}$$

Invoking (4.10.44) and Holder's inequality, we have

$$\eta_n^2(0) = \left(\int_0^\infty \sqrt{-\kappa'(s)} \frac{\eta_{n,s}(s)}{\sqrt{-\kappa'(s)}} \, ds \right)^2 \leq \kappa(0) \, \|\eta_n\|_{\mathscr{S}}^2,$$

that is,

$$\eta_n(0) \to 0. \tag{4.10.45}$$

According to (4.10.35), we have $f_n(0) \to 0$, which implies with (4.10.36) and (4.10.41) that

$$y_n(0) \to 0. \tag{4.10.46}$$

By virtue of (4.10.41), (4.10.45)–(4.10.46), and (4.10.39), it follows that

$$y_{n,x}(0) \to 0. \tag{4.10.47}$$

Next, we solve (4.10.36) for z_n and insert it into (4.10.37) to obtain an equation of y_n solely. Then, we multiply the resulting equation by $qy_{n,x}$, integrate by parts, use (4.10.42) together with (4.10.46)–(4.10.47) and the fact that $\|\Phi_n\|_{\mathscr{H}} = 1$, and we deduce

$$\begin{aligned}&\lambda^2 q(1)y_n^2(1) + a(1)q(1)y_{n,x}^2(1) - \int_0^1 (aq' - a'q)(x)y_{n,x}^2\, dx\\ &- \lambda^2 \int_0^1 q'(x)y_n^2\, dx = o(1).\end{aligned}$$

The latter implies, thanks to (4.10.34), that

$$y_n \to 0 \quad \text{in} \quad H_0^1(\Omega), \tag{4.10.48}$$

and

$$y_n(1) \to 0 \quad \text{and} \quad y_{n,x}(1) \to 0. \tag{4.10.49}$$

Whereupon, according to (4.10.36), (4.10.40), and (4.10.48)–(4.10.49), we get

$$\theta_n \to 0,$$

and

$$z_n \to 0 \quad \text{in} \quad L^2(\Omega).$$

Finally, we obtain $\Phi_n \to 0$ in $\mathscr{H}$, which contradicts the fact that $\|\Phi_n\|_{\mathscr{H}} = 1$ and the proof of the second assumption in (4.10.23) is achieved. □

As a direct consequence of Proposition 4.10.2, the operator $\mathscr{A}$ does not have purely imaginary eigenvalues, that is, the system (4.10.2) is strongly stable, i.e.,

$$\lim_{t \to +\infty} E(t) = 0,$$

where the energy $E(t)$ of (4.10.2) or equivalently (4.10.6) is given by

$$E(t) = \frac{1}{2}\int_0^1 (ay_x^2 + y_t^2)\,dx - \beta\int_0^\infty \frac{1}{\kappa'(s)}\eta_s^2(s)\,ds + \sigma y^2(0,t) + my_t^2(0,t) + My_t^2(1,t).$$

The following proposition shows that (4.6.29) holds for our operator $\mathscr{A}$.

Proposition 4.10.3 *In addition to the assumptions* A1 *and* A2, *assume that the conditions (4.1.5) and (4.10.14) are all verified. Then, the resolvent operator of* $\mathscr{A}$ *obeys the requirement* (2.3.17).

Proof Suppose that (2.3.17) is not true. Then, by the Banach-Steinhaus theorem (see Theorem 2.1.7), there exist a sequence of real numbers $\lambda_n \to \infty$ and a sequence of vectors $\Phi_n = (y_n, z_n, \eta_n, \mu_n, \theta_n)$ in $\mathscr{D}(\mathscr{A})$, with $\|\Phi_n\|_{\mathscr{H}} = 1$ such that

$$(i\lambda_n I - \mathscr{A})\Phi_n = (f_n, g_n, v_n, p_n, q_n) \to 0, \quad \text{in } \mathscr{H},$$

that is,

$$i\lambda_n y_n - z_n = f_n \to 0, \quad \text{in } H_0^1(\Omega), \tag{4.10.50}$$

$$i\lambda_n z_n - (ay_{n,x})_x + b(x)z_n = g_n \to 0, \quad \text{in } L^2(\Omega), \tag{4.10.51}$$

$$i\lambda_n \eta_n - \eta_{n,s} - \kappa(s)\mu_n = v_n \to 0, \quad \text{in } \mathscr{S},$$

$$i\lambda_n m\mu_n - (ay_{n,x})(0) + \sigma y_n(0) + \alpha\mu_n + \beta\eta_n(0) = mp_n \to 0, \quad \text{in } \mathbb{C}, \tag{4.10.52}$$

$$i\lambda_n M\theta_n + (ay_{n,x})(1) = Mq_n \to 0, \quad \text{in } \mathbb{C}. \tag{4.10.53}$$

Noting that $\Re((i\lambda_n I - \mathscr{A})\Phi_n, \Phi_n) \to 0$ and using (4.10.11), we deduce

$$\mu_n \to 0, \quad \frac{\eta_{n,x}(0)}{\kappa'(0)} \to 0, \tag{4.10.54}$$

$$\int_0^1 b(x)z_n^2\,dx \to 0, \tag{4.10.55}$$

and
$$\int_0^\infty \frac{\kappa''(s)}{(\kappa'(s))^2}\eta_{n,s}^2(s)\,ds \to 0.$$

Thanks to $0 \le -\varpi\int_0^\infty \frac{1}{\kappa'(s)}\eta_{n,s}^2(s)\,ds \le \int_0^\infty \frac{\kappa''(s)}{(\kappa'(s))^2}\eta_{n,s}^2(s)\,ds \to 0$, we get

$$\|\eta_n\|_{\mathscr{S}} \to 0,$$

which, together with Holder inequality, yields

$$\begin{aligned} \eta_n(0) &= \int_0^\infty \sqrt{-\kappa'(s)} \frac{\eta_{n,s}(s)}{\sqrt{-\kappa'(s)}} \, ds \\ &\le \left(\int_0^\infty -\kappa'(s) \, ds \right)^{\frac{1}{2}} \left(\int_0^\infty \frac{\eta_{n,s}^2}{-\kappa'(s)} \, ds \right)^{\frac{1}{2}} \\ &\le \sqrt{\kappa(0)} \, \|\eta_n\|_{\mathscr{S}} . \end{aligned} \tag{4.10.56}$$

Using (4.10.50) and the boundedness of z_n, we have

$$y_n = \frac{1}{i\lambda_n}(z_n + f_n) \to 0 \quad \text{in} \quad L^2(0, 1). \tag{4.10.57}$$

Recalling the fact that $(f_n, g_n, \eta_n, \mu_n, \theta_n) \to 0$ in $\mathscr{H}$, we obtain

$$f_n \to 0, \tag{4.10.58}$$

which, together with (4.10.50) and (4.10.54), implies that

$$i\lambda_n y_n(0) \to 0 \quad \text{and thus} \quad y_n(0) \to 0. \tag{4.10.59}$$

Combining (4.10.52) with (4.10.54), (4.10.56), and (4.10.59), it follows that

$$\frac{(y_n)_x(0)}{\lambda_n} \to 0. \tag{4.10.60}$$

On the other hand, multiplying (4.10.50) by $ay_{n,x}$ and integrating over (0, 1), we get

$$\lambda_n \int_0^\infty z_n y_{n,x} \, dx - \int_0^\infty (ay_{n,x})_x a y_{n,x} \, dx + \int_0^\infty b(x) z_n a y_{n,x} \, dx = \int_0^\infty g_n a y_{n,x} \, dx.$$

Using the fact that $\|\Phi_n\|_{\mathscr{H}} = 1$, $g_n \to 0$ in $L^2(0, 1)$, and (4.10.60) in the above identity, we deduce that $\frac{(ay_{n,x})(1)}{\lambda_n} \to 0$. The latter gives, thanks to (4.10.53), that

$$\theta_n \to 0, \tag{4.10.61}$$

and thus, $z_n(1) \to 0$. Then, combining (4.10.58) and

$$f_n(1) = \int_0^1 f_{n,x}(x) \, dx + f_n(0) \to 0,$$

we have

$$i\lambda_n y_n(1) = z_n(1) + f_n(1) \to 0 \quad \text{and hence} \quad y_n(1) \to 0. \tag{4.10.62}$$

Thereafter, we take the inner product in $L^2(0, 1)$ of (4.10.50) (resp. (4.10.51)) with z_n (resp. y_n) and then sum up. Integrating by parts and recalling $\|\Phi_n\|_{\mathscr{H}} = 1$, (4.10.55), (4.10.59), and (4.10.62), we obtain

$$\int_0^1 |z_n|^2\, dx - \int_0^1 a(x)|y_{n,x}|^2\, dx = o(1). \tag{4.10.63}$$

By virtue of (4.10.63) and the fact that $\|\Phi_n\|_{\mathscr{H}} = 1$, we get

$$\|z_n\|_{L^2(0,1)} \to \frac{1}{2} \quad \text{and} \quad \left\|a y_{n,x}\right\|_{L^2(0,1)} \to \frac{1}{2}. \tag{4.10.64}$$

On the other hand, substituting (4.10.50) in (4.10.51), multiplying the resulting equation by $\chi y_{n,x}$, where χ is a function of $C^1(0, 1; \mathbb{R})$ to be determined later, and integrating over (0, 1), we have

$$\begin{aligned}\int_0^\infty (-\lambda_n^2 y_n - (a y_{n,x})_x)\chi(x) y_{n,x}\, dx + \int_0^\infty b(x) z_n \chi(x) y_{n,x}\, dx \\ = \int_0^\infty (i\lambda_n f_n + g_n)\chi(x) y_{n,x}\, dx.\end{aligned} \tag{4.10.65}$$

Using the boundedness of y_n in $H^1(0, 1)$, (4.10.55), the convergence of (f_n, g_n) to zero in $H^1(0, 1) \times L^2(0, 1)$, and the convergence $f_n(0),\ f_n(1),\ \lambda_n y_n(0),\ \lambda_n y_n(1)$ again to zero in $\mathbb{C}$, a straightforward calculation shows that the second term of the left-hand side and the right-hand side of (4.10.65) converges to zero.

Next, in view of (4.10.59) and (4.10.62) and a direct computation based on simple integrations by parts, we obtain

$$\begin{aligned}\Re\left\{\int_0^1 -\lambda_n^2 \chi(x) y_n y_{n,x}\, dx\right\} &= \frac{1}{2}\int_0^1 \chi'(x)|\lambda_n y_n(x)|^2\, dx - \frac{\chi(1)}{2}|\lambda_n y_n(1)|^2 \\ &\quad + \frac{\chi(0)}{2}|\lambda_n y_n(0)|^2 \\ &= \frac{1}{2}\int_0^1 \chi'(x)|\lambda_n y_n(x)|^2\, dx + o(1),\end{aligned}$$

$$\begin{aligned}\Re\left\{-\int_0^1 \chi(x)\left(a(x) y_{n_x}\right)_x y_{n,x}\, dx\right\} &= \frac{1}{2}\int_0^1 (a\chi' - a'\chi)(x)|(y_n)_x|^2\, dx \\ &\quad - \frac{a(1)\chi(1)}{2}|y_{n,x}(1)|^2 \\ &\quad + \frac{a(0)\chi(0)}{2}|y_{n,x}(0)|^2.\end{aligned}$$

This, together with (4.10.65), leads to

$$\int_0^1 (a\chi' - a'\chi)(x)|y_{n,x}|^2\,dx + \int_0^1 \chi'(x)(x)|\lambda_n y_n|^2\,dx - a(1)\chi(1)|y_{n,x}(1)|^2 + a(0)\chi(0)|y_{n,x}(0)|^2 = o(1). \tag{4.10.66}$$

Thereafter, similarly to [167], let us choose the function χ as follows:

$$\chi(x) = \exp\left(\int_0^x \left|\frac{a'}{a}\right|(s)\,ds\right)\int_0^x \exp\left(-\int_0^s \left|\frac{a'}{a}\right|(\tau)\,d\tau\right)\,ds, \tag{4.10.67}$$

which satisfies

$$\begin{cases} \chi(0) = 0, \\ \chi(x) \geq 0, \quad \chi'(x) \geq 1, \quad \forall x \in [0,1], \\ (a\chi' - a'\chi)(x) \geq a(x),\ \forall x \in [0,1]. \end{cases} \tag{4.10.68}$$

Combining (4.10.66)–(4.10.68) and the fact that $\|\Phi_n\|_{\mathscr{H}} = 1$, it follows that $(y_{n,x}(1))$ must be bounded in $\mathbb{C}$ and so is $(y_{n,x}(0))$.

Now, let us pick up any smooth function ψ. Then, multiplying (4.10.51) by $(a\psi)_x y_n$ and integrating by parts, we obtain after using (4.10.55), (4.10.57), (4.10.59), (4.10.62) as well as the boundedness of $(y_{n,x}(0))$ and $(y_{n,x}(1))$:

$$\int_0^1 (a\psi)_x(x)|\lambda_n y_n|^2\,dx - \int_0^1 (a\psi)_x(x)a(x)|y_{n,x}|^2\,dx = o(1). \tag{4.10.69}$$

Thereafter, we go back to (4.10.66) and take χ so that $(a\chi' - a'\chi)(x) = (a(a\psi)_x)(x)$. This together with (4.10.69) implies that

$$a(1)\chi(1)|y_{n,x}(1)|^2 - a(0)\chi(0)|y_{n,x}(0)|^2 - \int_0^1 \left[\chi' + \left(\frac{\chi}{a}\right)_x a\right](x)|\lambda_n y_n|^2\,dx = o(1). \tag{4.10.70}$$

Now, we choose χ as

$$\chi(x) = \frac{\sqrt{a(x)}}{2}\int_0^x \frac{b(v)}{\sqrt{a(v)}}\,dv,$$

and hence,

$$\left(\chi' + \left(\frac{\chi}{a}\right)_x a\right)(x) = b(x), \quad \chi(0) = 0.$$

Then, using (4.10.70), it follows that

$$a(1)\chi(1)|y_{n,x}(1)|^2 - \int_0^1 b(x)|\lambda_n y_n|^2\, dx = o(1). \tag{4.10.71}$$

On the other hand, according to (4.10.50) and (4.10.55), we have

$$\int_0^1 b(x)|\lambda_n y_n|^2\, dx \to 0,$$

and therefore, (4.10.71) yields

$$y_{n,x}(1) = o(1). \tag{4.10.72}$$

Lastly, we turn to (4.10.66), where we again use the function χ defined in (4.10.67), and keeping in mind (4.10.72), we can deduce that

$$\int_0^1 a(x)|y_{n,x}|^2\, dx \quad \text{and} \quad \int_0^1 |\lambda_n y_n|^2\, dx \to 0, \tag{4.10.73}$$

which contradicts (4.10.64), and thus the proof is completed. □

Proof of Theorem 4.10.1 The two assumptions of Theorem 2.3.8 are being shown in Proposition 4.10.2 and Proposition 4.10.3, and the proof of Theorem 4.10.1 is thus achieved. □

4.10.3 Concluding Discussion

In this section, we were able to handle the presence of a boundary infinite memory term in the proposed feedback law for the exponential stabilization of the crane structure. To do so, a local interior damping control is introduced so that the system is well-posed and exponentially stable. The main ingredient of the proof is the utilization of the resolvent method, although there is a lack of the usual compactness property of the resolvent operator of the system's operator.

As in the previous section, it is highly recommended to weaken the conditions on the memory kernel so that a wide range of functions κ could be considered. Also, one could treat the case of time-dependent memory kernel.

4.11 Conclusion

This chapter has discussed the stabilization problem of the overhead crane subject to a boundary delay or memory. Several types of boundary and interior controls are designed to reach either polynomial or exponential stabilization outcomes. Each situation is thoroughly explored, and occasionally, numerical examples are provided to illustrate the theoretical findings. Additionally, for each problem tackled, we gave some concluding remarks and discussed future research directions.

Bibliography

1. Z. Abbas, K. Ammari and D. Mercier, Remarks on stabilization of second order evolution equations by unbounded dynamic feedbacks, *Journal of Evolution Equations,* 16 (2016), 95–130.
2. Z. Abbas and S. Nicaise, The multidimensional wave equation with generalized acoustic boundary conditions I: Strong stability, *SIAM J. Control Optim.*, 53 (2015), 2558–2581.
3. E. M. Abdel-Rahman, A. H. Nayfeh and Z. N. Masoud, Dynamics and control of cranes: A review, *J. Vibration and Control*, 9 (2003), 863–908.
4. R. A. Adams, *Sobolev spaces,*, Pure and Applied Mathematics, Vol. 65. Academic Press, New York-London, 1975.
5. E. M. Ait Ben Hassi, K. Ammari, S. Boulite and L. Maniar, Feedback stabilization of a class of evolution equations with delay, *J. Evol. Eq.*, 1 (2009), 103–121.
6. ———, Exponential energy decay of some coupled second order systems, *Semigroup Forum*, 86 (2013), 362–382.
7. F. Alabau-Boussouira, Convexity and weighted integral inequalities for energy decay rates of nonlinear dissipative hyperbolic systems, *Appl. Math. Optim.*, 51 (2005), 61–105.
8. F. Alabau-Boussouira, A unified approach via convexity for optimal energy decay rates of finite and infinite dimensional vibrating damped systems with applications to semi-discretized vibrating damped systems, *J. Differ. Eqs.*, 248 (2010), 1473–1517.
9. F. Alabau-Boussouira and K. Ammari, Sharp energy estimates for nonlinearly locally damped PDEs via observability for the associated undamped system, *J. Funct. Anal.*, 260 (2011), 2424–2450.
10. F. Alabau-Boussouira, On some recent advances on stabilization for hyperbolic equations. In Control of partial differential equations, volume 2048 of Lecture Notes in Math., pages 1–100. Springer, Heidelberg, 2012.
11. F. Alabau-Boussouira, P. Cannarsa and G. Leugering, Control and stabilization of degenerate wave equations, *SIAM Journal on Control and Optimization*, 55 (2017), 2052–2087.
12. F. Al-Musallam, K. Ammari and B. Chentouf, Asymptotic behavior of a 2D overhead crane with input delays in the boundary control, *Zeitschrift fur Angewandte Mathematik und Mechanik*, 98 (2018), 1103–1122.
13. G. Amendola, M. Fabrizio, and J. M. Golden, *Thermodynamics of Materials with Memory: Theory and Applications*, Springer, Berlin, 2012.
14. K. Ammari, A. Bchatnia and K. EL Mufti, Stabilization of the nonlinear damped wave equation via linear weak observability, *NoDEA Nonlinear Differential Equations and Applications,* 2 (2016), 23:6.

K. Ammari, B. Chentouf, *Stabilization of Wave-Type Equations with Delay or Memory*, Advances in Mechanics and Mathematics 57,
https://doi.org/10.1007/978-3-032-28496-9

15. K. Ammari and B. Chentouf, Asymptotic behavior of a delayed wave equation without displacement term, *Z. Angew. Math. Phys.*, 68 (2017), no. 5, Art. 117, 13 pp.
16. ———, On the exponential and polynomial convergence for a delayed wave equation without displacement, *Applied Mathematics Letters*, 86 (2018), 126–133.
17. ———, Further results on the long-time behavior of a 2D overhead crane with a boundary delay: Exponential convergence, *Applied Math. and Computation*, 365 (2020), 365, Article 124698, 17 pages.
18. K. Ammari and S. Gerbi, Interior feedback stabilization of wave equations with dynamic boundary delay, *Zeitschrift für Analysis und Ihre Anwendungen*, 36 (2017), 297–327.
19. K. Ammari and S. Nicaise, Stabilization of elastic systems by collocated feedback, *Lecture Notes in Mathematics*, 2124, Springer, Cham, 2015.
20. K. Ammari, S. Nicaise, and C. Pignotti, Feedback boundary stabilization of wave equations with interior delay, *Syst. Control Lett.*, 59 (2010), 623–628.
21. ———, Stabilization by switching time-delay, *Asymptotic Analysis*, 83 (2013), 263–283.
22. ———, Stability of an abstract-wave equation with delay and a Kelvin–Voigt damping, *Asymptotic Analysis*, 95 (2015), 21–38.
23. K. Ammari and M. Tucsnak, Stabilization of second order evolution equations by a class of unbounded feedbacks, *ESAIM Control Optim. Calc. Var.*, 6 (2001), 361–386.
24. B. d'Andréa-Novel, F. Boustany and B. Rao, Feedback stabilization of a hybrid PDE-ODE system: Application to an overhead crane, *Math. Control. Signals Systems.*, 7 (1994), 1–22.
25. B. d'Andréa-Novel and J. M. Coron, Exponential stabilization of an overhead crane with flexible cable via a back-stepping approach, *Automatica,* 36 (2000), 587–593.
26. B. d'Andréa-Novel, I. Moyano and L. Rosier, Finite-time stabilization of an overhead crane with a flexible cable, *Math. Control. Signals Systems.*, 31 (2019), 1–19.
27. K. T. Andrews, K. L. Kuttler, and M. Shillor, Second order evolution equations with dynamic boundary conditions, *J. Math. Anal. Appl.*, 197 (1996), 781–795.
28. V. I. Arnold, *Mathematical Methods of Classical Mechanics*, Springer-Verlag, New York, 1989.
29. B. Augner, Well-posedness and stability of infinite-dimensional linear Port-Hamiltonian systems with nonlinear boundary feedback, *SIAM Journal on Control and Optimization*, 57 (2019), 1818–1844.
30. J. Auriol, U. J. F. Aarsnes, P. Martin and F. D. Meglio, Delay-robust control design for two heterodirectional linear coupled hyperbolic PDEs, *IEEE Transactions on Automatic Control*, 63 (2018), 3551–3557.
31. J. Auriol, F. Bribiesca-Argomedo, D. Bou Saba, M. D. Loreto, F. D. Meglio, Delay-robust stabilization of a hyperbolic PDE-ODE system, *Automatica*, 95 (2018), 494–502.
32. C. Bardos, G. Lebau and J. Rauch, Sharp sufficient conditions for the observation, controllability ans stabilization of waves from the boundary, *SIAM J. Control Optim.*, 30 (1992), 1024–1064.
33. A. Bátkai and S. Piazzera, *Semigroups for delay equations,* Research Notes in Mathematics 10, A. K. Peters, Wellesley MA (2005).
34. C. J. K. Batty and T. Duyckaerts, Non-uniform stability for bounded semi-groups on Banach spaces, *J. Evol. Eqs.,* 8 (2008), 765–780.
35. L. Baudouin, E. Crépeau, J. Valein, Two approaches for the stabilization of nonlinear KdV equation with boundary time-delay feedback, *IEEE Trans. Automat. Control.*, 64 (2019), 1403–1414.
36. J. T. Beale, *Spectral properties of an acoustic boundary condition*, Indiana Univ. Math. J., 25 (1976), 895–917.
37. A. Benaissa, A. Benaissa and S. A. Messaoudi, Global existence and energy decay of solutions for the wave equation with a time varying delay term in the weakly nonlinear internal feedbacks, *J. Math. Phys.,* 53 (2012), 123514, https://doi.org/10.1063/1.4765046.
38. A. Borichev and Y. Tomilov, Optimal polynomial decay of functions and operator semigroups, *Math. Ann.,* 347 (2010), 455–478.

39. H. Brezis, *Opérateurs Maximaux Monotones et Semi-groupes de Contractions dans les Espaces de Hilbert*, North-Holland, Mathematics Studies, 1973.
40. H. Brezis, Functional Analysis, Sobolev Spaces and Partial Differential Equations, Universitex, Springer, 2011.
41. B. M. Budak, A. A. Samarskii, and A. N. Tikhonov, *A collection of problems on mathematical physics*, Translated by A. R. M. Robson, The Macmillan Co., New York, 1964.
42. N. Burq, Décroissance de l'énergie locale de l'équation des ondes pour le problème extérieur et absence de résonance au voisinage du réel, *Acta Math.,* 180 (1998), 1–29.
43. R. W. Carroll and R. E. Showalter, *Singular and degenerate Cauchy problems*, Academic Press [Harcourt Brace Jovanovich Publishers], New York, 1976. Mathematics in Science and Engineering, Vol. 127.
44. C. Castro and E. Zuazua, *Boundary controllability of a hybrid system consisting in two flexible beams connected by a point mass.*, SIAM J. Control Optimization, 36 (1998), 1576–1595.
45. M. M. Cavalcanti, V. D. Cavalcanti, and I. Lasiecka, Well-posedness and optimal decay rates for the wave equation with nonlinear boundary damping-source interaction, *J. Differ. Equations*, 236 (2007), 407–459.
46. G. Chen, Energy decay estimates and exact boundary value controllability for the wave equation in a bounded domain, *J. Math. Pures Appl.*, 58 (1979), 249–274.
47. G. Chen, Control and stabilization for the wave equation in a bounded domain, Part I, *SIAM J. Control Optim.*, 17 (1979), 66–81.
48. G. Chen, Control and stabilization for the wave equation in a bounded domain, Part II, *SIAM J. Control Optim.*, 19 (1981), 114–122.
49. G. Chen, A note on boundary stabilization of the wave equations, *SIAM J. Control Optim.,* 19 (1981), 106–113.
50. S. Chen, K. Liu, and Z. Liu, *Spectrum and stability for elastic systems with global or local kelvin-voigt damping.*, SIAM J. Appl. Math., 59 (1999), 651–668.
51. B. Chentouf, Stabilization of memory type for a rotating disk-beam system, *Applied Math. Comput.*, 258 (2015), 227–236.
52. ———, A minimal state approach to dynamic stabilization of the rotating disk-beam system with infinite memory, IEEE Trans. Automat. Control, 61 (2016), 3700–3706.
53. ———, Compensation of the interior delay effect for a rotating disk-beam system, *IMA Journal of Math. Control and Information*, 33 (2016), 963–978.
54. ———, Stabilization of a non-linear rotating flexible structure under the presence of time-dependent delay term in a dynamic boundary control, *IMA J. Math. Control Info.*, 33 (2016), 349–363.
55. ———, Effect compensation of the presence of a time-dependent interior delay on the stabilization of the rotating disk-beam system. *Nonlinear Dynamics*, 84 (2016), 977–990.
56. ———, Well-posedness and exponential stability results for a nonlinear Kuramoto-Sivashinsky equation with a boundary time-delay, *Analysis and Mathematical Physics*, 11, 144 (2021).
57. B. Chentouf and M. S. Boudellioua, A new approach to stabilization of the wave equation with boundary damping control, Sultan Qaboos University. Science, vol. 9, 7 pages, 2004.
58. ———, On the stabilization of the wave equation with dynamical control. In: Proc. 16th International Symposium on Mathematical Theory of Networks and Systems, 2004, Leuven, Belgium, 6 pages.
59. B. Chentouf and B. Feng, On the stabilization of a flexible structure via a nonlinear delayed boundary control, *Discrete and Continuous Dynamical Systems-B*, 27 (2022), 7043–7063.
60. Chentouf and A. Guesmia, Neumann-boundary stabilization of a non-homogeneous n-dimensional wave equation with damping control. In: Proc. *Fifth Asian Mathematical Conf.*, Kuala Lampur, Malaysia. June 22–26, 2009, 6 pages.
61. ———, Neumann-boundary stabilization of the wave equation with damping control and applications. *Communications in Applied Analysis*, 14 (2010), 541–566.

62. ———, Well-posedness and asymptotic behavior of a wave equation with distributed time-delay and Neumann boundary conditions, *Mathematical Methods in the Applied Sciences*, 42 (2019), 4584–4605.
63. B. Chentouf and Z. Han, On the stabilization of an overhead crane system with dynamic and delayed boundary conditions, *IEEE Transactions on Automatic Control*, 65 (2020), 4273–4280.
64. B. Chentouf and S. Mansouri, Exponential decay rate for the energy of a flexible structure with dynamic delayed boundary conditions and a local interior damping, *Applied Math. Letters*, 103 (2020), no. 106185.
65. ———, On the exponential stabilization of a flexible structure with dynamic delayed boundary conditions via one boundary control only. *Journal of The Franklin Institute*, 358 (2021) 934–962.
66. ———, Boundary stabilization of a flexible structure with dynamic boundary conditions via one time-dependent delayed boundary control, *Discrete & Continuous Dynamical Systems–S*, 15 (2022), 1127–1141.
67. ———, Stabilization of an overhead crane system with infinite memory in the boundary control, *IEEE Transactions on Automatic Control*, 68 (2023), 4344–4350.
68. ———, Exponential stabilization of a flexible structure with a local interior control and under the presence of a boundary infinite memory, *Mathematical Methods in the Applied Sciences*, 46 (2023), 2955–2971.
69. B. Chentouf and N. Smaoui, Exponential stabilization of a non-uniform rotating disk-beam system via a torque control and a finite memory type dynamic boundary control, *Journal of The Franklin Institute*, 356 (2019), 11318–11344.
70. B. Chentouf, C. Z. Xu and G. Sallet, On the stabilization of a vibrating equation, *Nonlinear Analysis: Theory, Methods Applications,* 39 (2000), 537–558.
71. M. Cherkaoui and F. Conrad, Stabilisation d'un bras fléxible en torsion, *Research report INRIA,* 1805 (1992).
72. C. Chicone and Y. Latushkin, *Evolution semigroups in dynamical systems and differential equations*, Mathematical Surveys and Monographs, vol. 70. Providence,(1999), RI: American Mathematical Society.
73. E. A. Coddington and N. Levinson, *Theory of Ordinary Differential Equations*, Tata McGraw-Hill, New Delhi, 1977.
74. F. Conrad and A. Mifdal, Strong stability of a model of an overhead crane, *Control Cybernet.*, 27 (1998) 363–394.
75. F. Conrad and Ö. Morgül, *On the stabilization of a flexible beam with a tip mass*, SIAM J. Control Optim., 36 (1998), 1962–1986 (electronic).
76. F. Conrad, G. O'Dowd and F. Z. Saouri, Asymptotic behavior for a model of flexible cable with tip masses, *Asymptot. Anal.*, 30 (2002) 313–330.
77. F. Conrad and B. Rao, Decay of solutions of the wave equation in a star-shaped domain with nonlinear boundary feedback, *Asymptotic Analysis,* 7 (1993) 159–177.
78. R. F. Curtain and H. J. Zwart, *An Introduction to Infinite-Dimensinal Linear Systems*. Texts in Applied Mathematics No. 21, Springer-Verlag, 1995.
79. C. M. Dafermos, Asymptotic stability in viscoelasticity, *Arch. Rational Mech. Anal.,* 37 (1970) 297–308.
80. C. M. Dafermos and M. Slemrod, Asymptotic behavior of nonlinear contraction semigroups, *J. Func. Anal.*, 13 (1973), 97–106
81. R. Datko, Not all feedback stabilized hyperbolic systems are robust with respect to small time delays in their feedbacks, *SIAM J. Control Optim.*, 26 (1988), 697–713.
82. ———, *Two questions concerning the boundary control of certain elastic systems*, J. Differential Equations, 92 (1991), 27–44.
83. ———, Two examples of ill-posedness with respect to time delays revisited, *IEEE Trans. Autom. Control.,* 42 (1997), 511–515.
84. R. Datko, J. Lagnese and M. P. Polis, An example on the effect of time delays in boundary feedback stabilization of wave equations, *SIAM J. Control Optim.*, 24 (1986), 152–156.

85. F. Dell'Oro and V. Pata, Lack of exponential stability in Timoshenko systems with flat memory kernels, *Appl. Math. Optim.*, 71 (2015), 79–93.
86. J. Deutscher and J. Gabriel, Fredholm Backstepping Control of Coupled Linear Parabolic PDEs With Input and Output Delays,*in IEEE Transactions on Automatic Control*, 65 (2020), 3128–3135.
87. G. Doronin, N. Larkin, and A. Souza, *A hyperbolic problem with nonlinear second-order boundary damping.*, Electron. J. Differ. Equ. 1998, paper 28, (1998), 1–10.
88. G. Doronin and N. A. Larkin, *Global solvability for the quasilinear damped wave equation with nonlinear second-order boundary conditions.*, Nonlinear Anal., Theory Methods Appl., 8 (2002), 1119–1134.
89. M. Dreher, R. Quintanilla, and R. Racke, Ill-posed problems in thermomechanics, *Appl. Math. Lett.*, 22 (2009), 1374–1379.
90. A. K. Drame, D. Dochain, and J. J. Winkin, Dynamical Analysis of a Biochemical Reactor Distributed Parameter Model with Time Delay, Proceedings of the European Control Conference 2007, Kos, Greece, July 2–5, 2007, 5003–5007.
91. A.K. Drame, D. Dochain, and J. J. Winkin, P.R. Wolenski, Periodic trajectories of distributed parameter biochemical systems with time delay, *Applied Mathematics and Computation*, 218 (2012), 7395–7405.
92. A. Elharfi, Exponential stabilization and motion planning of an overhead crane system, *IMA J. Math. Control Info.*, dnw026. https://doi.org/10.1093/imamci/dnw026.
93. A. Elharfi, Control design of an overhead crane system from the perspective of stabilizing undesired oscillations, *IMA J. Math. Control Info.*, 28 (2011), 267–278.
94. A. Elharfi, Exponential stabilization and motion planning of an overhead crane system, *IMA J. Math. Control Info.*, 34 (2017), 1299–1321.
95. A. Elharfi, Exponential stabilization of a class of 1-D hyperbolic PDEs, *J. Evol. Equ.*, 16 (2016), 665–679.
96. K. J. Engel and R. Nagel, *One-Parameter Semigroups for Linear Evolution Equations,* Graduate Texts in Mathematics No. 194, Springer-Verlag, New York, 2000.
97. K. J. Engel and R. Nagel, *A Short Course on Operator Semigroups,* Universitext, Springer, New York, 2006.
98. M. Fabrizio, C. Giorgi, and V. Pata, A new approach to equations with memory, *Arch. Ration. Mech. Anal.*, 198 (2010), 189–232.
99. E. E. Fridman, S. Nicaise and J. Valein, Stabilization of second order evolution equations with unbounded feedback with time-dependent delay, *SIAM J. Control Optim.,* 48 (2010), 5028–5052.
100. S. S. Ge, W. He, B. Voon, and Y. S. Choo, Boundary control of a coupled nonlinear flexible marine riser, *IEEE Trans. Contr. Syst. Technol.*, 18 (2010), 1080–1091.
101. M. Grobbelaar-Van Dalsen, On fractional powers of a closed pair of operators and a damped wave equation with dynamic boundary conditions, *Appl. Anal.*, 53 (1994), 41–54.
102. ———, On the solvability of the boundary-value problem for the elastic beam with attached load, *Math. Models Meth. Appl. Sci.*, 4 (1994), 89–105.
103. ———, On the initial-boundary-value problem for the extensible beam with attached load, *Math. Methods Appl. Sci.*, 19 (1996), 943–957.
104. ———, Uniform stabilization of a one-dimensional hybrid thermo-elastic structure, *Math. Methods Appl. Sci.*, 26 (2003), 1223–1240.
105. M. Grobbelaar-Van Dalsen and A. Van Der Merwe, Boundary stabilization for the extensible beam with attached load, *Math. Models Methods Appl. Sci.*, 9 (1999), 379–394.
106. A. Guesmia, Well-posedness and exponential stability of an abstract evolution equation with infinite memory and time-delay, *IMA J. Mathematical Control and Information*, 30 (2013), 507–526.
107. A. Guesmia and A. Soufyane, On the stability of Timoshenko-type systems with internal frictional dampings and discrete time delays, Applicable Analysis, 96 (2017), 2075–2101.

108. A. Guesmia and N. Tatar, Some well-posedness and stability results for abstract hyperbolic equations with infinite memory and distributed time-delay, *Comm. Pure. Appl. Anal.*, 14 (2015), 457–491.
109. M. Gueye, Exact boundary controllability of 1-d parabolic and hyperbolic degenerate equations, *SIAM J. Control Optim.*, 52 (2014), 2037–2054.
110. M. Gugat, Boundary feedback stabilization by time delay for one-dimensional wave equations, *IMA J. Math. Control Inform.*, 27 (2010), 189–203.
111. B. Z. Guo and K. Y. Yang, Output feedback stabilization of a one-dimensional Schrödinger equation by boundary observation with time delay, *IEEE Trans. Autom. Control*, 55 (2010), 1226–1232.
112. B. Z. Guo, C. Z. Xu, and H. Hammouri, Output feedback stabilization of a one-dimensional wave equation with an arbitrary time delay in boundary observation, *ESAIM: Control, Optimization and Calculus of Variations*, 18 (2012), 22–35.
113. P. Guzmáan, S. Marx and E. Cerpa, Stabilization of the linear Kuramoto–Sivashinsky equation with a delayed boundary control, *IFAC Papers On Line*, 52-2 (2019) 70–75.
114. A. Haraux, *Systèmes Dynamiques Dissipatifs et Applications,* Collection Recherche en Mathématiques Appliqueés, Masson, Paris, 1991.
115. G. H. Hardy, J. E. Littlewood and G. Pólya, *Inequalities*, 2nd ed. Cambridge, England: Cambridge University Press, 1988.
116. A. Hastir, F. Califano and H. Zwart, Well-posedness of infinite-dimensional linear systems with nonlinear feedback, *Systems and Control Letters*, 128 (2019), 19–25.
117. W. He, S. Zhang, and S. S. Ge, Boundary control of a flexible riser with the application to marine installation, *IEEE Trans. Industr. Electronics*, 60 (2013), 5802–5810.
118. F. Huang, Characteristic conditions for exponential stability of linear dynamical systems in Hilbert space, *Ann. Differential Equations,* 1 (1985), 43–56.
119. B. Jacob and H. Zwart, *Linear Port-Hamiltonian Systems on Infinite-dimensional Spaces*, Operator Theory: Advances and Applications, 223, Birkhäuser, 2012.
120. P. M. Jordan, W. Dai and R. E. Mickens, A note on the delayed heat equation: Instability with respect to initial data, *Mech. Res. Comm.*, 35 (2008), 414–420.
121. T. Kato, *Perturbation theory of linear Operators,* Springer, 1976.
122. ———, Linear and quasilinear equations of evolution of hyperbolic type, C.I.E.M., II (1976), 125–191.
123. J. B. Kennedy, *On the isoperimetric problem for the Laplacian with Robin and Wentzell boundary bonditions*, PhD thesis, School of Mathematics and Statistics, University of Sydney, 2010.
124. M. Kirane, Blow-up for some equations with semilinear dynamical boundary conditions of parabolic and hyperbolic type, *Hokkaido Math. J.*, 21 (1992), 221–229.
125. V. Kolmanovskii and A. Myshkis, *Applied theory of functional differential equations*, Kluwer, 1999.
126. V. Komornik and E. Zuazua, A direct dethod for boundary stabilization of the wave equation, *Journal de Mathématiques Pures et Appliqués*, 69 (1990), 33–54.
127. M. Krstic, *Delay Compensation for Nonlinear, Adaptive and PDE Systems*. Boston: Birkhäuser, 2009.
128. J. Lagnese, Decay of solutions of the wave equation in a bounded region with boundary dissipation, *J. Diff. Equ.,* 50 (1983), 163–182.
129. ———, Note on boundary stabilization of wave equations, *SIAM J. Control Optim.,* 26 (1988), 1250–1256.
130. I. Lasiecka and D. Tataru, Uniform boundary stabilization of semilinear wave equations with nonlinear boundary damping, *Differential Integral Equations,* 6 (1993), 507–533.
131. P. D. Lax, C. S. Morawetz and R. S. Philips, Exponential decay of solutions of the wave equation in the exterior of a star-shaped obstacle, *Comm. Pure Appl. Math.*, 16 (1963), 447–486.

132. G. Lebeau, Equation des ondes amorties, Boutet de Monvel, Anne (ed.) et al., *Algebraic and geometric methods in mathematical physics*. Kluwer Academic Publishers. Math. Phys. Stud., 19 (1996), 73–109.
133. G. Lebeau and L. Robbiano, Stabilisation de l'équation des ondes par le bord, *Duke Math. J.*, 86 (1997), 465–491.
134. J. L. Lions, Exact controllability, stabilization and perturbations for distributed systems, *SIAM Review*, 30 (1988), 1–68.
135. ———, *Contrôlabilité exacte et stabilisation de systèmes ditribués*, vol. 1, Masson, Paris, 1988.
136. W. Littman and L. Markus, Stabilization of a hybrid system of elasticity by feedback boundary damping, *Ann. Mat. Pura Appl., IV. Ser.* , 152 (1988), 281–330.
137. K. Liu and Z. Liu, Exponential decay of energy of the Euler-Bernoulli beam with locally distributed Kelvin–Voigt damping, *SIAM J. Control Optimization*, 36 (1998), 1086–1098.
138. ———, *Semigroups associated with dissipative systems*, Chapman & Hall/CRC, New York, 1999.
139. ———, *Exponential decay of energy of vibrating strings with local viscoelasticity.*, Z. Angew. Math. Phys., 53 (2002), 265–280.
140. Z. Liu and B. Rao, Characterization of polynomial decay rate for the solution of linear evolution equation, *Z. Angew. Math. Phys.*, 56 (2005), 630–644.
141. Z. Liu and S. Zheng, *Semigroups Associated with Dissipative Systems*, Chapman & Hall/CRC, Boca Raton, 1999.
142. W. J. Liu and E. Zuazua, Decay rates for dissipative wave equations, *Ricerche Mat.*, 48 (1999), 61–75.
143. Z. H. Luo, B. Z. Guo, and O. Morgul, *Stability and Stabilization of Infinite Dimensional Systems with Applications,* Springer-Verlag, London, 1999.
144. A. Majda, Disappearing solutions for the dissipative wave equations, *Indiana Univ. Math. J.*, 24 (1975), 1119–1133.
145. A. Mifdal, Stabilisation uniforme d'un système hybride, *C. R. Acad. Sci. Paris. Série I.*, 324 (1997), 37–42.
146. O. Morgul, B. Rao and F. Conrad, On the stabilization of a cable with a tip mass, *IEEE Transactions on Automatic Control*, 39 (1994), 440–454.
147. C. S. Morawetz, Decay of solutions of the exterior problem for the wave equations, *Comm. Pure Appl. Math.*, 28 (1975), 229–264.
148. A. Mifdal, Stabilisation uniforme d'un système hybride, *C. R. Acad. Sci. Paris. Série I.*, 324 (1997) 37–42.
149. S. Nicaise and C. Pignotti, Stability and instability results of the wave equation with a delay term in the boundary or internal feedbacks, *SIAM J. Control Optim.*, 45 (2006), 1561–1585.
150. ———, Stabilization of the wave equation with variable coefficients and boundary condition of memory type, *Asymptotic Analysis*, 50 (2006), 31–67.
151. ———, Stabilization of the wave equation with boundary or internal distributed delay., *Diff. Int. Eqs.*, 21 (2008), 935–958.
152. ———, Exponential stability of second-order evolution equations with structural damping and dynamic boundary delay feedback, *IMA J. Math. Control Inform.*, 28 (2011), 417–446.
153. S. Nicaise, C. Pignotti and J. Valein, *Exponential stability of the wave equations with boundary time-varying delays*, Discrete Contin. Dyn. Syst. Ser. S, 4, 693–722 (2011).
154. S. Nicaise and J. Valein, Stabilization of second order evolution equations with unbounded feedback with delay, *ESAIM Control Optim. Calc. Var.*, 16 (2010), 420–456.
155. S. Nicaise, J. Valein and E. Fridman, Stability of the heat and of the wave equations with boundary time-varying delays, *Discrete Contin. Dyn. Syst. Ser. S*, 2 (2009), 559–581.
156. S. Nicaise, C. Pignotti and J. Valein, Exponential stability of the wave equations with boundary time-varying delays, *Discrete Contin. Dyn. Syst. Ser. S*, 4 (2011), 693–722.
157. K. Ono, On global existence, asymptotic stability and blowing up of solutions for some degenerate nonlinear wave equations of Kirchhoff type, *Math. Methods Appl. Sci.*, 20 (1997), 151–177.

158. A. Pazy, *Semigroups of Linear Operators and Applications to Partial Differential Equations*, Springer-Verlag, New York, 1983.
159. M. Pellicer, Large time dynamics of a nonlinear spring-mass-damper model, *Nonlin. Anal.*, 69 (2008), 3110–3127.
160. M. Pellicer and J. Solà-Morales, Analysis of a viscoelastic spring-mass model, *J. Math. Anal. Appl.*, 294 (2004), 687–698.
161. ———, Spectral analysis and limit behaviours in a spring-mass system, *Commun. Pure Appl. Anal.*, 7 (2008), 563–577.
162. K. D. Phung, Polynomial decay rate for the dissipative wave equation, *J. Differential Equations.*, 240 (2007), 92–124.
163. J. Prüss, *On the spectrum of* C_0-semigroups, *Trans. Amer. Math. Soc.*, 284 (1984), 847–857.
164. J. Prüss, *Evolutionary integral equations and applications*, Monograhs Mathematics, 87, Birkhäuser Verlag, Basel, 1993.
165. J. P. Quinn and D. L. Russell, Asymptotic stability and energy decay rate for solutions of hyperbolic equations with boundary damping, *Proc. Roy. Soc. Edinburgh, Sect. A,* 77 (1977), 97–127.
166. R. Racke, *Instability of coupled systems with delay,* Commun. Pure Appl. Anal., 11 (2012), 1753–1773.
167. B. Rao, Decay estimate of solution for hybrid system of flexible structures, *Euro. J. Appl. Math.,* 4 (1993), 303–319.
168. J. Rauch and M. Taylor, Exponential decay of solutions to hyperbolic equations in bounded domain, *Indiana. Univ. Math. J.*, 24 (1974), 79–86.
169. G. Ruiz Goldstein, Derivation and physical interpretation of general boundary conditions, *Adv. Differential Equations*, 11 (2006), 457–480.
170. D. Russell, Decay rates for weakly damped systems in Hilbert space obtained with control-theoretic methods, *J. Differential. Equations,* 19 (1975) 344–370.
171. ———, Controllability and stabilizability theory for inear partial differential equations : Recent progress and open questions, *SIAM Review*, 20 (1978), 639–739.
172. H. Sano, Boundary stabilization of hyperbolic systems related to overhead cranes, *IMA J Math Control Info*, 25 (2008), 353–366.
173. Y. F. Shang, G. Q. Xu and Y. L. Chen, Stability analysis of Euler-Bernoulli beam with input delay in the boundary control, *Asian J. Control.,* 14 (2012), 186–196.
174. J. Simon , Compact sets in the space $L^p(0; T; B)$. *Ann. Mat. Pura Appl.,* 146 (1987), 65–96.
175. R. Stahn, Optimal decay rate for the wave equation on a square with constant damping on a strip, *Z. Angew. Math. Phys,* 68 (2017), no. 2, Art. 36, 10 pp.
176. O. Solomon and E. Fridman, New stability conditions for systems with distributed delays, *Automatica*, 49 (2013), 3467–3475.
177. ———, Stability and passivity analysis of semilinear diffusion PDEs with time-delays, *International Journal of Control*, 88 (2015), 180–192.
178. L. N. Trefethen, *Spectral methods in Matlab*. PA: SIAM, Philadelphia (2000).
179. R. Triggiani, Wave equation on a bounded domain with boundary dissipation: an operator approach, *J. Math. Anal. Appl.*, 137 (1989), 438–461.
180. M. Tucsnak and G. Weiss, *Observation and control for operator semigroups.* Birkhäuser Advanced Texts, Birkhäuser Verlag, Basel, 2009.
181. J. Wang, M. Krstic, Delay-compensated event-triggered boundary control of hyperbolic PDEs for deep-sea construction, *Automatica*, 138 (2022), 110137.
182. ———, Delay-compensated control of sandwiched ODE-PDE-ODE hyperbolic systems for oil drilling and disaster relief, *Automatica*, 120 (2020), 109131.
183. F. Wolf, On the essential spectrum of partial differential boundary problems, *Comm. Pure Appl. Math.*, 12 (1959), 211–228.
184. G. Q. Xu, S. P. Yung, and L. K. Li, Stabilization of wave systems with input delay in the boundary control, *ESAIM Control Optim. Calc. Var.*, 12 (2006), 770–785.
185. G. Q. Xu, Stabilization of string system with linear boundary feedback, *Nonlinear Analysis: Hybrid Systems,* 1 (2007), 383–397.

186. W. H. Young, On classes of summable functions and their Fourier series, *Proc. Roy. Soc. London Ser. A*, 87 (1912), 225–229.
187. A. Zeidler, *Nonlinear Functional Analysis and its Applications*, vol. 2, Springer Verlag, New York, 1986.
188. H. Zhang and Q. Hu, Energy decay for a nonlinear viscoelastic rod equations with dynamic boundary conditions, *Math. Methods Appl. Sci.*, 30 (2007), 249–256.
189. H. C. Zhou and B. Z. Guo, Stabilization of ODE with hyperbolic equation actuator subject to boundary control matched disturbance, *International J. Control,* 92 (2019), 12–26.
190. E. Zuazua, Uniform stabilization of the wave equation by nonlinear boundary feedback, *SIAM J. Control Optim.*, 2 (1990), 466–477.
191. E. Zuazua, Exponential decay for the semilinear wave equation with locally distributed damping, *Comm. Partial. Diff. Eq.*, 15 (1990), 205–235.

Index

K. Ammari, B. Chentouf, *Stabilization of Wave-Type Equations with Delay or Memory*, Advances in Mechanics and Mathematics 57,
https://doi.org/10.1007/978-3-032-28496-9

www.ingramcontent.com/pod-product-compliance
Ingram Content Group UK Ltd.
Pitfield, Milton Keynes, MK11 3LW, UK
UKHW020934180926
475692UK00003B/25
9783032284952